Springer Tracts in Mechanical Engineering

Springer Tracts in Mechanical Engineering (STME) publishes the latest developments in Mechanical Engineering - quickly, informally and with high quality. The intent is to cover all the main branches of mechanical engineering, both theoretical and applied, including:

- Engineering Design
- Machinery and Machine Elements
- Mechanical Structures and Stress Analysis
- Automotive Engineering
- Engine Technology
- Aerospace Technology and Astronautics
- Nanotechnology and Microengineering
- Control, Robotics, Mechatronics
- MEMS
- Theoretical and Applied Mechanics
- Dynamical Systems, Control
- Fluids Mechanics
- Engineering Thermodynamics, Heat and Mass Transfer
- Manufacturing
- Precision Engineering, Instrumentation, Measurement
- Materials Engineering
- Tribology and Surface Technology

Within the scope of the series are monographs, professional books or graduate textbooks, edited volumes as well as outstanding PhD theses and books purposely devoted to support education in mechanical engineering at graduate and post-graduate levels.

Indexed by SCOPUS. The books of the series are submitted for indexing to Web of Science.

Please check our Lecture Notes in Mechanical Engineering at http://www.springer.com/series/11236 if you are interested in conference proceedings.

To submit a proposal or for further inquiries, please contact the Springer Editor **in your country**:

Dr. Mengchu Huang (China)
Email: mengchu.Huang@springer.com
Priya Vyas (India)
Email: priya.vyas@springer.com
Dr. Leontina Di Cecco (All other countries)
Email: leontina.dicecco@springer.com

More information about this series at http://www.springer.com/series/11693

Marco Gambini · Michela Vellini

Turbomachinery

Fundamentals, Selection and Preliminary Design

Springer

Marco Gambini
Department of Industrial Engineering
University of Rome Tor Vergata
Rome, Italy

Michela Vellini
Department of Industrial Engineering
University of Rome Tor Vergata
Rome, Italy

ISSN 2195-9862 ISSN 2195-9870 (electronic)
Springer Tracts in Mechanical Engineering
ISBN 978-3-030-51301-6 ISBN 978-3-030-51299-6 (eBook)
https://doi.org/10.1007/978-3-030-51299-6

This Springer imprint is published by the registered company Springer Nature Switzerland AG
The registered company address is: Gewerbestrasse 11, 6330 Cham, Switzerland

To our students.
Their enthusiasm and quest for learning are the energy that fuels our commitment and passion in transferring knowledge.

Preface

Turbomachines are extremely important from a technological point of view: they are the basis of most energy conversion systems for electricity generation (steam cycles, gas turbines, combined steam-gas cycles, as well as hydroelectric and wind power plants) and for heat and power production (cogeneration plants), for aeronautical propulsion (gas turbines), for fluid transport and compression (compressors, fans and pumps) in both industrial and civil applications.

For several decades, academic and industrial studies have extensively investigated turbomachines, favoring their formidable technological evolution. One example of this technological evolution is gas turbines for aeronautical and stationary applications: aeronautical gas turbines are the most technologically sophisticated and best performing engines that the market can offer. Similarly, also the technological evolution of gas turbines for stationary applications has been remarkable: together with the most advanced steam turbines, they realize nowadays combined steam-gas cycles that attain efficiencies of above 60%, becoming the most efficient and clean energy conversion systems available on the market.

The design of a turbomachinery is complex and laborious: it starts with the choice of the appropriate turbomachinery stage (radial, mixed flow, axial), including rotational speed and number of stages; it continues with the preliminary design of each stage and ends with the detailing design, conducted with three-dimensional computational fluid-dynamic analysis (CFD) techniques, including sophisticated structural, acoustic and mechanical analyses.

In the preliminary design, a one-dimensional performance evaluation of each single stage is carried out (mean-line analysis): in this phase, kinematic (velocity triangles), thermodynamic (fluid conditions upstream and downstream of the stator and rotor) and geometric parameters (inlet/outlet diameters, stator and rotor length, blade heights, chord, pitch and solidity, etc.) are defined. If these parameters are calculated in the most effective way, by using the most advanced loss correlations, then the turbomachinery efficiency can be predicted with reasonable accuracy and, in most cases, in good agreement with the CFD calculations. This means that the subsequent detailed aerodynamic design phase, conducted with advanced calculation tools that require a significant commitment of resources in economic and temporal terms, can be significantly reduced.

Precisely for these reasons, to reduce design costs and time, it is essential to perform these first design phases (selection and preliminary design of turbomachinery) correctly and realistically. In the subsequent phases, indeed, although greater levels of details are reached, any changes to the basic configuration would be particularly complicated and expensive, also taking into account that many other disciplines are involved in the detailed design of turbomachinery (structural, aerodynamic, acoustic, mechanical).

For over thirty years, the authors of this book have lectured, investigated and designed turbomachines and energy systems in which they are installed: conventional steam power plants, gas turbines and combined cycles for power production, cogeneration plants, advanced power plants with low CO_2 emissions, etc. In recent years, significant changes have been taking place in this sector, especially in order to limit CO_2 emissions and to develop new conversion systems integrated with renewable energy sources: new thermodynamic cycles integrated with pre-combustion CO_2 removal systems or post-combustion ones, new thermodynamic cycles using supercritical CO_2 (sCO_2), mixtures of CO_2-based fluids and other non-conventional fluids to be integrated in CSP plants (concentrating solar power), Organic Rankine Cycle (ORCs) coupled with renewable sources (solar, geothermal) and with low temperature thermal sources, to name the most promising. In essence, these are mostly turbomachinery-based conversion systems operated with unconventional fluids (different from air, exhaust gases, water and water vapor); these fluids cannot be treated with the schematization valid for perfect gases or with the classic equations of state valid for water–water vapor. Historically, however, correlations used in the preliminary design of a turbomachinery are referred to conventional fluids (gas turbines and steam turbines, compressors operating with gases similar to perfect gases).

Based on the above, this book aims to provide users with a selection process and a preliminary design procedure of turbomachinery with the following features:

- optimize the stage kinematics, thermodynamics and geometry already in the preliminary design phase, using geometric correlations and loss models, proposed in the most recent literature of the sector, able to predict the turbomachine efficiency with good accuracy and, in most cases, in excellent agreement with CFD calculations;
- be applicable to both conventional and non-conventional fluids;
- analyze homogeneously all the turbomachine types taken into consideration (axial and radial turbines, axial and centrifugal compressors, centrifugal pumps); this procedure calculates the kinematics, thermodynamics, geometry and stage losses in sequence, after identifying exactly the input parameters necessary to develop these numerical calculations;
- guide the user in choosing the sets of input parameters according to the most recent design developments.

Given its peculiarities, this book is addressed to bachelor's and master's degree students in industrial, mechanical and energy engineering, and to researchers, professionals and technicians of the energy systems and turbomachinery sector. For all these users, this book can become an instrument to guide them effectively "step by step" in the first sizing of turbomachines and in the verification of their technological feasibility when these turbomachines are conceived for new conversion systems operated with unconventional fluids.

The book is divided into eight chapters.

Chapter 1 is an introductory chapter and is a synthesis of the book titled *Lecture notes on Fluid Machines—M. Gambini, M. Vellini (in Italian)*, 2007, adopted for the Turbomachinery and Energy Systems courses for students of the bachelor's Degree in Mechanical and Energy Engineering of the University of Rome Tor Vergata. This chapter intends to provide a summary of the basic principles of thermodynamics and fluid-dynamics applied to turbomachinery necessary to define all the performance parameters (work transfer, isentropic and polytropic efficiency, nozzles and diffusers efficiency, degree of reaction and so on) used in the proposed procedures of turbomachinery selection and design.

Chapter 2 concerns the selection process of the turbomachine configuration (radial, mixed flow, axial) to be used in a specific application defined by the operating conditions (type of fluid, mass flow rate, pressure and inlet temperature, outlet pressure). This selection process is based on the application of the *similitude theory*, which allows to "capitalize" all previous experience in the turbomachinery sector by transferring the results obtained for an existing machine, considered as a model, to another machine, to be designed, which is "similar" to the model one, used as a reference.

The following chapters, from Chaps. 3 to 7, explain the preliminary design procedures of the five turbomachinery configurations here considered: axial turbines (Chap. 3), axial compressors (Chap. 4), radial turbines (Chap. 5), centrifugal compressors (Chap. 6) and centrifugal pumps (Chap. 7).

The book ends with a chapter (Chap. 8) entirely dedicated to numerical applications of the proposed procedures for the selection and preliminary design of turbomachines. In particular, the concentrating solar power (CSP) sector has been chosen as a case study, for which several unconventional conversion cycles are proposed: closed Brayton cycles operated with helium, argon, supercritical CO_2 (sCO_2), also in combined arrangements with bottoming cycles operated with organic fluids (ORC). When defining the thermodynamic cycle of each power block, preliminary hypotheses on the turbomachinery efficiencies must be made. These hypotheses must necessarily be verified through the calculation of the turbomachine losses, which, in turn, requires assessing of the kinematics, thermodynamics and geometry of the turbomachine (Chaps. 3–7). In fact, only if these hypotheses are validated, the reliability of the calculation of the cycles' performance is ensured.

Therefore, the organization of the book is the following:

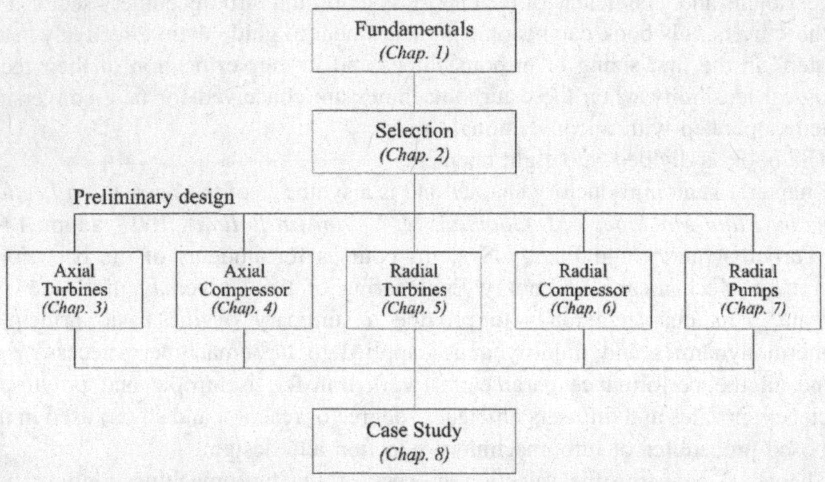

It is evident that the preliminary design procedures (Chaps. 3–7) are developed in parallel. So, some repetitions of similar topics for each type of turbomachine can occur, but the preliminary design procedure becomes homogeneous for all turbomachines and exhaustive for each of them; therefore, the reader can study in depth only one type of turbomachine without having to analyze all the other ones. For example, one interested in radial compressor can read through Chaps. 1–2–6–8 only.

The authors hope that the efforts made in the elaboration of an organic and homogeneous textbook on turbomachinery principles and design will be beneficial for the cultural and professional enrichment of a wide audience of users who, whether for study, research or work, will investigate new energy systems and their related turbomachines.

Rome, Italy
May 2020

Marco Gambini
Michela Vellini

Acknowledgements

The authors would like to express their appreciation and fond gratitudc to Mattco Sammarco, young and talented engineer, for his precious contribution in rendering the writing and graphics of this textbook as clear and effective as possible.

Moreover, the authors would like to thank all the numerous scholars from the academic, industrial and professional communities, whose research provided the decisive advancements included in this textbook. Our special thanks go to:

- Aungier, for his books on turbines and compressors,
- Casey, and his team for their papers on centrifugal compressors,
- Denton, for his extensive work on energy losses in turbomachines,
- Gulich, for his book on centrifugal pumps,
- Tournier and El Genk for their comprehensive treatise of axial stages of turbines and compressors.

Contents

Symbols

A	Flow area (m^2)
A_w	Cross-sectional area (m^2)
AR	Blade aspect ratio (-)
A_R	Area ratio (-)
b	Blade height (radial stage) (m)
$\vec{b}$	Angular momentum (kgm^2/s)
BL	Blade loading parameter (-)
c	Absolute velocity (m/s)
c_L	Blades lift coefficient (-)
c_p	Specific heat at constant pressure (kJ/kgK)
c_v	Specific heat at constant volume (kJ/kgK)
c_l	Liquid specific heat (kJ/kgK)
c_s	Speed of sound (m/s)
C	Blade chord length (m)
C_a	Blade axial chord (m)
D	Diameter (m)
D_s	Specific diameter (-)
DF	Diffusion coefficient (-)
DF_{eq}	Equivalent diffusion ratio (-)
e	Blade maximum camber (axial stage) (m)
	Specific energy (Chap. 1) (kJ/kg)
E	Specific energy (kJ/kg)
f	Friction factor (-)
$\vec{F}$	Vector sum of the forces exerted on the control volume (N)
$\vec{F}_W$	Vector sum of the forces exerted on the duct walls (N)
F_{AR}	Aspect ratio factor(-)
F_t	Tangential loading parameter (-)
g	Gravitational acceleration (m/s^2)
h	Specific enthalpy (kJ/kg)
h_B	Blade height (axial stage) (m)

H	Head (m)
H_{TE}	Boundary layer trailing-edge shape factor (-)
i	Incidence angle (°)
I	Rothalpy (kJ/kg)
IF	Incidence factor (-)
k	Ratio of specific heats (-)
k_s	Surface roughness (i.e. equivalent sandgrain roughness) (m)
K	Empirical factors (-)
l	Position of maximum camber (axial stage) (m)
L	Length (m)
m	Mass flow rate (kg/s)
M	Mass (kg)
$\vec{M}$	Vector sum of the moments of the forces exerted on the control volume (Nm)
M_a	Moment on rotating axis
Ma	Mach number (-)
Mu	Peripheral Mach number (or blade Mach number)
N_B	Blade number (-)
n	Rotational speed (rev/min)
	Polytropic exponent (Chap. 1) (-)
O	Blade distance (m)
p	Pressure (bar)
P	Power (kW)
P_w	wetted perimeter (m)
$\vec{q}$	Momentum (kg·m/s)
Q	heat transfer (kJ/kg)
R	degree of reaction (-)
Re	Reynolds number
R_g	specific gas constant (kJ/kgK)
R_0	universal gas constant (kJ/kgK)
r	Radius (m)
s	specific entropy (kJ/kgK)
S	blade pitch (m)
SF	slip factor (-)
SP	size parameter (m)
t	blade thickness (m)
T	Temperature (K)
U	internal energy (kJ/kg)
u	blade speed (or peripheral speed) (m/s)
v	specific volume (m^3/kg)
V	volume flow rate (m^3/s)
W	work transfer (kJ/kg)
w	relative velocity (m/s)
Y	pressure loss coefficient

z	number of stages (-)
	coordinate in vertical direction (Chap. 1) (m)
Z_H	head loss (m)
Z_{TE}	spanwise penetration depth between primary and secondary loss regions (m)
ZW	Zweifel coefficient (-)

Greek Symbols

α	Absolute flow angle (°)
α_B	Stator blade angle (°)
β	Relative flow angle (°)
β	Pressure ratio (Chap. 1, Sects. 1.3 and 1.4)
β_B	Rotor blade angle (°)
γ	Blade stagger angle (°)
Γ	Blade circulation parameter (-)
δ	Boundary layer thickness (m)
δ^*	Boundary layer displacement thickness (m)
δ_b	Rotor outlet width to diameter ratio (-)
δ_c	Deviation angle at the stator blade trailing edge (°)
δ_h	Rotor hub diameter ratio (-)
δ_M	Rotor mean diameter ratio
δ_R	Rotor deflection angle (°)
δ_S	Stator deflection angle (°)
δ_t	Rotor tip diameter ratio (-)
δ_w	Deviation angle at the rotor blade trailing edge (°)
ε	Isentropic exponent (-)
η	Efficiency (-)
η_p	Polytropic efficiency (-)
η_{is}	Isentropic efficiency (-)
θ	Blade camber angle (°)
θ	Boundary-layer momentum thickness at the blade outlet (m)
θ_C	Divergence angle (°)
λ	Hub-to-tip diameter ratio (axial stage)
μ	Dynamic viscosity (Pa·s)
ν_s	Isentropic velocity ratio (-)
ν_N	Nozzle velocity coefficient(-)
ξ	Rotor meridional velocity ratio (-)
ρ	Density (kg/m^3)
σ	Solidity (-)
τ	Blade tip clearance (axial stage) (m)
τ_a	Axial clearance (radial stage) (m)
τ_r	Radial clearance (radial stage) (m)

τ_b Rotor backface clearance (radial stage) (m)
φ Flow coefficient (-)
Φ Flow factor (-)
χ Blade angle measured from the chord line (°)
ψ Work coefficient (or stage loading coefficient) (-)
Ψ Work factor (-)
ω Speed of rotation (rad/s)
ω_s Specific speed (-)

Subscripts

0 Nozzle inlet
1 Rotor inlet
Inlet of machine (Chap. 1, Sects. 1.1–1.5)
2 Rotor outlet
Outlet of machine (Chap. 1, Sects. 1.1–1.5)
3 Diffuser outlet
a Axial
ad Adiabatic
AM Ainley and Mathieson
B Blade
c Kinetic
CL Clearance
diss Dissipated
ext External
EW End wall
h Hub value
hyd Hydraulic
in Inlet
inc Incidence
int Internal
irr Irreversible
is Isentropic
m Meridional (referred to kinematic and geometric parameters)
Mechanical (referred to energy and power)
max Maximum
min Minimum
mix Mixing
M Mean
opt Optimum
out Outlet
ov Overall

p	Pressure
pol	Polytropic
rec	Recovered
rev	Reversible
R	Rotor
S	Stator
sf	Skin friction
t	Total state (referred to thermodynamic parameters)
	Tip value (referred to kinematic and geometric parameters)
tr	Total relative state
TE	Trailing edge
TC	Tip clearance
TS	Total-to-static
TT	Total-to-total
u	Tangential
W	Related to work transfer W
z	Potential
ω	Centrifugal forces

Chapter 1
Fundamentals of Thermodynamics and Fluid Dynamics of Turbomachinery

Abstract This introductory Chapter is a synthesis of the text titled *"Lecture notes on Fluid Machines—Gambini and Vellini" (in Italian)*, 2007, adopted for the Turbomachinery and Energy Systems courses for students of the bachelor's Degree in Mechanical and Energy Engineering of the University of Rome Tor Vergata. This Chapter intends to provide a summary of the basic principles of thermodynamics and fluid-dynamics applied to turbomachinery in order to define all the performance parameters (work transfer, isentropic and polytropic efficiency, nozzles and diffusers efficiency, degree of reaction, etc.) used in the proposed procedures for turbomachinery selection and preliminary design (Chaps. 2–7). After introducing the basics of the First and Second Principles of Thermodynamics (Sect. 1.1), the energy equation is deduced in the mechanical and thermodynamic form (Sect. 1.2). Then, fluid properties are introduced (Sect. 1.3). Next, we analyze work transfer in turbomachines handling compressible (Sect. 1.4) and incompressible (Sect. 1.5) fluids, and we define isentropic, polytropic (Sect. 1.4), and hydraulic efficiency (Sect. 1.5). Then, the fluid flow in turbomachinery is studied through cardinal equations, and the Euler equation is deduced (Sect. 1.6). The principles of turbomachinery operation are therefore illustrated, highlighting stator and rotor configurations and the relative energy transfer processes (Sect. 1.7). Afterwards, an energy analysis of both axial and radial turbine and compressor stages is carried out (Sect. 1.8), aimed at defining all the performance parameters (work transfer, efficiency, degree of reaction, power, etc.). Finally (Sect. 1.9), we analyze losses in turbomachinery stages by highlighting the loss sources and providing a classification of these losses for both axial and radial stages. Throughout the discussion, real fluids are considered, illustrating only a few particular applications for perfect gases.

1.1 Review of Basic Thermodynamics of Turbomachinery

The First Law of Thermodynamics can be derived by applying two different approaches*: the Lagrangian approach* and *the Eulerian approach.*

M. Gambini and M. Vellini, *Turbomachinery*, Springer Tracts in Mechanical Engineering, https://doi.org/10.1007/978-3-030-51299-6_1

The Lagrangian approach follows over time the thermodynamic evolutions of a well-defined and constant mass of a homogeneous fluid in thermodynamic equilibrium (system status identifiable univocally through two thermodynamic parameters, for example, pressure and temperature).

The Eulerian approach instead provides the history of the fluid properties in a well-defined region of space, occupied by the fluid, without reference to the fact that the fluid is continually renewed in it. The surface that delimits this region, through which fluid can flow continuously, is called *control surface.* This approach is particularly effective in steady-state processes, that is when the properties of the fluid can vary from point to point in the space but in any point do not vary in time.

In other words, the Lagrangian approach studies the evolutions of a *closed system*, that is a portion of fluid undergoing changes in energy and momentum but not in mass; the Eulerian approach studies a portion of space, called *control volume*, that is an arbitrary volume across whose borders transfers of fluid can occur, as well as transfers of energy and momentum. This control volume represents an *open system* whose mass is therefore not necessarily constant. Only if the process is steady, in fact, the mass contained within the control volume remains constant, even though the open system is continuously crossed by the fluid.

For the purposes of this book, the Eulerian approach is extremely more suitable and effective, especially under the assumption of steady flow processes; by using this approach it is possible to delimit a turbomachine or a portion of it (in any case, a portion of the space occupied by the fluid) with a control surface through which energy/mass flux occurs.

Then, following this approach, let us consider an open system (Fig. 1.1), that is a system in which fluid is continuously introduced and extracted and heat and work transfers occur.

The *principle of conservation of energy* applied to this system allows to correlate the energy transfers through the control surface during a time interval dt to the energy variation of the system, dE^*(kJ), during the same time interval. In a fixed frame of reference:

$$dE^* = \sum_i dE_i^* = dE_1^* - dE_2^* + dQ^* + dW^*$$

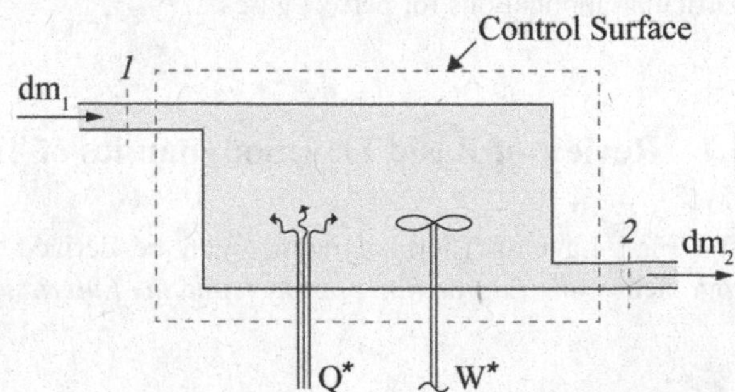

Fig. 1.1 Open system with heat and work transfers

where:

- dE^* is the variation of the system energy (kJ) during the time interval dt;
- dQ^* and dW^* are respectively the heat (kJ) exchanged across the control surface and the work through the control surface transmitted (kJ) by components connected to the shafts, during the time interval dt. They are both considered positive if supplied to the system;
- dE_1^* and dE_2^* are the variation of the system energy (kJ) associated to the fluid mass entering (dm_1) and exiting (dm_2) the system during the time interval dt.

So it is necessary to express the terms dE_1^* and dE_2^*.

Let us first consider the term dE_1^*. When the fluid element of mass dm_1(kg), and volume dV_1 (m^3), enters the system (Fig. 1.1), the system energy enriches itself with the energy of this fluid element and the energy correlated with the work of the pressure forces in the inlet station 1 (p_1) to put the volume dV_1 into the system:

$$dE_1^* = e_1 \cdot dm_1 + p_1 \cdot dV_1 = dm_1 \cdot \left(e_1 + p_1 \cdot \frac{dV_1}{dm_1} \right) = dm_1 \cdot (e_1 + p_1 \cdot v_1)$$

where:

- e_1 is the specific energy (kJ/kg) of the entering fluid element;
- p_1 and v_1 are respectively the pressure (kPa) and the specific volume (m^3/kg) of the entering fluid element;
- $p_1 \cdot v_1$ represents the work done by the fluid just upstream of the inlet to move the fluid ahead of it into the control volume. This work is also called *flow work*.

In thermo-mechanic systems, the specific energy, e_1 (kJ/kg), of the fluid element is composed of three terms: kinetic and potential energy and internal energy:

$$e_1 = \frac{c_1^2}{2} + g \cdot z_1 + U_1$$

where:

- c_1 (m/s) is the velocity of the fluid element;
- z_1 (m) is the height of the fluid element in a fixed frame of reference;
- g (m/s^2) is the gravitational acceleration;
- U_1 (kJ/kg) is the *internal energy* of the fluid element. This energy is a *state function* and represents the energy associated with the thermodynamic state of the fluid element.

Combining the expressions of dE_1^* and e_1, we get:

$$dE_1^* = dm_1 \cdot \left(\frac{c_1^2}{2} + g \cdot z_1 + U_1 + p_1 \cdot v_1 \right)$$

Thermodynamics defines the *state function* called *enthalpy* as:

$$h = U + p \cdot v$$

and hence we infer:

$$dE_1^* = dm_1 \cdot \left(\frac{c_1^2}{2} + g \cdot z_1 + h_1\right)$$

Analogously, the term dE_2^* is expressed as:

$$dE_2^* = dm_2 \cdot \left(\frac{c_2^2}{2} + g \cdot z_2 + h_2\right)$$

Hence, the energy variation of the system, dE^*(kJ), becomes:

$$dE^* = dm_1\left[\frac{c_1^2}{2} + gz_1 + h_1\right] - dm_2\left[\frac{c_2^2}{2} + gz_2 + h_2\right] + dQ^* + dW^*$$

The above equation allows the study of a generic open system even in unsteady operation. No hypotheses have been made on the process reversibility and so it is valid even in irreversible processes.

In the case of *steady-state processes*, i.e. when the state parameters in each point of the system are constant in time (this does not imply that they remain constant from one point to another one of the system), the energy of the system does not change ($dE^* = 0$) and its mass remains constant ($dm_1 = dm_2 = dm$). In this case, by dividing the heat and work transfers by the mass dm that enters and leaves the system (Q and W expressed in kJ/kg), *the principle of conservation of energy becomes*:

$$\left[\frac{c_2^2}{2} + gz_2 + h_2\right] - \left[\frac{c_1^2}{2} + gz_1 + h_1\right] = Q + W$$

$$\frac{\Delta c^2}{2} + g\Delta z + \Delta h = Q + W$$

and in *differential form*:

$$cdc + gdz + dh = dQ + dW \tag{1.1}$$

according to the following specification: Q and W are not state functions because their magnitudes depend on the path followed during a process as well as the initial and final states. Indeed, they are also called *path functions*. For this reason dQ and dW are inexact differential and hence they represent infinitesimal amounts of heat and work and, when integrated, they give finite amounts Q and W, respectively. In the technical literature the notations δQ and δW are often used; however, in this book we use the notation dQ and dW, remembering that they are infinitesimal amounts and not exact differentials.

However, since the first member of Eq. (1.1) is a state function, i.e. an exact differential, the sum of dQ and dW becomes an exact differential. Hence, during an adiabatic process ($dQ = 0$), i.e. the reference process for turbomachines, dW becomes an exact differential.

The Eq. (1.1) represents the *First Law of Thermodynamics based on the Eulerian approach* applied to an open system in steady-state and in a fixed frame of reference.

As the *state function enthalpy*, h, summarizes the First Law of Thermodynamics (Eq. 1.1), the *state function entropy*, s, is at the basis of the formulation of the *Second Law of Thermodynamics*. Entropy is very important in the identification and quantification of the irreversibilities that occur in fluid machines and energy systems.

Addressing to specialized thermodynamic texts for the in-depth formulation of the Second Law of Thermodynamics, the expression of the entropy is reported below:

$$ds = \frac{dQ}{T} + ds_{irr} \tag{1.2}$$

where the term ds_{irr} (entropy generation by irreversibilities), always greater than or equal to zero, takes into account the different types of irreversibilities:

$$ds_{irr} \geq 0$$

Equation (1.2) shows that in adiabatic processes (dQ = 0), such as those occurring in turbomachines, the entropy remains constant (isentropic process) only if the processes are ideal; in real processes, entropy increases, and this increase is equal to the entropy generated by irreversibilities.

Considering again the system represented in Fig. 1.1, during reversible processes, work done on the fluid increases its mechanical energy (in kinetic, potential and pressure terms) from station 1 to station 2 of the system:

$$dW_{rev} = dE_m = dE_c + dE_z + dE_p = cdc + gdz + vdp$$

In real (irreversible) processes, the same energy increase of the fluid flow requires greater work to compensate for energy dissipations due to irreversibilities:

$$dW = dE_m + dE_{diss} = cdc + gdz + vdp + dE_{diss}$$

If in a system irreversibilities are only those connected with adiabatic fluid flow (absence of heat exchanges, chemical reactions and other types of irreversibility), we have:

$$dE_{diss} = Tds_{irr} \tag{1.3}$$

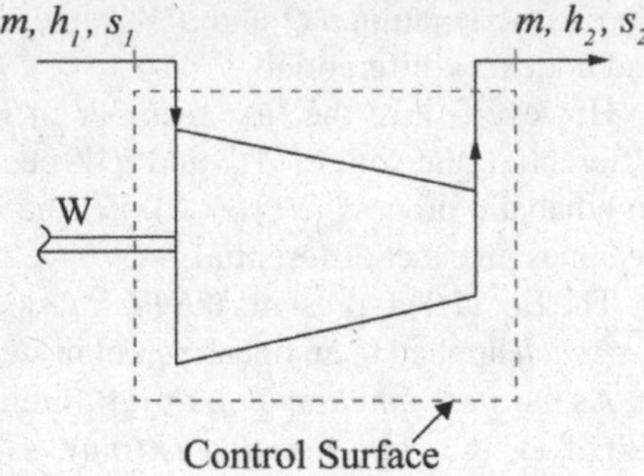

Fig. 1.2 Open system with work transfer

Then, considering a turbomachine (Fig. 1.2), where the process can be considered adiabatic (dQ = 0), assuming negligible variations of kinetic and potential energy between stations 1 and 2, the application of Eq. (1.1) allows to obtain the work transfer:

$$W = h_2 - h_1 = \Delta h$$

but this expression provides no information about the dissipations that occur in this process. Hence, on the basis of the Second Law of Thermodynamics, Eq. (1.2), we obtain:

$$\begin{aligned} Tds &= Tds_{irr} = dE_{diss} \\ \Delta s_{irr} &= \Delta s = s_2 - s_1 \end{aligned}$$

These equations clearly show that entropy increases between stations 1 and 2 because of irreversibilities; therefore, the greater this entropy increase, the greater the dissipated energy.

1.2 The Energy Equation in Mechanical and Thermodynamic Form

Now it is possible to formulate the *energy equation* in expressions, valid also for real (irreversible) processes, which simultaneously take into account the I and II Law of Thermodynamics.

For open systems, in steady operations, we have (Sect. 1.1):

$$\begin{aligned} cdc + gdz + dh &= dQ + dW \\ dW &= cdc + gdz + vdp + Tds_{irr} \\ dQ &= Tds - Tds_{irr} \end{aligned}$$

so:

$$dh = Tds + vdp \tag{1.4}$$

Equation (1.1) expresses the enthalpy change as a function of energy transfers across the control surface, while Eq. (1.4) expresses this variation as a function of the state parameters of the fluid flow.

Equation (1.1) represents the *energy equation in thermodynamic form*:

$$cdc + gdz + dh = dQ + dW$$

Combining Eqs. (1.1)–(1.4), we obtain the *energy equation in mechanical form*:

$$cdc + gdz + vdp = dW - dE_{diss} \tag{1.5}$$

The Eq. (1.5) is equivalent to Eq. (1.1), but expressed with explicit reference only to the mechanical terms of the fluid energy.

Equations (1.1) and (1.5) are extremely important in the study of turbomachinery. In particular, Eq. (1.1) allows calculating work and heat transfers across the control surface knowing the fluid state (velocity, position, enthalpy) at inlet and outlet stations of the system, without requiring any information on the particular process undergone by the fluid inside the system (in other words, the system can be seen as a "black box"). As proof of what has been stated, the integration of Eq. (1.1), for any type of fluid, depends only on the fluid state in the inlet (1) and outlet (2) stations and not on the process undergone by the fluid within the system:

$$\frac{\Delta c^2}{2} + g\Delta z + \Delta h = \frac{c_2^2 - c_1^2}{2} + g \cdot (z_2 - z_1) + (h_2 - h_1) = Q + W \tag{1.6}$$

Equation (1.5) instead correlates the work transfer only to the mechanical parameters of the fluid. The term vdp however presupposes to know the process within the system. Indeed, its integration provides:

$$\frac{\Delta c^2}{2} + g\Delta z + \int \frac{dp}{\rho} = \frac{c_2^2 - c_1^2}{2} + g \cdot (z_2 - z_1) + \int_1^2 \frac{dp}{\rho} = W - E_{diss} \tag{1.7}$$

which requires knowing how the term vdp varies and therefore the process itself.

1.3 Fluid Properties

For the practical application of the formulae deduced in Sects. 1.1–1.2, knowledge of the thermodynamic properties (p, T, v, ρ, h, s, and so on) of fluids handled by turbomachines is fundamental. In particular, in turbomachinery, work transfer is related to enthalpy variations while irreversibilities are directly related to entropy variations. The adiabatic process is of reference.

The thermodynamic state of any fluid is defined by the knowledge of three parameters: pressure (p), temperature (T), and specific volume (v, or, also its inverse, the density ρ), also called thermodynamic properties or state properties. There is always a correlation among these parameters, called the *equation of state*, which can be expressed by:

$$f(p, T, v) = 0$$

Through in this equation, the number of independent parameters to define the thermodynamic state of the system is reduced, at most, to a pair (for example p and T, or T and v, and so on).

Working fluids in turbomachinery are:

- compressible fluids (gases, gas mixtures, vapors);
- incompressible fluids (liquids).

Before tackling the problem of evaluating the properties of these fluids, it is advisable to refer to the ideal schematization of the behavior of compressible and incompressible fluids, namely the perfect gas and the perfect liquid.

1.3.1 Perfect Gases

State equation

The equation of state of a perfect gas is:

$$p \cdot v = R_g \cdot T \tag{1.8}$$

where R_g is the gas constant:

$$R_g = \frac{R_0}{PM}$$

R_0 is the universal gas constant (equal to 8.314 kJ/kmole K) and PM is the molecular weight of the gas.

For a perfect gas, internal energy (U) and enthalpy (h) are functions of the temperature only:

$$\left(\frac{\partial U}{\partial T}\right)_v = \frac{dU}{dT} \quad and \quad \left(\frac{\partial h}{\partial T}\right)_p = \frac{dh}{dT}$$

and therefore, defining the specific heats at constant volume and pressure as:

$$c_v = \left(\frac{\partial U}{\partial T}\right)_v \quad \text{and} \quad c_p = \left(\frac{\partial h}{\partial T}\right)_p$$

we obtain:

$$dU = c_v \cdot dT \tag{1.9}$$

$$dh = c_p \cdot dT \tag{1.10}$$

From the thermodynamic definition of enthalpy (Sect. 1.1):

$$h = U + p \cdot v$$

applying the *equation of state* (1.8) and the Eqs. (1.9)–(1.10) yields:

$$dh = c_p dT = dU + d(pv) = c_v dT + R_g dT$$

and then:

$$R_g = c_p - c_v = const$$

On the basis of the last equation, an ideal gas is a gas for which not only R_g but also the specific heats, c_p and c_v, are constant.

For perfect gases, therefore, specific heats can vary with the temperature as long as they keep their difference (R_g) constant.

Combining Eqs. (1.4), (1.8) and (1.10), we derive the expression of the entropy of a perfect gas:

$$Tds = dh - vdp = c_p dT - R_g T \frac{dp}{p}$$

so that:

$$ds = c_p \cdot \frac{dT}{T} - R_g \cdot \frac{dp}{p} \tag{1.11}$$

The expression (1.11) shows that entropy, unlike enthalpy and internal energy that depend only on temperature, is a function of two state parameters, p and T.

Thermodynamic processes

A general expression can summarize all the possible reversible processes of perfect gases. Indeed, assuming constant specific heats, we obtain:

$$pv^n = const \tag{1.12}$$

where:

$$n = \frac{c_p - c}{c_v - c}$$

and c is the specific heat (constant) of the particular process considered.

The Eq. (1.12) describes the polytropic process, that is any reversible process where specific heats are constant. By defining the specific heat of the process, the Eq. (1.12) can describe all the processes of perfect gases.

By assuming an isentropic (reversible adiabatic) process, the following expression is obtained:

$$pv^k = const \tag{1.13}$$

where:

$$c = 0$$
$$n = \frac{c_p - c}{c_v - c} = \frac{c_p}{c_v} = k$$

By combining the polytropic Eq. (1.12) with the equation of state (1.8), it is possible to correlate the initial and final state of the fluid during a polytropic process. The polytropic relation between temperature and pressure for a perfect gas is:

$$\frac{T_2}{T_1} = \left(\frac{p_2}{p_1}\right)^{\frac{n-1}{n}} = \beta^{\frac{n-1}{n}}$$

where β is the pressure ratio.

For an isentropic process, Eq. (1.13), we obtain:

$$\frac{T_2}{T_1} = \beta^{\frac{k-1}{k}} = \beta^{\varepsilon}$$

where:

$$\varepsilon = \frac{k-1}{k} = \frac{R}{c_p}$$

is the isentropic coefficient.

Another important process is the isenthalpic one (typically this process occurs in throttle valves), which in this case coincides with an isothermal process ($dh = c_p \cdot dT = 0$).

1.3.2 *Perfect Liquids*

Equation of state

The equation of state of a perfect liquid is:

$$v = \frac{1}{\rho} = const \tag{1.14}$$

Based on the Eq. (1.14), since all processes occur at constant specific volume, a single specific heat of the liquid can be defined and it coincides precisely with that at constant volume:

$$c_l = \left(\frac{\partial U}{\partial T}\right)_v = \frac{dU}{dT}$$

and thus:

$$dU = c_l \cdot dT \tag{1.15}$$

Also for perfect liquids, therefore, the internal energy is a function only of temperature: $U = U(T)$.

By applying Eq. (1.4), the thermodynamic definition of enthalpy and the equation of state (1.14), we obtain the expression of entropy; in fact:

$$dh = dU + vdp + pdv = Tds + vdp$$

and thus:

$$ds = \frac{dU}{T} = c_l \frac{dT}{T} \tag{1.16}$$

and:

$$dh = c_l dT + vdp \tag{1.17}$$

Therefore entropy, Eq. (1.16), as well as internal energy, Eq. (1.15), is a function only of temperature, while enthalpy, Eq. (1.17), is a function of two thermodynamic parameters.

The two Eqs. (1.16) and (1.17) demonstrate that, except for irreversibilities, the temperature variations of a perfect liquid are exclusively connected with heat exchange, since isentropic work has no effect on the temperature.

Thermodynamic processes

By definition, any process of a perfect liquid is isochoric, i.e. with constant volume.

Entropic diagrams give fewer information about liquid processes compared to compressible fluid processes. The T-s diagram becomes a single curve: indeed, for all processes, the relation between temperature and entropy for a perfect liquid is given by Eq. (1.16):

$$ds = \frac{dU}{T} = c_l \frac{dT}{T}$$

This curve is, therefore, representative of all isobars: the vertical distance between two isobars in fact is null since the isentropic process is also isothermal. Isentropic and isothermal processes, therefore, degenerate in one point.

The situation is similar also in the h-s diagram, even if in this case the isobars, although very close to each other, are not coincident. Their vertical distance (isentropic process) is given, Eq. (1.17), by the term $v \cdot (p_2 - p_1)$.

In the case of perfect liquids, the isothermal process does not coincide with the isenthalpic one, as in perfect gases, but with the isentropic one.

1.3.3 Real Fluids

Although several fluids operating in turbomachinery in the usual pressure and temperature ranges can be schematized as perfect gases (for example air, combustion gases) or perfect liquids (for example water), these schematizations cannot be used for vapors (for example steam, organic fluid vapors, and so on) nor for other types of fluids which, by their nature or for operating conditions close to the saturation lines, have properties significantly different from those of perfect fluids.

Since one of the purposes of this text is to design turbomachinery operating also with unconventional fluids, in Chaps. 2–7 we will refer to real fluids studied through their state functions (for example those supplied by NIST libraries).

Choosing any pair of thermodynamic parameters, for example, pressure and temperature, these libraries allow to calculate all the other parameters through specific functions:

$$\begin{aligned} h &= f_h(p, T) \\ s &= f_s(p, T) \\ U &= f_U(p, T) \\ \rho &= f_\rho(p, T) \\ \mu &= f_\mu(p, T) \end{aligned}$$

However, the properties of perfect gases and perfect liquids, briefly recalled in Sects. 1.3.1 and 1.3.2, allow to express analytically, in a very effective form, some key performance indicators (work transfer, isentropic and polytropic efficiency, for example) and their correlations with operating parameters of turbomachinery (for example, with pressure ratio and so on).

1.4 Compressible Fluids: Work Transfer and Efficiencies

In turbomachinery, the high flow velocity and the machine compactness allow to neglect the heat exchanged between the fluid and the surroundings. For this reason turbomachinery flow processes can be assumed adiabatic ($dQ = 0$), except for the first stages of gas turbines where gas temperatures can go above 1000 °C thanks to the employment of blade cooling systems.

First, adiabatic compression and expansion of real fluids will be analyzed; then, in order to highlight some characteristic correlations, these processes will be specialized for perfect gases. In the following analysis, the fluid velocity in inlet and outlet stations of turbomachine (Sections 1 and 2 in Fig. 1.2) will be the same. In between these stations, the variation of potential energy will be neglected.

1.4.1 Compression Process

Let us first consider the reversible adiabatic (isentropic) process 1–2_{is} on h-s diagram of Fig. 1.3. By applying Eq. (1.1) and neglecting the variation of kinetic and potential energy of the fluid between the inlet and outlet stations of the machine, as already mentioned, the isentropic compression work is:

$$dW_{is} = dh_{is}$$

and therefore:

$$W_{is} = \Delta h_{is} = h_{2is} - h_1 \qquad (1.18)$$

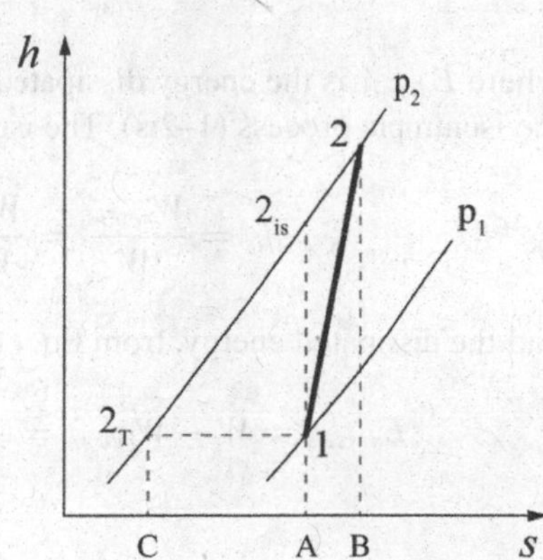

Fig. 1.3 Compression process on the h-s diagram

Then, from Eqs. (1.2) and (1.4), we deduce that:

$$W_{is} = \int_{1}^{2_{is}} v \cdot dp$$

The real process can not be drawn as a line on the thermodynamic diagram since it is an irreversible process; in fact, only reversible processes can be drawn as a line on thermodynamic diagrams because any point of this line is a point in thermodynamic equilibrium. However, since the real compression endpoint 2 can be identified exactly, it is possible to draw a polytropic line that is equivalent to the real process.

Since the real process is adiabatic, the adiabatic compression work done on the fluid is:

$$W = \Delta h_{ad} = h_2 - h_1 \tag{1.19}$$

On the basis of the energy equation in mechanical form, Eqs. (1.5) and (1.7), we can write:

$$W = W_{rev} + E_{diss}$$

Dissipations during work transfer can be taken into account by the compression efficiency, defined as:

$$\eta = \frac{W_{rev}}{W}$$

but it is necessary to specify the reference reversible process. If we assume the isentropic process as the reference reversible process, the previous equation becomes:

$$W = W_{rev,is} + E_{diss,is} = \int_{1}^{2_{is}} v \cdot dp + E_{diss,is} = (h_{2is} - h_1) + E_{diss,is} \tag{1.20}$$

where $E_{diss,is}$ is the energy dissipated in the adiabatic process (1–2) with respect to the isentropic process (1–2is). The isentropic efficiency is, therefore:

$$\eta_{is} = \frac{W_{rev,is}}{W} = \frac{W_{is}}{W} = \frac{\Delta h_{is}}{\Delta h_{ad}} = 1 - \frac{E_{diss,is}}{\Delta h_{ad}} \tag{1.21}$$

and the dissipated energy, from Eq. (1.20), is:

$$E_{diss,is} = W - W_{rev,is} = (h_2 - h_1) - (h_{2is} - h_1) = h_2 - h_{2is}$$

In order to calculate the enthalpy increase between 2is and 2, Eq. (1.4) can be applied along the isobar p_2:

$$(dh)_p = (T \cdot ds)_p$$

$$E_{diss,is} = h_2 - h_{2is} = \int_{2is}^{2} T \cdot ds \tag{1.22}$$

The expression of the isentropic efficiency, plugging (1.22) in (1.21), becomes:

$$\eta_{is} = \frac{W_{rev,is}}{W} = \frac{\Delta h_{is}}{\Delta h_{ad}} = 1 - \frac{\int_{2is}^{2} T \cdot ds}{\Delta h_{ad}} \tag{1.23}$$

This energy dissipation (1.22) can be displayed graphically on T-s diagram (Fig. 1.4) because it is the area underlying the isobar p_2 between 2is and 2:

$$E_{diss,is} = area(A2_{is}2B)$$

Instead, if the reference reversible process is the polytropic process (1–2), that is a non-adiabatic process, the polytropic work becomes (Eq. 1.1):

$$W_{pol} = W_{rev,pol} = \Delta h_{pol} - Q_{pol} = h_2 - h_1 - \int_{1}^{2} T \cdot ds \tag{1.24}$$

The polytropic efficiency is, therefore:

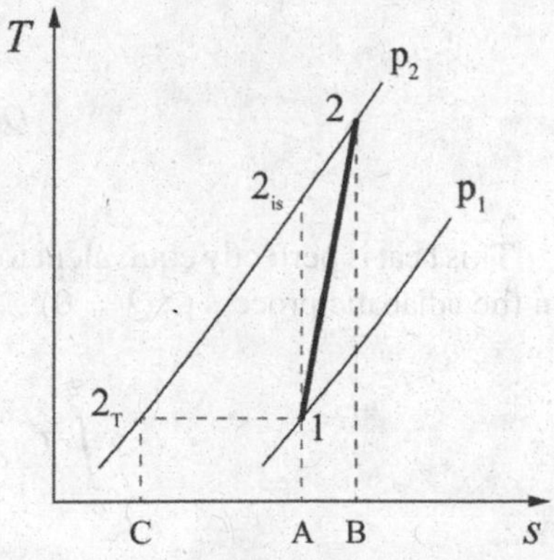

Fig. 1.4 Compression process on the T-s diagram

$$\eta_p = \frac{W_{rev,pol}}{W} = \frac{\Delta h_{ad} - \int_1^2 T \cdot ds}{\Delta h_{ad}} = 1 - \frac{\int_1^2 T \cdot ds}{\Delta h_{ad}} \tag{1.25}$$

Now, by applying the energy equation in mechanical form, Eqs. (1.5) and (1.7), given that polytropic work is:

$$W_{rev,pol} = \int_1^2 v \cdot dp$$

, we obtain:

$$\begin{aligned} W = W_{rev,pol} + E_{diss,pol} &= \int_1^2 v \cdot dp + E_{diss,pol} \\ &= (h_2 - h_1) - \int_1^2 T \cdot ds + E_{diss,pol} \end{aligned} \tag{1.26}$$

and therefore, using (1.19) we infer:

$$E_{diss,pol} = \int_1^2 T \cdot ds \tag{1.27}$$

It is worthwhile to investigate the meaning of this dissipated energy ($E_{diss,pol}$); it is the energy dissipated in the adiabatic process with respect to the reversible polytropic process (1–2). By applying the Second Law of Thermodynamics (Eq. 1.2), the heat exchanged along a reversible polytropic process ($T \cdot ds_{irr} = 0$) is:

$$Q = \int_1^2 T \cdot ds$$

This heat is perfectly equivalent to the heat generated internally by irreversibilities in the adiabatic process ($dQ = 0$):

$$\int_1^2 T \cdot ds = \int_1^2 T \cdot ds_{irr}$$

The term $E_{diss,pol}$, therefore, represents precisely mechanical-fluid-dynamic irreversibilities (viscous dissipations in the boundary layers, mixing in wakes, vortices, and so on, see Sect. 1.9):

$$E_{diss,pol} = \int_1^2 T \cdot ds_{irr}$$

This energy dissipation can be also displayed graphically on T–s diagram (Fig. 1.4) because it is equal to the area underlying the reversible polytropic process 1–2:

$$E_{diss,pol} = area(A12B)$$

and therefore:

$$E_{diss,is} = \int_{2is}^2 T \cdot ds > E_{diss,pol} = \int_1^2 T \cdot ds \tag{1.28}$$

It is important to understand the physical meaning of this result ($E_{diss,is} > E_{diss,pol}$): along the adiabatic process, mechanical-fluid-dynamic dissipations degrade mechanical energy into heat, heating the fluid; this heating increases the specific volume of the fluid and so more work is necessary to perform the fluid compression. Therefore, we can identify a term, E'_{diss}, to take into account these further dissipations; this term can be viewed on the T–s diagram (Fig. 1.4) as the mixtilinear triangle area 1–2is-2:

$$E_{diss,is} = E_{diss,pol} + E'_{diss} = \int_1^2 T \cdot ds + area(12_{is}2) = area(A12_{is}2B)$$

For this reason, along with an adiabatic compression we have:

$$\eta_p > \eta_{is}$$

This result can also be explained by considering average temperatures along the isobaric process 2is-2 and the polytropic process 1–2:

$$\int_1^2 T \cdot ds = \bar{T}_{1-2} \cdot \Delta s_{1-2}$$

$$\int_{2is}^2 T \cdot ds = \bar{T}_{2is-2} \cdot \Delta s_{2is-2} = \bar{T}_{2is-2} \cdot \Delta s_{1-2}$$

As it always is:

$$\bar{T}_{2is-2} > \bar{T}_{1-2}$$

it follows $\eta_p > \eta_{is}$.

Casey (2007) provided an interesting correlation between polytropic and isentropic efficiency for a real fluid. For an infinitesimal compression, we can write Eq. (1.25) as:

$$\eta_p = 1 - \frac{T \cdot ds}{dh}$$
$$1 - \eta_p = \frac{T \cdot ds}{dh}$$

For an isobar process, we know that (Eq. 1.4):

$$dh = T \cdot ds_p$$

and therefore, considering the same temperatures along with the polytropic and isobaric processes, we have:

$$1 - \eta_p = \frac{T \cdot ds}{T \cdot ds_p} = \frac{ds}{ds_p}$$

By integrating the last expression for a process where the polytropic efficiency is constant, we obtain:

$$1 - \eta_p = \frac{\Delta s}{\Delta s_p} = \frac{s_2 - s_1}{s_{1h} - s_1}$$

and therefore the polytropic efficiency can also be expressed as a function of entropy:

$$\eta_p = \frac{s_{1h} - s_2}{s_{1h} - s_1} \tag{1.29}$$

in this expression, 1h is on isobar p_1 (Fig. 1.5) and its enthalpy is equal to h_2 ($h_{1h} = h_2$).

In Fig. 1.5 the polytropic efficiency can be visualized as a ratio between two horizontal lengths (c/d) and the isentropic one as a ratio between two vertical lengths (a/b). By knowing, for example, the isentropic efficiency, we can calculate enthalpy at the compression end and then, through the properties of the fluid, entropy s_2 ($s_2 = f(h_2, p_2)$) and entropy s_{1h} ($s_{1h} = f(h_2, p_1)$); so we obtain the value of the polytropic efficiency of this compression process.

Specialization for perfect gases

For perfect gases polytropic and isentropic efficiencies are a function of pressure ratio. The isentropic work, Eq. (1.18), is:

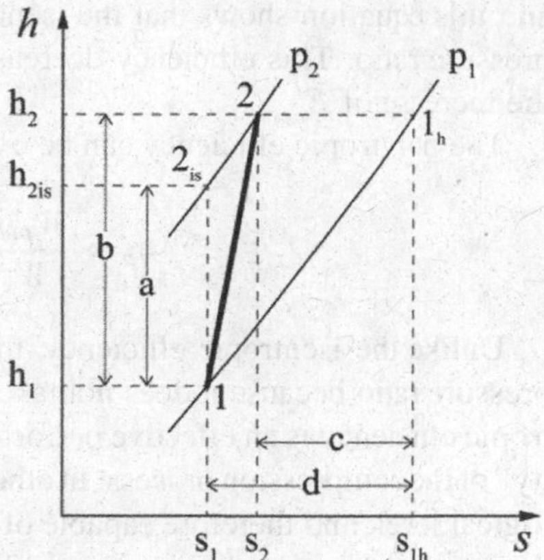

Fig. 1.5 Polytropic and isentropic efficiency for a compression

$$W_{is} = \Delta h_{is} = c_p \cdot T_1 \cdot \left(\frac{T_{2is}}{T_1} - 1\right) = c_p \cdot T_1 \cdot \left(\left(\frac{p_2}{p_1}\right)^{\frac{k-1}{k}} - 1\right)$$
$$= \frac{k}{k-1} \cdot R_g \cdot T_1 \cdot (\beta^{\varepsilon} - 1)$$

The adiabatic work, Eq. (1.19), is:

$$W = \Delta h_{ad} = c_p \cdot T_1 \cdot \left(\frac{T_2}{T_1} - 1\right) = c_p \cdot T_1 \cdot \left(\left(\frac{p_2}{p_1}\right)^{\frac{n-1}{n}} - 1\right)$$
$$= \frac{k}{k-1} \cdot R_g \cdot T_1 \cdot \left(\beta^{\frac{n-1}{n}} - 1\right)$$

The polytropic work, Eq. (1.24), is:

$$W_{pol} = c_p \cdot (T_2 - T_1) - c \cdot (T_2 - T_1) = \left(c_p - c\right) \cdot (T_2 - T_1)$$

As:

$$c_p - c = n \cdot (c_v - c) = \frac{n}{n-1} \cdot R_g$$

we obtain:

$$W_{pol} = \frac{n}{n-1} \cdot R_g \cdot T_1 \cdot \left(\beta^{\frac{n-1}{n}} - 1\right)$$

The isentropic efficiency is, therefore:

$$\eta_{is} = \frac{W_{is}}{W} = \frac{\beta^{\varepsilon} - 1}{\beta^{\frac{n-1}{n}} - 1} \tag{1.30}$$

and this equation shows that the isentropic compression efficiency depends on the pressure ratio. This efficiency decreases as the pressure ratio increases because of the increase of E'_{diss}.

The polytropic efficiency can be expressed as:

$$\eta_p = \frac{W_{pol}}{W} = \frac{n}{n-1} \cdot \frac{k-1}{k} \tag{1.31}$$

Unlike the isentropic efficiency, the polytropic efficiency is independent of the pressure ratio because it does not take into account E'_{diss}. For this reason, the polytropic efficiency is an effective performance indicator because it evaluates the "quality" of the compression process. In other words, turbomachines with the same technological level, and therefore capable of the same compression quality (as mentioned, the same mechanical-fluid-dynamic dissipations), even if characterized by different pressure ratios, can be studied assuming the same polytropic efficiency.

Hence, polytropic efficiency is very effective in the analysis of gas turbine cycles:

- in the hypothesis of maintaining the same technological level of compressors, we can analyze gas cycles with different pressure ratios assuming constant polytropic efficiency;
- we can choose the polytropic efficiency of a compressor, assuming it equal to that of another compressor already built of the same technology, even if operating at a different pressure ratio and different operating conditions (for example different inlet pressure and temperature).

On the basis of the previous considerations, it results particularly effective to express the adiabatic temperature ratio along the compression process as a function of polytropic efficiency:

$$\frac{T_2}{T_1} = \beta^{\frac{n-1}{n}} = \beta^{\frac{k-1}{k} \cdot \frac{1}{\eta_p}} = \beta^{\frac{\varepsilon}{\eta_p}}$$

Indeed, by using this equation, we can express the polytropic efficiency as a function of the characteristic temperatures of the compression process:

$$\eta_p = \frac{\ln \frac{T_{2is}}{T_1}}{\ln \frac{T_2}{T_1}}$$

The isentropic efficiency for perfect gases, as a function of these temperatures, is:

$$\eta_{is} = \frac{T_{2is} - T_1}{T_2 - T_1}$$

The last two equations establish a direct correlation between isentropic and polytropic efficiency and are particularly effective for calculating η_p and η_{is} knowing the pressure p_2 and the temperature T_2 (for example, from measurements) and T_{2is}

(calculated on the basis of the pressure ratio and the exponent of the isentropic process): this is the case of turbomachinery efficiency testing.

1.4.2 Expansion Process

The discussion is entirely analogous to that of compression process even if, in this case, we will see that part of energy dissipations can be recovered.

In the previous section, work done on fluid was considered positive. Therefore, in the case of expanders (turbines), as we want to write a positive work, we will change the sign of the terms in the energy equation.

Regarding the expansion processes in Fig. 1.6, initially we consider the reversible (isentropic) 1–2is and the adiabatic process. By applying Eq. (1.1) and neglecting the variation of kinetic and potential energy between inlet and outlet stations of the machine, the isentropic expansion work is:

$$dW_{is} = dh_{is}$$

and therefore:

$$W_{is} = \Delta h_{is} = h_1 - h_{2is} \tag{1.32}$$

Then, from expressions (1.2) and (1.4), we deduce that:

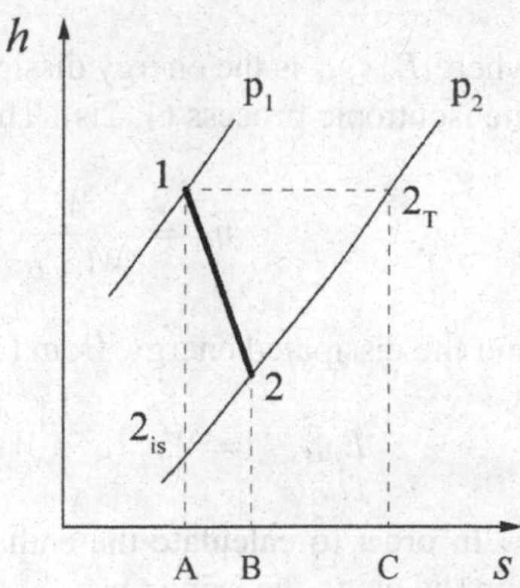

Fig. 1.6 Expansion process on the h-s diagram

$$W_{is} = \left| \int_{1}^{2_{is}} v \cdot dp \right|$$

The real process cannot be drawn as a line on the thermodynamic diagram since it is an irreversible process; only reversible processes can be drawn as a line on thermodynamic diagrams because any point of this line is a point in thermodynamic equilibrium. However, since the real expansion endpoint 2 can be identified exactly, it is possible to draw a polytropic line that is equivalent to the real process (Fig. 1.6).

Since the real process is adiabatic, the adiabatic expansion work done by the fluid is:

$$W = \Delta h_{ad} = h_1 - h_2 \tag{1.33}$$

On the basis of the energy equation in mechanical form, Eqs. (1.5) and (1.7), we can write:

$$W = W_{rev} - E_{diss}$$

Dissipations during work transfer can be taken into account by the expansion efficiency, defined as:

$$\eta = \frac{W}{W_{rev}}$$

but it is necessary to specify the reference reversible process. If we assume the isentropic process as the reference reversible process, the previous equation becomes:

$$W = W_{rev,is} - E_{diss,is} = \left| \int_{1}^{2_{is}} v \cdot dp \right| - E_{diss,is} = (h_1 - h_{2is}) - E_{diss,is} \tag{1.34}$$

where $E_{diss,is}$ is the energy dissipated in the adiabatic process (1–2) with respect to the isentropic process (1–2is). The isentropic efficiency is, therefore:

$$\eta_{is} = \frac{W}{W_{rev,is}} = \frac{W}{W_{is}} = \frac{\Delta h_{ad}}{\Delta h_{is}} = 1 - \frac{E_{diss,is}}{\Delta h_{is}} \tag{1.35}$$

and the dissipated energy, from (1.34), is:

$$E_{diss,is} = W_{rev,is} - W = (h_1 - h_{2is}) - (h_1 - h_2) = h_2 - h_{2is}$$

In order to calculate the enthalpy increase between 2is and 2, Eq. (1.4) can be applied along the isobar p_2:

$$(dh)_p = (T \cdot ds)_p$$

$$E_{diss,is} = h_2 - h_{2is} = \int_{2is}^{2} T \cdot ds \tag{1.36}$$

The expression of the isentropic efficiency, plugging (1.36) in (1.35), becomes:

$$\eta_{is} = \frac{W}{W_{is}} = \frac{\Delta h_{ad}}{\Delta h_{is}} = 1 - \frac{\int_{2is}^{2} T \cdot ds}{\Delta h_{is}} \tag{1.37}$$

This energy dissipation (1.36) can be displayed graphically on the T-s diagram (Fig. 1.7) because it is the area underlying the isobar p_2 between 2is and 2:

$$E_{diss,is} = area(A2_{is}2B)$$

Instead, if the reference reversible process is the polytropic process (1–2), which is a non-adiabatic process, the polytropic work becomes (Eq. 1.1):

$$W_{pol} = W_{rev,pol} = \Delta h_{pol} + Q_{pol} = h_1 - h_2 + \int_{1}^{2} T \cdot ds \tag{1.38}$$

The polytropic efficiency is, therefore:

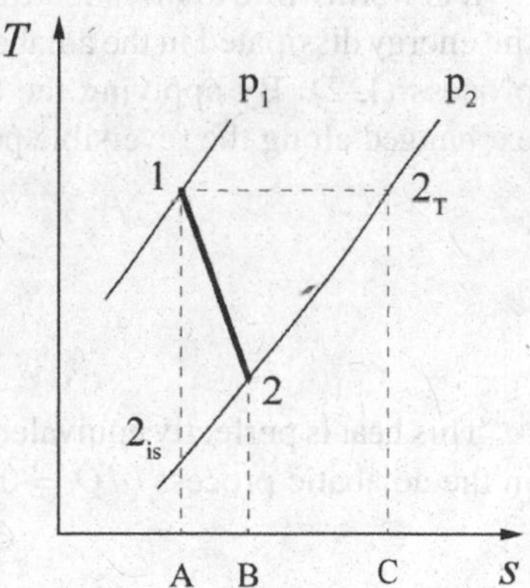

Fig. 1.7 Expansion process on the T-s diagram

$$\eta_p = \frac{W}{W_{pol}} = \frac{W_{pol} - \int_1^2 T \cdot ds}{W_{pol}} = 1 - \frac{\int_1^2 T \cdot ds}{W_{pol}} \tag{1.39}$$

Now, by applying the energy equation in mechanical form, Eqs. (1.5) and (1.7), given that polytropic work is:

$$W_{rev,pol} = \left| \int_1^2 v \cdot dp \right|$$

we obtain:

$$W = W_{rev,pol} - E_{diss,pol} = \left| \int_1^2 v \cdot dp \right| - E_{diss,pol}$$

$$= (h_1 - h_2) + \int_1^2 T \cdot ds - E_{diss,pol} \tag{1.40}$$

and therefore, using (1.33), we obtain:

$$E_{diss,pol} = \int_1^2 T \cdot ds \tag{1.41}$$

It is worthwhile to investigate the meaning of the dissipated energy ($E_{diss,pol}$); it is the energy dissipated in the adiabatic process with respect to the reversible polytropic process (1–2). By applying the Second Law of Thermodynamics, Eq. (1.2), heat exchanged along the reversible polytropic process ($T \cdot ds_{irr} = 0$) is:

$$Q = \int_1^2 T \cdot ds$$

This heat is perfectly equivalent to the heat generated internally by irreversibilities in the adiabatic process ($dQ = 0$):

$$\int_1^2 T \cdot ds = \int_1^2 T \cdot ds_{irr}$$

The term $E_{diss,pol}$, therefore, represents precisely the mechanical-fluid-dynamic irreversibilities (viscous dissipations in the boundary layers, mixing in wakes, vortices, and so on, see Sect. 1.9):

$$E_{diss,pol} = \int_1^2 T \cdot ds_{irr}$$

This energy dissipation can be also displayed graphically on the T–s diagram (Fig. 1.7) because it is equal to the area underlying the reversible polytropic process 1–2:

$$E_{diss,pol} = area(A12B)$$

and therefore:

$$E_{diss,is} = \int_{2is}^2 T \cdot ds < E_{diss,pol} = \int_1^2 T \cdot ds \tag{1.42}$$

It is important to understand the physical meaning of this result ($E_{diss,is} < E_{diss,pol}$): during an adiabatic process, mechanical-fluid-dynamic dissipations degrade mechanical energy into heat, heating the fluid; this heating increases the specific volume of the fluid, so part of the work loss is recovered. Therefore we can identify a term, E''_{diss}, to take into account this partial energy dissipation recovery; this term can be viewed on the T–s diagram of Fig. 1.7 as the mixtilinear triangle area 1–2is-2:

$$E_{diss,is} = E_{diss,pol} - E''_{diss} = \int_1^2 T \cdot ds - area(12_{is}2) = area(A2_{is}2B)$$

For this reason, along an adiabatic expansion we have:

$$\eta_{is} > \eta_p$$

This result can also be explained by considering average temperatures along the isobaric process 2is-2 and the polytropic process 1–2:

$$\int_1^2 T \cdot ds = \bar{T}_{1-2} \cdot \Delta s_{1-2}$$

$$\int_{2is}^2 T \cdot ds = \bar{T}_{2is-2} \cdot \Delta s_{2is-2} = \bar{T}_{2is-2} \cdot \Delta s_{1-2}$$

As it always is:

$$\bar{T}_{2is-2} < \bar{T}_{1-2}$$

it follows $\eta_{is} > \eta_p$.

As for the compression process, the correlation between polytropic and isentropic efficiency for a real fluid can be derived from Casey's dissertation (2007). For an infinitesimal expansion, we can write Eq. (1.39) as:

$$\eta_p = \frac{dh}{dh + T \cdot ds}$$
$$\frac{1}{\eta_p} - 1 = \frac{T \cdot ds}{dh}$$

For an isobar process, we know that, Eq. (1.4):

$$dh = T \cdot ds_p$$

and therefore, considering the same temperatures along the polytropic and isobaric processes, we have:

$$\frac{1}{\eta_p} - 1 = \frac{T \cdot ds}{T \cdot ds_p} = \frac{ds}{ds_p}$$

By integrating the last expression for a process where the polytropic efficiency is constant, we obtain:

$$\frac{1}{\eta_p} - 1 = \frac{\Delta s}{\Delta s_p} = \frac{s_2 - s_1}{s_1 - s_{1h}}$$

and therefore the polytropic efficiency can also be expressed as a function of entropy:

$$\eta_p = \frac{s_1 - s_{1h}}{s_2 - s_{1h}} \tag{1.43}$$

in this expression, the point 1h is on isobar p_1 (Fig. 1.8) and its enthalpy is equal to h_2 ($h_{1h} = h_2$).

In Fig. 1.8 the polytropic efficiency can be visualized as a ratio between two horizontal lengths (c/d) and the isentropic one as the ratio between two vertical lengths (a/b). By knowing, for example, the isentropic efficiency, we can calculate enthalpy at the expansion end and then, through the properties of the fluid, entropy s_2 ($s_2 = f(h_2, p_2)$) and entropy s_{1h} ($s_{1h} = f(h_2, p_1)$); so we obtain the value of the polytropic efficiency of this expansion process.

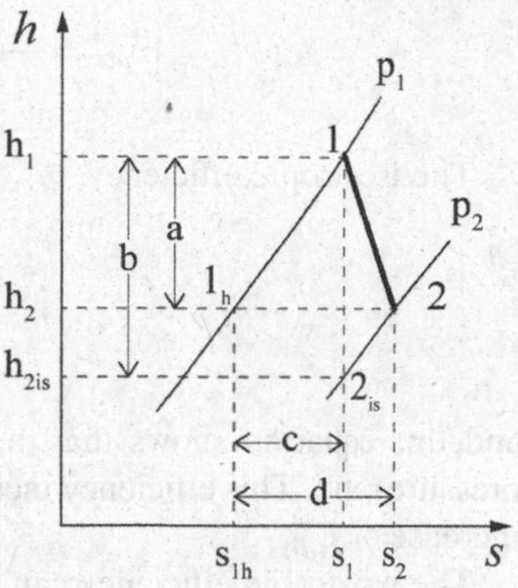

Fig. 1.8 Polytropic and isentropic efficiency for an expansion

Specialization for perfect gases

For perfect gases polytropic and isentropic efficiencies are a function of the pressure ratio. The isentropic work, Eq. (1.32), is:

$$W_{is} = \Delta h_{is} = c_p \cdot T_1 \cdot \left(1 - \frac{T_{2is}}{T_1}\right) = c_p \cdot T_1 \cdot \left(1 - \left(\frac{p_2}{p_1}\right)^{\frac{k-1}{k}}\right)$$

$$= \frac{k}{k-1} \cdot R_g \cdot T_1 \cdot \left(1 - \frac{1}{\beta^{\varepsilon}}\right)$$

The adiabatic work, Eq. (1.33), is:

$$W = \Delta h_{ad} = c_p \cdot T_1 \cdot \left(1 - \frac{T_2}{T_1}\right) = c_p \cdot T_1 \cdot \left(1 - \left(\frac{p_2}{p_1}\right)^{\frac{n-1}{n}}\right)$$

$$= \frac{k}{k-1} \cdot R_g \cdot T_1 \cdot \left(1 - \frac{1}{\beta^{\frac{n-1}{n}}}\right)$$

The polytropic work, Eq. (1.38), is:

$$W_{pol} = c_p \cdot (T_1 - T_2) - c \cdot (T_1 - T_2) = \left(c_p - c\right) \cdot (T_1 - T_2)$$

As

$$c_p - c = n \cdot (c_v - c) = \frac{n}{n-1} \cdot R_g$$

we obtain

$$W_{pol} = \frac{n}{n-1} \cdot R_g \cdot T_1 \cdot \left(1 - \frac{1}{\beta^{\frac{n-1}{n}}}\right)$$

The isentropic efficiency, η_{is}, is, therefore:

$$\eta_{is} = \frac{W}{W_{is}} = \frac{1 - \frac{1}{\beta^{\frac{n-1}{n}}}}{1 - \frac{1}{\beta^{\varepsilon}}} \tag{1.44}$$

and this equation shows that the isentropic expansion efficiency depends on the pressure ratio. This efficiency increases as the pressure ratio increases because of the increase of E''_{diss}.

The polytropic efficiency can be expressed as:

$$\eta_p = \frac{W}{W_{pol}} = \frac{k}{k-1} \cdot \frac{n-1}{n} \tag{1.45}$$

Based on the aforementioned correlations, the final considerations are similar to those made for the compression process and they regard the effectiveness of the use of polytropic efficiency in gas turbine cycles analysis.

Finally, we can deduce the expression of the adiabatic temperature ratio along the expansion as a function of polytropic efficiency; it is:

$$\frac{T_1}{T_2} = \beta^{\frac{n-1}{n}} = \beta^{\frac{k-1}{k} \cdot \eta_p} = \beta^{\varepsilon \cdot \eta_p}$$

The above equation also allows to express the polytropic efficiency as a function of the characteristic temperatures of the expansion:

$$\eta_p = \frac{\ln \frac{T_1}{T_2}}{\ln \frac{T_1}{T_{2is}}}$$

The isentropic efficiency for perfect gases, as a function of these temperatures, is:

$$\eta_{is} = \frac{T_1 - T_2}{T_1 - T_{2is}}$$

1.4.3 Synthetic Comparison Between Compression and Expansion Processes

The above considerations about the efficiency of compression and expansion processes can be summarized as follows:

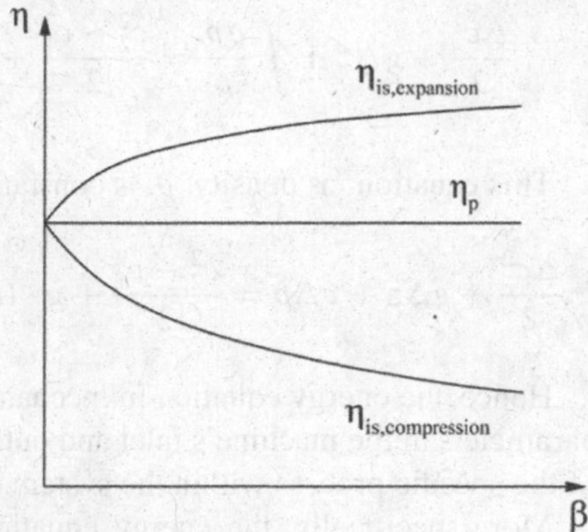

Fig. 1.9 Polytropic and isentropic efficiency for expansion and compression

- in an adiabatic compression, the additional energy dissipation, E'_{diss}, implies $\eta_{is,compression} < \eta_p$;
- in an adiabatic expansion, the partial recovery of energy dissipation, E''_{diss}, implies $\eta_{is,expansion} > \eta_p$;
- the isentropic compression efficiency, $\eta_{is,compression}$, decreases as the pressure ratio, β, increases;
- the isentropic expansion efficiency, $\eta_{is,expansion}$, increases as the pressure ratio, β, increases;
- the polytropic compression and expansion efficiencies, η_p, are substantially independent of the pressure ratio;
- for turbines and compressors with the same polytropic efficiency and pressure ratio, the isentropic efficiency of turbines, $\eta_{is,expansion}$, is greater than that of compressors, $\eta_{is,compression}$.

These considerations, therefore, justify the qualitative representation of the variation of these efficiencies with the pressure ratio, on the diagram in Fig. 1.9.

1.5 Incompressible Fluids: Work Transfer and Hydraulic Efficiency

In this case, the discussion is considerably simplified compared to compressible fluids since, as already illustrated above, the work transfer entails irrelevant thermal effects; therefore the work analysis on the thermodynamic diagrams is scarcely meaningful.

Since for incompressible fluids the specific volume is nearly constant (Sect. 1.3.2), work is analyzed using almost exclusively the energy equation in the mechanical form (1.7):

$$\frac{\Delta c^2}{2} + g\Delta z + \int \frac{dp}{\rho} = \frac{c_2^2 - c_1^2}{2} + g \cdot (z_2 - z_1) + \int_1^2 \frac{dp}{\rho} = W - E_{diss}$$

This equation, as density, ρ, is constant, becomes:

$$\frac{\Delta c^2}{2} + g\Delta z + v\Delta p = \frac{c_2^2 - c_1^2}{2} + g \cdot (z_2 - z_1) + \frac{p_2 - p_1}{\rho} = W - E_{diss} \quad (1.46)$$

Hence, the energy equation in mechanical form (1.46) depends only on the fluid parameters in the machine's inlet and outlet stations, and therefore it is independent of the specific process within the system.

More specifically, the energy equation expressed as a height (meters) of the working fluid column is used. In this regard, the concept of *head* is introduced; it represents the mechanical energy of the working fluid expressed in meters:

$$H = \frac{E}{g} = \frac{c^2}{2 \cdot g} + z + \frac{p}{\rho \cdot g} \quad (1.47)$$

Expressing also energy dissipations as a head loss:

$$Z_H = \frac{E_{diss}}{g} \quad (1.48)$$

the energy Eq. (1.46), by using Eqs. (1.47) and (1.48), becomes:

$$W = g \cdot \Delta H + g \cdot Z_H \quad (1.49)$$

From now on, we will analyze the compression process (pumps) for which ΔH is the *head rise across the pump* and, therefore, it is the net mechanical energy transferred to the fluid by the machine. In the following, ΔH is simply called *head*.

Assuming to be negligible the variations of kinetic and potential energy between pump inlet and outlet, we obtain:

$$\Delta H = \frac{\Delta p}{\rho \cdot g} = \frac{p_2 - p_1}{\rho \cdot g} \quad (1.50)$$

$$W_{rev} = g \cdot \Delta H \quad (1.51)$$

$$W = W_{rev} + E_{diss} = \frac{p_2 - p_1}{\rho} + g \cdot Z_H \quad (1.52)$$

As the dissipations due to irreversibility in an adiabatic process are (Eqs. (1.2), (1.3), (1.15) and (1.16)):

$$\Delta U = c_l \cdot \Delta T = \int T ds - \int p dv = \int T ds_{irr} = E_{diss}$$

we can write Eq. (1.48) as:

$$E_{diss} = g \cdot Z_H = c_l \cdot \Delta T \tag{1.53}$$

and therefore, Eq. (1.52) becomes:

$$W = \frac{\Delta p}{\rho} + c_l \cdot \Delta T \tag{1.54}$$

The two Eqs. (1.53) and (1.54) show that work transferred to incompressible fluids involves very low thermal effects (temperature variations) compared to compressible fluids; these effects are due only to irreversibilities (the reversible compression is isothermal, Sect. 1.3.2). For this reason, machines operating with compressible fluids are also called *thermal machines*, while machines operating with incompressible fluids are also called *hydraulic machines.*

Unlike thermal machines where in the definition of efficiency different reversible processes can be chosen (for example isentropic or polytropic processes, Sect. 1.4), in hydraulic machines, there is only one reference reversible process and therefore the definition of efficiency is univocal.

This univocal efficiency is called *hydraulic efficiency* and, with reference to pumps, is expressed, by using Eqs. (1.48–1.52), through:

$$\eta_{hyd} = \frac{W_{rev}}{W} = \frac{W_{rev}}{W_{rev} + E_{diss}} = \frac{\Delta H}{\Delta H + Z_H} \tag{1.55}$$

1.6 Basic Fluid Mechanics. The Euler Equation

The analyses carried out so far allow us to evaluate energy transfers by using parameters of working fluids in machine inlet and outlet stations, without knowing the internal configuration of the machine itself.

Now it is necessary to analyze this internal configuration; in other words, we will analyze how such energy transfers take place in practice.

In turbomachinery, the work transfer occurs between a flowing fluid and moving parts of the machine (rotating ducts, formed by two adjacent blades, crossed by the fluid) in continuous rotary motion around an axis.

The blade channels directly connected to the machine shaft are called *rotor channels*. However there are others blade channels connected to the casing of the machine, which are called *stator channels*.

The stator transforms the fluid energy into the most convenient form: in *driving turbomachines* (they produce power by expanding fluid to a lower pressure), it is placed upstream of the rotor in order to properly prepare the fluid flow for work transfer; in *driven turbomachines* (they absorb power to increase the fluid pressure), it is placed downstream of the rotor in order to transform the fluid energy (kinetic energy) into the most convenient form (pressure energy).

Furthermore, if these channels are arranged in an axial direction, the machine is called *axial*, while if the direction is radial it is called *radial* (centrifugal or centripetal).

To describe the energy transfer between a fluid flow and machines, it is necessary to study the fluid flow through stator and rotor channels.

Following the Eulerian approach, we must know in each point of the control volume the fluid velocity vector and three thermodynamic parameters, which define the physical state of the fluid: pressure, temperature, and density.

It is therefore necessary to define the following system of four equations (Caputo 1994):

- n. 3 scalar equations
- n. 1 vector equation.

together with two fundamental hypotheses: steady and one-dimensional flow. Under this last hypothesis, the physical and kinematic parameters of the fluid in each duct section are uniform; this means that the fluid state can be defined on each duct section by a single value of velocity, pressure, temperature, and so on. And again, under this last hypothesis absolute or relative velocity (respectively in the stator and rotor) can be considered tangential to the axis of the duct. Under these assumptions, simple expressions will describe the operating principles and the general conformation of turbomachines.

The first scalar equation is the state equation of the fluid, already widely discussed in Sect. 1.3.

The second scalar equation is the process equation; this process, as already mentioned, can always be considered adiabatic; this process can be considered isentropic only for ideal machines. Also in this case the adiabatic equations have already been widely discussed in Sects. 1.4–1.5.

The third scalar equation is the *mass conservation law*:

$$\dot{m} = \frac{dm}{dt} = const \tag{1.56}$$

where $\dot{m}$ indicates the mass flow rate. In the following, for short hand notation we will refer to $\dot{m}$ with m.

Indicating with 1 and 2 the inlet and outlet sections of the generic duct (Fig. 1.10), under the two hypotheses of steady and one-dimensional flow, the mass conservation law is expressed simply as:

$$m = \rho \cdot c \cdot A = \rho_1 \cdot c_1 \cdot A_1 = \rho_2 \cdot c_2 \cdot A_2 = const$$

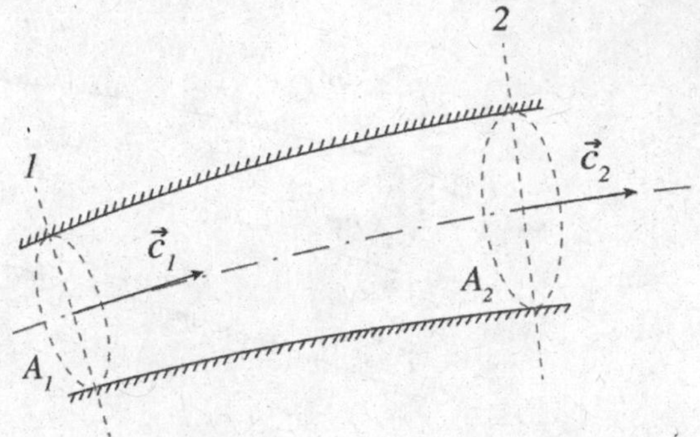

Fig. 1.10 One-dimensional flow in a duct

where A is the cross-section area of the duct, and c the velocity component perpendicular to section A.

In differential form, the mass conservation law (1.56) is expressed as:

$$\frac{dc}{c} + \frac{d\rho}{\rho} + \frac{dA}{A} = 0 \tag{1.57}$$

The volumetric flow rate is expressed by:

$$V = \frac{m}{\rho} = c \cdot A \tag{1.58}$$

and, unlike the mass flow rate, it generally varies also for a steady flow following the density variation of the fluid (compressible fluids), with the exception of incompressible fluids for which the density can be considered constant along the process.

The fourth vector equation is the *momentum equation*. The result of the application of this equation will be obtained also starting from the energy equation, as proof of the existence of a single general principle which governs all the thermo-mechanical phenomena, expressed in different forms.

Newton's second law states that:

$$\vec{F} = \frac{d\vec{q}}{dt} \tag{1.59}$$

i.e. the sum of the forces exerted on the control volume of a system is equal to the change of the system momentum per unit time.

Considering the duct shown in Fig. 1.11, the change of the system momentum per unit time is, under the usual hypotheses of steady and one-dimensional flow:

$$\frac{d\vec{q}}{dt} = \vec{c}_2 \cdot \frac{dm_2}{dt} - \vec{c}_1 \cdot \frac{dm_1}{dt} = \frac{dm}{dt} \cdot (\vec{c}_2 - \vec{c}_1) = m \cdot (\vec{c}_2 - \vec{c}_1) \tag{1.60}$$

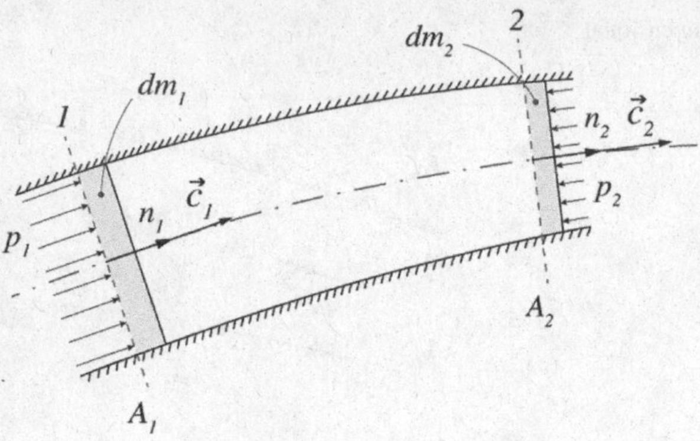

Fig. 1.11 Generic duct for momentum balance

The sum of forces exerted on the system control volume is composed of:

- the resultant force, $\vec{F}_W$, exerted by the duct walls on the fluid as reaction to the pressure and friction forces transmitted by the fluid;
- the resultant pressure force transmitted to the system through the inlet A_1 and outlet A_2 sections by the contiguous fluid. Indicating with $\vec{n}$ the versor normal to these sections and directed as the flow, it is $p_1 \cdot A_1 \cdot \overrightarrow{n_1} - p_2 \cdot A_2 \cdot \overrightarrow{n_2}$;
- the resultant friction force transmitted to the system through the inlet A_1 and outlet A_2 sections by the contiguous fluid. Under the hypothesis of one-dimensional flow, these forces are null because there is no velocity tangential components in these sections;
- the resultant gravity forces, $\vec{G}$, applied to the fluid mass in the duct; generally, in turbomachinery, this force can be neglected in comparison with the other forces. In other cases, it can, however, be easily taken into account.

Hence, we can write Eq. (1.59), by using (1.60), as:

$$\overrightarrow{F_w} + p_1 \cdot A_1 \cdot \overrightarrow{n_1} - p_2 \cdot A_2 \cdot \overrightarrow{n_2} + \vec{G} = m \cdot \left(\overrightarrow{c_2} - \overrightarrow{c_1}\right) \tag{1.61}$$

This Eq. (1.61) gains considerable importance for calculating the force exerted by the duct walls on the fluid flow.

If this equation is applied to stator channels, the work transfer is zero and this equation allows to evaluate the reactions exerted on the duct constraints.

1.6.1 The Euler Equation from the Mechanics Equation

In turbomachinery the work transfer takes place in rotor channels. As these channels rotate around the axis of the machine, it is more convenient to apply the equation of mechanics in terms of the variation of moment of momentum.

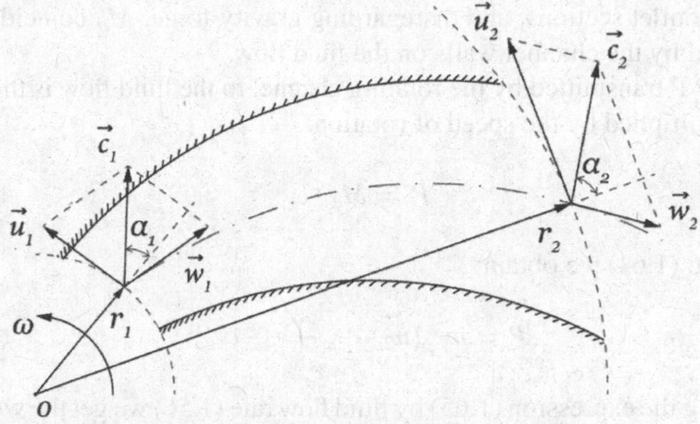

Fig. 1.12 Rotating channel for angular momentum balance

Considering a generic channel rotating respect to the point O (Fig. 1.12), the second law of dynamics expressed in terms of the *moment of momentum* is:

$$\vec{M} = \frac{d\vec{b}}{dt} \tag{1.62}$$

where $\vec{M}$ is the torque, with respect to the rotation center O, of the forces applied to the system and $\vec{b}$ is the moment of system momentum with respect to the center O, also called *angular momentum.*

Always under the usual hypotheses of steady and one-dimensional flow, the variation of the angular momentum per unit time is expressed as:

$$\frac{d\vec{b}}{dt} = \vec{r}_2 \times \vec{c}_2 \cdot \frac{dm_2}{dt} - \vec{r}_1 \times \vec{c}_1 \cdot \frac{dm_1}{dt} = m \cdot (\vec{r}_2 \times \vec{c}_2 - \vec{r}_1 \times \vec{c}_1) \tag{1.63}$$

Projecting this vector Eq. (1.63) onto the machine rotation axis, we have the value of the axial torque M_a (a scalar value):

$$M_a = m \cdot (r_2 \cdot c_{2u} - r_1 \cdot c_{1u}) = m \cdot (r_2 \cdot c_2 \cdot sin\alpha_2 - r_1 \cdot c_1 \cdot sin\alpha_1) \tag{1.64}$$

where c_u indicates the component of the fluid absolute velocity on the *blade speed* $(u = \omega \cdot r)$ vector; its direction is perpendicular to the distance, r, of the section from the rotation axis. This velocity component is called *tangential component.*

Returning now to the expression of the axial torque M_a, in this case, the moment of the pressure forces about the machine axis is necessarily zero since the inlet and outlet sections of the channel are surfaces of revolution around this axis. So, under the hypothesis of one-dimensional flow, which allows to neglect viscous forces on the inlet and outlet sections, and disregarding gravity force, M_a coincides with the torque exerted by the channel walls on the fluid flow.

The power P transmitted by the rotating channel to the fluid flow is therefore this torque M_a multiplied by the speed of rotation:

$$P = M_a \cdot \omega$$

Then, from (1.64) we obtain:

$$P = m \cdot (u_2 \cdot c_{2u} - u_1 \cdot c_{1u}) \tag{1.65}$$

By dividing the expression (1.65) by fluid flow rate (1.56) we get the work transfer per mass unit:

$$W = \frac{P}{m}$$

If we consider a driven turbomachine, we have:

$$W = u_2 \cdot c_{2u} - u_1 \cdot c_{1u} \tag{1.66}$$

This Eq. (1.66), also called *Euler's equation*, is a fundamental equation for the study of turbomachinery, even if some simplifying hypotheses were made in obtaining it.

This equation clearly shows how work is related to the tangential component of the absolute velocity (c_u) in the inlet and outlet channel sections and to the blade speed (u). The absolute velocity can, therefore, be subdivided into two components: one, c_u, tangential to the sections and responsible for work transfer, and the other one, c_m, perpendicular to these sections (called the *meridional component* of the velocity, which is parallel to the rotation axis in axial machines and radial in radial ones) directly *related to fluid flow rate* handled by the machine.

Equation (1.66) was obtained by assuming positive forces and moments if exerted by the channel walls on the fluid flow. This equation, therefore, gives positive work for driven turbomachinery. To have positive work also in driving machines, it is necessary to change the sign of the second member terms in (1.66), obtaining the following equation:

$$W = u_1 \cdot c_{1u} - u_2 \cdot c_{2u} \tag{1.67}$$

Many times it is useful to study the fluid flow in a rotating frame of reference, placed on the machine rotor. Hence, an observer positioned on the rotor sees the fluid moving with relative velocity, $\vec{w}$, tangent to the channel axis under the hypothesis of one-dimensional flow. In a fixed channel, instead, the absolute velocity, $\vec{c}$, is tangent to this channel axis.

These three velocities ($\vec{w}$, $\vec{c}$ and blade speed $\vec{u}$) are linked by the following vector equation:

$$\vec{c} = \vec{w} + \vec{u}$$

This means that these three velocity vectors form a triangle, called *velocity triangle* (Fig. 1.13); these triangles, at inlet and outlet sections of the rotor, are fundamental for the study of turbomachinery.

By applying Carnot's theorem to the generic velocity triangle of Fig. 1.13, we obtain:

$$w^2 = u^2 + c^2 - 2 \cdot u \cdot c \cdot sin\alpha$$

where α is the angle between the absolute velocity, $\vec{c}$, and the perpendicular to the blade speed, $\vec{u}$. From this equation we obtain:

$$u \cdot c \cdot sin\alpha = u \cdot c_u = \frac{1}{2} \cdot \left(u^2 + c^2 - w^2\right)$$

The Euler's equation, (1.66) and (1.67), can, therefore, be expressed, with reference to the rotor inlet and outlet velocities, as:

$$\text{driven turbomachines} \quad W = \frac{c_2^2 - c_1^2}{2} + \frac{u_2^2 - u_1^2}{2} + \frac{w_1^2 - w_2^2}{2} \tag{1.68a}$$

$$\text{driving turbomachines} \quad W = \frac{c_1^2 - c_2^2}{2} + \frac{u_1^2 - u_2^2}{2} + \frac{w_2^2 - w_1^2}{2} \tag{1.68b}$$

In differential form, Eqs. (1.68a) and (1.68b) become:

$$dW = cdc + udu - wdw \tag{1.69}$$

Fig. 1.13 Velocity triangle

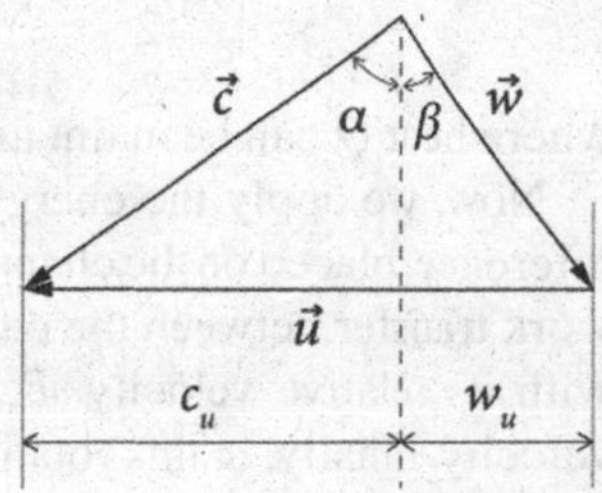

To deepen the meaning of energy contributions in (1.69), we can use the energy equation in mechanical form (Eq. (1.5)):

$$cdc + gdz + vdp = dW - dE_{diss}$$

which shows that work transfer takes place by variation of kinetic and pressure energy of the fluid and, even if it is generally negligible, by variation of fluid potential energy. While the term related to the variation of kinetic energy is straightforward:

$$dE_c = cdc \tag{1.70}$$

the combination of the Euler's Eq. (1.69) with the energy equation in mechanical form (1.5), assuming the potential energy variation to be negligible, gives on explicit correlation among pressure energy of the fluid, its relative velocity, the blade speed and dissipations. Indeed we have:

$$dE_p = vdp = udu - wdw - dE_{diss} \tag{1.71}$$

This Eq. (1.71) links the fluid pressure change to the centrifugal forces' work and to the fluid relative velocity change in the rotating channels.

1.6.2 The Euler Equation from the Energy Equation

As proof of the uniqueness of the general principles of mechanics, now the Euler's equation will be directly obtained by applying the energy equation.

For this purpose, let us consider the energy equation in thermodynamic form (1.1):

$$cdc + gdz + dh = dQ + dW$$

As previously noted, integration of this equation, for any type of fluid, depends only on the fluid conditions at the inlet (1) and outlet (2) stations of the control surface. So, by integrating (1.1) between the inlet and outlet sections of the rotating channel (Fig. 1.12), we obtain Eq. (1.6):

$$\frac{c_2^2 - c_1^2}{2} + g \cdot (z_2 - z_1) + (h_2 - h_1) = Q + W$$

where heat Q can be maintained even if the turbomachines are considered adiabatic.

Now, we apply the energy Eq. (1.6) to the rotor channel in a rotating frame of reference, placed on the channel. In this rotating frame, since the channel is stationary, work transfer between the channel and the fluid is zero, $W = 0$, and the fluid moves with its relative velocity $\vec{w}$, so kinetic energy must be expressed in terms of this velocity. Finally, in this rotating frame, the energy contribution due to the centrifugal forces (ΔE_ω) appears. In analogy with the gravitational field, in the centrifugal force field, the energy variation of the fluid element is related to the work necessary to move it from one equipotential surface to another, or simply to vary its distance from the rotation center. Considering work positive if done on fluid, we obtain:

$$\Delta E_\omega = -\int_1^2 \omega^2 \cdot r \cdot dr = -\frac{1}{2}\omega^2\left(r_2^2 - r_1^2\right)$$

and therefore:

$$\Delta E_\omega = -\frac{u_2^2 - u_1^2}{2} \tag{1.72}$$

The last Eq. (1.72) demonstrates that if the inlet and outlet sections of turbomachine channels are placed at different distances from the rotation axis (radial machines), the centrifugal forces contribute to work transfer. In axial machines, instead, as $r_1 = r_2$ and therefore $u_1 = u_2$, this contribution does not appear.

On the basis of these considerations, the energy Eq. (1.6), applied to a rotating channel where the contribution (1.72) occurs, provides:

$$\frac{w_2^2 - w_1^2}{2} + g \cdot (z_2 - z_1) + (h_2 - h_1) - \frac{u_2^2 - u_1^2}{2} = Q$$

Then, by subtracting the last equation from (1.6), we obtain:

$$\frac{c_2^2 - c_1^2}{2} + \frac{u_2^2 - u_1^2}{2} + \frac{w_1^2 - w_2^2}{2} = W$$

The last equation is precisely the Euler's equation, Eq. (1.68a–1.68b), and can be also rearranged by using the tangential components of the fluid absolute velocity and the blade speed to obtain Eqs. (1.66) and (1.67).

1.7 Working Principles of Turbomachinery

Now, the basic elements of thermodynamics and fluid-dynamics so far exposed will be applied to turbomachinery channels in order to highlight their working principles.

We will mainly refer to thermal turbomachines (handling compressible fluids) considering that the basic working principles are the same as hydraulic turbomachinery and also that the discussion extension to the latter is a simpler particular case.

1.7.1 General Information on Turbomachinery Stage

The turbomachine channels are generally formed by two adjacent blades; there are rows of stationary blades (stators) and rows of rotating blades (rotors). These channels can be nozzles (generally convergent channels or convergent-divergent channels in case of supersonic flows), diffusers (divergent channels), or deviators (channels with approximately constant sections) where there is no fluid pressure change.

A row of stator channels and a row of rotor channels together constitute the *stage* of a turbomachine. Depending on work transfer, turbomachines can be single-stage (for example all hydraulic turbines, most of the radial turbines, some radial pumps, and compressors), or multistage (generally steam and gas axial turbines, axial compressors, some pumps and radial compressors).

In order to study the fluid flow in turbomachine channels, we will refer to two main classes: axial and radial stages.

For axial stages, where stator and rotor channels, and fluid flow, are approximately parallel to the rotation axis, fluid flow is studied on a cylindrical meridional section of blades developed in a plane (see the example of axial stage shown in Fig. 1.14).

For radial stages, on the other hand, where the stator and rotor channels, and fluid flow, are radial, fluid flow is studied on two stage sections: one perpendicular and one containing the rotation axis (see the example of a radial stage shown in Fig. 1.15).

As already mentioned (Sect. 1.6), axial and radial turbomachines can be driving (they produce power by expanding fluid to a lower pressure) or driven turbomachines (they absorb power to increase the fluid pressure). Driving turbomachines are called turbines (hydraulic and thermal turbines handling respectively incompressible and compressible fluids) while driven turbomachines are called pumps (handling incompressible fluids) or compressors (handling incompressible fluids).

The energy equation, both in the thermodynamic (1.1) and mechanical (1.5) forms, will be applied between input and output sections of the stage, of the rotor and the stator; the variation of potential energy will be always considered negligible.

The hypothesis of one-dimensional flow cannot be applied to the blades, and therefore channels, of considerable height. In this case, it is necessary to study the fluid flow in different sections along the blade (for example, for rotor blades of axial stages, we will consider the flow near the hub, at the mean radius and near the casing). This procedure will be described in Sect. 1.7.4 and in the Chapters dedicated to the turbomachinery stage design (Chaps. 3–7).

1.7.2 Stator

Stator channels can be divided into two classes: nozzles, which accelerate fluid flow by transforming pressure energy into kinetic energy, and diffusers, which decelerate fluid by transforming kinetic energy in pressure energy.

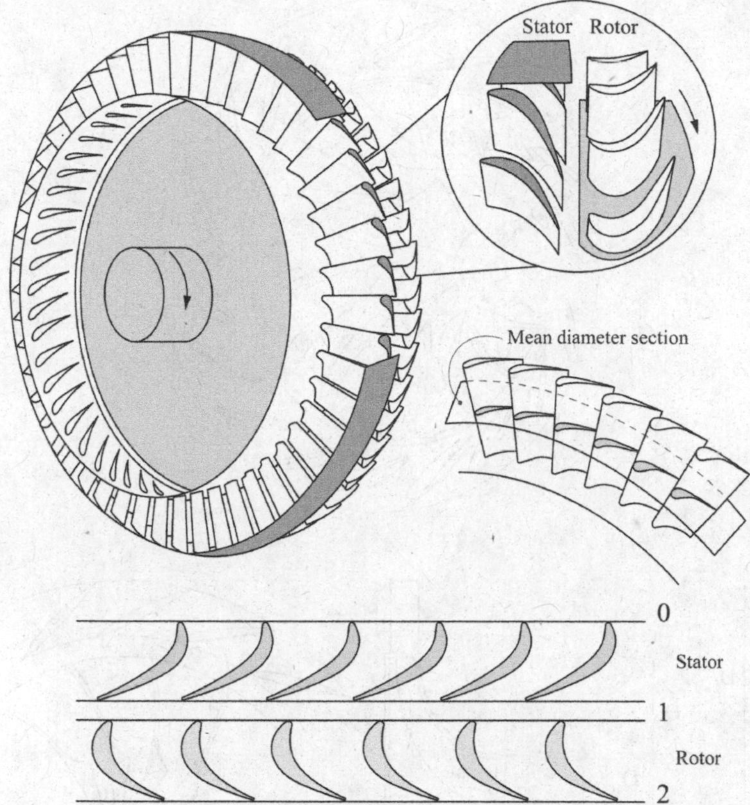

Fig. 1.14 Axial turbine: stator and rotor representation

Their purpose in turbomachinery is illustrated by Euler's Eqs. (1.66)–(1.71):

- in driving turbomachines work transfer (Eq. (1.67)) requires a large tangential component (c_{1u}) of the absolute fluid velocity at the rotor inlet, that is high kinetic energy (Eq. (1.68b)), and therefore the stator is placed upstream of the rotor and must produce this high tangential component of the fluid velocity by exploiting its pressure energy, which is transformed into kinetic energy (nozzles in driving turbomachines);
- in driven turbomachines work done on fluid (Eq. (1.66)) produces, at the rotor exit, a tangential component of the absolute fluid velocity (c_{2u}) greater than that at the rotor inlet (c_{1u}) and increases fluid kinetic energy (Eq. (1.68a)). If the machine must provide energy to the fluid in the form of pressure energy (pumps and compressors), it is necessary to reduce this velocity downstream of the rotor

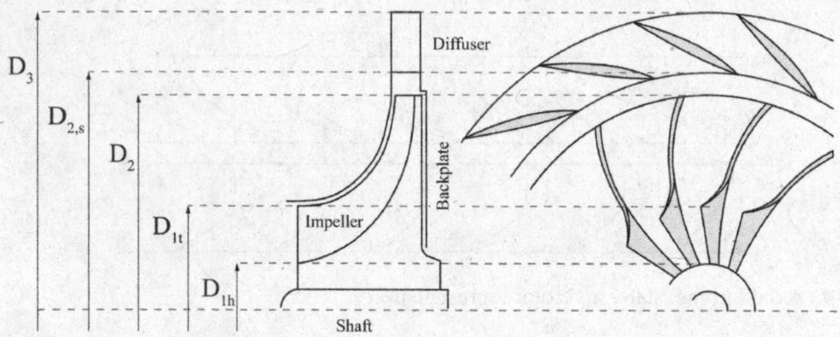

Fig. 1.15 Radial compressor: stator and rotor representation

by transforming part of this kinetic energy into pressure energy. In this case, stator channels are diffusers, arranged downstream of the rotor.

The equations illustrated in Sect. 1.6 allow studying the fluid flow in the stator. We must determine the channel configuration to obtain the desired fluid state in each section.

For incompressible fluids, a combination of the equation of state (ρ = cost, Eq. (1.14)) and the continuity Eq. (1.57) establishes a unique correlation between section and velocity:

$$\text{incompressible fluids} \quad \frac{dc}{c} + \frac{dA}{A} = 0 \quad \Rightarrow \quad \frac{dc}{c} = -\frac{dA}{A}$$

This equation demonstrates that if a channel has decreasing sections along the direction of fluid flow (converging channel, Fig. 1.16a), liquid progressively increases its velocity; vice versa in a channel with increasing sections (divergent channel, Fig. 1.16b). The energy equation in the mechanical form (1.7) highlights that the kinetic energy variation, taking into account dissipations, is opposite to the pressure energy variation and that the fluid mechanical energy decreases because of dissipations:

$$\frac{c_2^2 - c_1^2}{2} + \frac{p_2 - p_1}{\rho} = -E_{diss}$$

and in terms of total head (Sect. 1.5):

$$H_2 = H_1 - \frac{E_{diss}}{g}$$

The study of compressible fluid flow is more complex since the continuity Eq. (1.57) relates fluid velocity not only to channel section but also to fluid density:

$$\text{compressible fluids} \quad \frac{dc}{c} + \frac{dA}{A} + \frac{d\rho}{\rho} = 0 \quad \Rightarrow \quad \frac{dc}{c} = -\frac{dA}{A} - \frac{d\rho}{\rho}$$

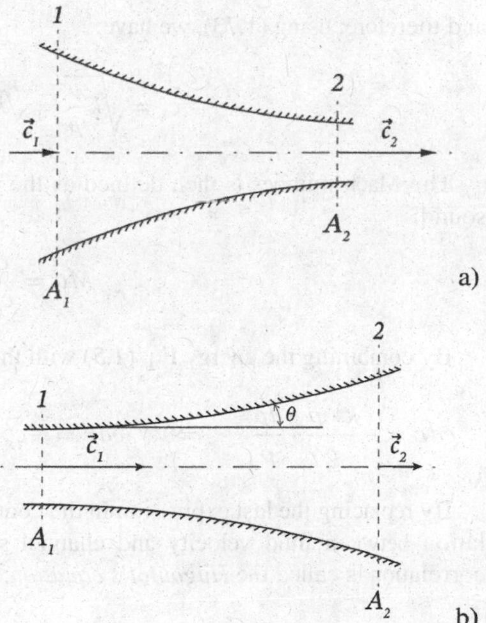

Fig. 1.16 Nozzle (**a**) and diffuser (**b**)

For this reason, it is necessary to evaluate the influence of density variations to establish a velocity-section correlation.

Let us consider an isentropic flow of a perfect gas in a stationary channel (or in any case in a rotating frame placed on the channel).

In this case, we have the following three cardinal Eqs. (1.5), (1.8) and (1.13):

$$cdc + \frac{dp}{\rho} = 0$$

$$\frac{p}{\rho} = R_g \cdot T \quad \Rightarrow \quad dp = R_g \cdot \rho \cdot dT + R_g \cdot T \cdot d\rho$$

$$\frac{p}{\rho^k} = const \quad \Rightarrow \quad \frac{dp}{\rho^k} - kp\frac{d\rho}{\rho^{k+1}} = 0 \quad \Rightarrow \quad \frac{d\rho}{\rho} = \frac{1}{k}\frac{dp}{p}$$

It is now necessary to make some considerations on the meaning of the term $dp/d\rho$, which, according to the previous equations, can be expressed as:

$$\frac{dp}{d\rho} = \frac{k \cdot p}{\rho} = k \cdot R_g \cdot T \tag{1.73}$$

Let us introduce the speed of sound, c_s, which represents the speed of small pressure perturbations propagating in a fluid:

$$c_s = \sqrt{\left(\frac{\partial p}{\partial \rho}\right)_{s=const}}$$

and therefore, using (1.73), we have:

$$c_s = \sqrt{k\frac{p}{\rho}} = \sqrt{k \cdot R_g \cdot T} \tag{1.74}$$

The Mach number is then defined as the ratio of fluid velocity to the speed of sound:

$$Ma = \frac{c}{c_s} \tag{1.75}$$

By combining the energy Eq. (1.5) with the process Eq. (1.13), we obtain:

$$cdc = -\frac{k \cdot p}{\rho} \cdot \frac{d\rho}{\rho} \quad \Rightarrow \quad cdc = -c_s^2 \cdot \frac{d\rho}{\rho} \quad \Rightarrow \quad \frac{d\rho}{\rho} = -Ma^2 \cdot \frac{dc}{c}$$

By replacing the last expression in the continuity Eq. (1.57), we obtain the correlation between fluid velocity and channel section for a compressible fluid; this correlation is called the *Hugoniot's equation*:

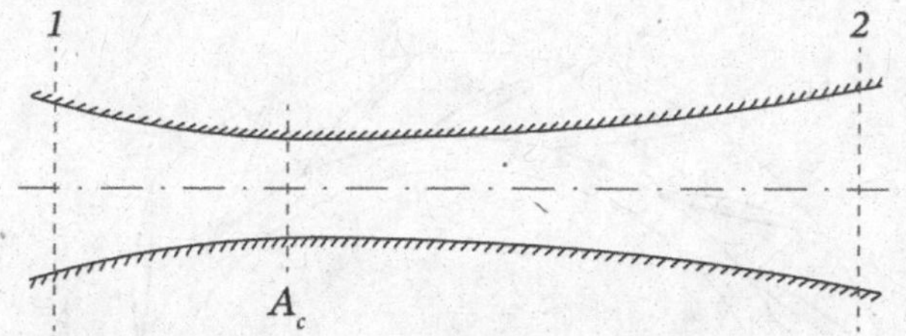

Fig. 1.17 Convergent-divergent nozzle

$$\frac{dc}{c} \cdot \left(1 - Ma^2\right) = -\frac{dA}{A} \tag{1.76}$$

This equation is of fundamental importance for discussing compressible fluid flow within channels.

In this regard, it is necessary to distinguish two cases: Ma < 1 and Ma > 1.

When Ma < 1 (subsonic fluid flow), the situation is similar to that of incompressible fluids: a channel with decreasing section along the fluid flow produces a fluid acceleration (expansion), and therefore the nozzle is a simply converging channel; conversely, a channel with increasing section along the fluid flow produces a fluid deceleration (compression), and therefore the diffuser is a divergent channel.

When Ma > 1 (supersonic fluid flow), the situation is completely reversed: in order to further accelerate the fluid (expansion), it is necessary to increase the passage areas (divergent channel).

This means that if we want to accelerate a fluid up to supersonic velocities, starting from subsonic ones, we need to have first a convergent channel to get the speed of sound (Ma = 1) and then a divergent channel to continue fluid acceleration up to supersonic velocity (Ma > 1); in other words, a convergent-divergent nozzle is necessary (Fig. 1.17) where in the minimum flow area section A_c (throat section) the sonic conditions are established (Ma = 1, $c = c_s$).

Another consequence is that in a simply convergent nozzle a fluid can be accelerated to the local velocity of sound at the most.

Fig. 1.18 Nozzles of axial turbines

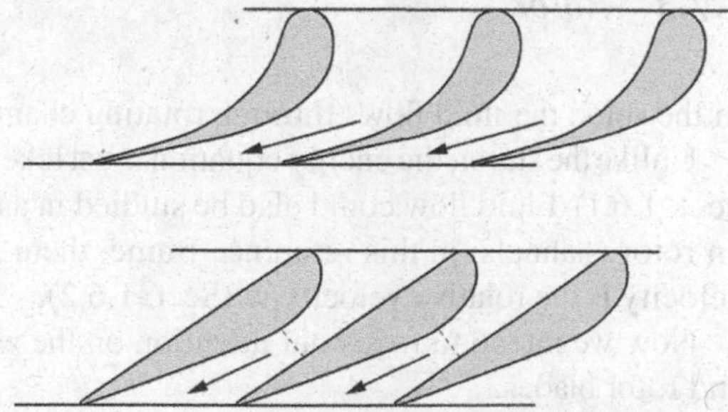

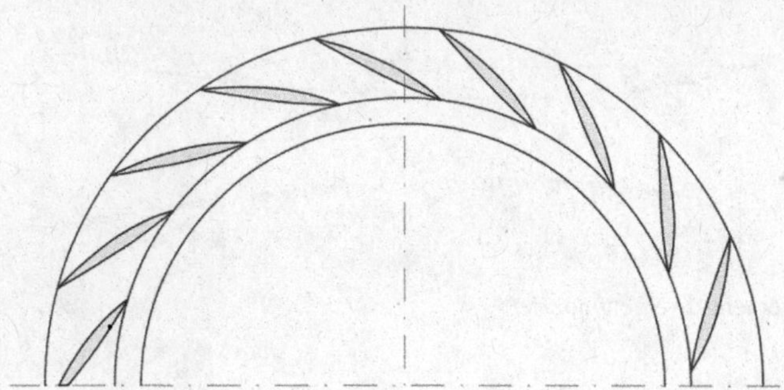

Fig. 1.19 Diffuser of a radial driven machine

On the basis of what has been illustrated, the stator of a turbine stage, which is placed upstream of the rotor, is composed of convergent nozzles, or convergent-divergent nozzles in case of fluid accelerations up to supersonic speeds (Fig. 1.18). In any case, in driving turbomachines, the stator channel axis has a strong inclination precisely to produce the desired tangential component of the fluid velocity (c_{1u}) at rotor inlet, maintaining a meridional component of the fluid velocity (perpendicular to passage areas, and therefore axial in axial machines and radial in radial machines) compatible with the fluid flow rate handled by the machine.

In driven turbomachine (compressors and pumps) stages, the stator is placed downstream of the rotor and it is composed of diffusers (divergent channels) that perform fluid compression while slowing it down. In these channels, it is necessary to keep in mind the limitations on the opening angle of the diffuser (Fig. 1.16): if it is too high, it can cause a "detachment" of the fluid from the channel walls (channel walls are not "wetted" by fluid), energy dissipations occur due to vortex creation and no compression is carried out anymore. By limiting this angle to about 10°, the diffuser becomes long and energy dissipations increase.

In Fig. 1.19 a stator of a radial driven stage is shown.

1.7.3 Rotor

In the rotor, the fluid flows through rotating channels, formed by adjacent blades.

Unlike the stator, the energy equation must take work into account (Euler equation, Sect. 1.6.1). Fluid flow could also be studied in a rotating frame of reference, placed on rotor channels: in this reference frame, there is no work exchange and the fluid velocity is the relative velocity w (Sect. 1.6.2).

Now we intend to focus our attention on the work exchanged between fluid flow and rotor blades.

As evident from the correlations deduced in Sect. 1.6.1 (Euler's equation), it is essential to know the velocity triangles at the rotor inlet and outlet. Figures 1.20 and

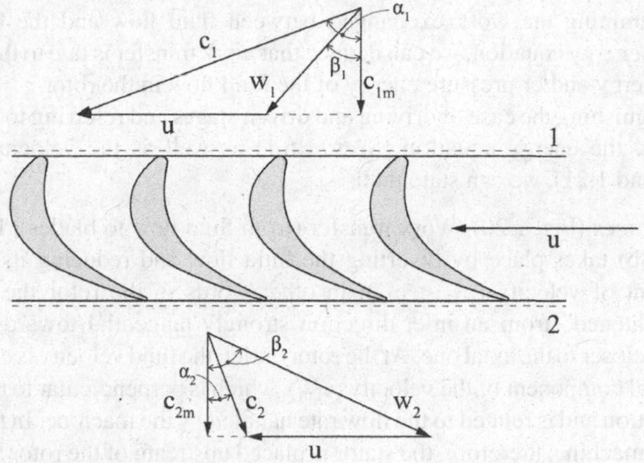

Fig. 1.20 Velocity triangle in an axial turbine rotor

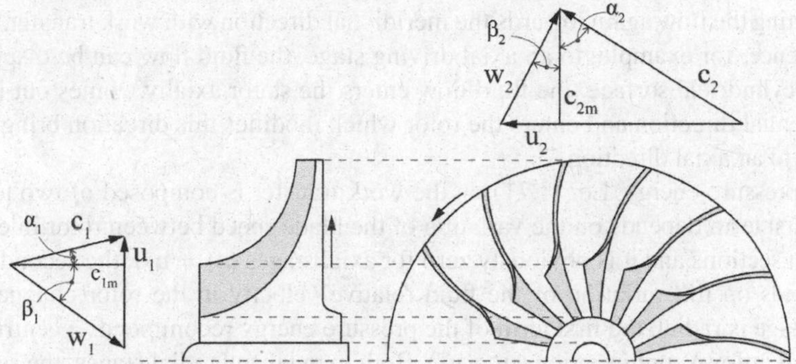

Fig. 1.21 Velocity triangle in a radial compressor rotor

1.21 show two examples of velocity triangles: the first one is referred to the rotor of an axial driving stage and the second one to the rotor of a radial driven machine. In these Figs. 1.1 and 1.2 respectively indicate the section just upstream and downstream of the rotor.

The blade profile influences the direction of the fluid relative velocity: assuming the absence of an angle of incidence at rotor inlet, here the blade profile is tangent to the direction of w_1 (this velocity is the vector difference of fluid absolute velocity c_1 and blade velocity u_1) and, assuming perfectly guided fluid flow, the direction of w_2 is parallel to the blade exit profile.

This hypothesis of perfectly guided fluid flow can be conceived only by assuming an infinite number of blades and therefore channels of infinitesimal passage area. In reality, with a finite number of blades, the relative velocity direction of the fluid flow has a deviation with respect to the direction of the trailing edge of the blade (Chaps. 3–7).

Now, examining the work exchanged between fluid flow and the blades and applying the energy equation, we can deduce that work transfer is due to the variation of kinetic energy and/or pressure energy of the fluid flow in the rotor.

By distinguishing the case of driving and driven stages and referring to the Euler's equation and the energy equation (Sect. 1.6.1) as well as the velocity triangles (Figs. 1.20 and 1.21), we can state that:

- *driving stages* (Fig. 1.20). Work transfer (from fluid flow to blades), Eqs. (1.67) and (1.68b) takes place by diverting the fluid flow and reducing its tangential component of velocity ($c_{2u} < c_{1u}$); in other words in the rotor the fluid flow is "straightened" from an inlet direction strongly tangential towards an outlet direction closer to the axial one. At the rotor outlet, the fluid velocity is closer to the meridional component of the velocity (c_{2m}), which is perpendicular to the channel outlet section and is related to the flow rate handled by the machine. In the stage of a driving machine, therefore, the stator is placed upstream of the rotor and has the task of diverting the fluid from a direction close to the meridional direction to a strong tangential direction, thanks to the fluid acceleration, Eq. (1.70), produced in stator nozzles; the rotor "reabsorbs" the tangential component of fluid velocity by diverting the flow again towards the meridional direction with work transfer. With reference, for example, to an axial driving stage, the fluid flow can be displayed on a cylindrical surface: the fluid flow enters the stator axially, comes out in the tangential direction and enters the rotor which modifies this direction bringing it back to an axial direction.
  The pressure energy, Eq. (1.71), in the work transfer is composed of two terms: the first term depends on the variation of the blade speed between rotor inlet and outlet sections and it is obviously zero for axial stages ($u_1 = u_2$); the second term depends on the variation of the fluid relative velocity in the rotor channels. If the stage is radial, the first term of the pressure energy recommends a centripetal (radial inflow) configuration ($u_1 > u_2$). Both in axial and radial stages, the second term suggests a further fluid expansion ($w_2 > w_1$) in the rotor.

- *driven stages* (Fig. 1.21). Briefly, the first blade row of the stage is the rotor and the work, Eqs. (1.66) and (1.68a) is done on the fluid; it is diverted in a tangential direction ($c_{2u} > c_{1u}$) and therefore acquires kinetic energy ($c_2 > c_1$), Eq. (1.70), and pressure energy, Eq. (1.71), thanks to the deceleration in the relative flow ($w_2 < w_1$) and, where possible (radial stages), to the action of centrifugal forces ($u_2 > u_1$). Precisely in this regard, unlike driving stages, a radial driven stage has a centrifugal configuration.
  Then, the fluid flow enters the stator where it undergoes a deviation in the meridional direction; stator reduces the tangential component of the fluid velocity and part of the kinetic energy, conferred by the rotor, is recovered in pressure energy. Even in the case of driven axial stages, the fluid flow can be displayed on a cylindrical surface, but in this case, the deviation from axial to tangential direction is given by the rotor and the subsequent "straightening" of the fluid flow, again in the axial direction, is accomplished by the stator.

1.7.4 The Radial Equilibrium. The Free Vortex Law

So far the fluid flow has been studied on planes intersecting the blades (planes on which the velocity triangles are drawn) through the two velocity components: the meridional velocity, c_m, correlated to the mass flow rate of the working fluid, and the tangential velocity, c_u, correlated to the work transfer. In fact, as illustrated in Sect. 1.7.1, in axial stages, fluid flow is studied on a cylindrical meridional section of blades developed in a plane while in radial stages fluid flow is studied on a stage section perpendicular to the axis of rotation.

In turbomachinery, the fluid flow is three-dimensional (Sect. 1.9) so it is necessary to extend the study in the direction perpendicular to the aforementioned planes, that is, along the blade span. By indicating with z this direction, there will therefore exist a third component of the velocity vector along this direction, c_z. In the study of turbomachinery (Dixon and Hall 2014), this velocity component is considered only within the blade channels because outside them it can be considered null. This means that in stations just upstream and downstream of the bladed channels, each element of fluid is in *radial equilibrium* in the z-direction, that is, *the centrifugal force, generated by the velocity component c_u, is balanced by the pressure forces.*

So, with reference to an axial stage, let us consider the fluid flow in a plane perpendicular to the axis of rotation in a section upstream or downstream of the blades (Fig. 1.22), that is, the fluid flow in the spaces between the blade rows. In this case, the z-direction is the radial direction and the element of fluid has a velocity c_u perpendicular to this direction (the velocity component c_m is perpendicular to this plane).

Referring to a unit depth, the pressure force exerted on this element of fluid is:

$$dF_p = (p + dp) \cdot (r + dr) \cdot d\theta - p \cdot r \cdot d\theta - 2 \cdot \left(p + \frac{dp}{2}\right) \cdot dr \cdot sin\frac{d\theta}{2}$$

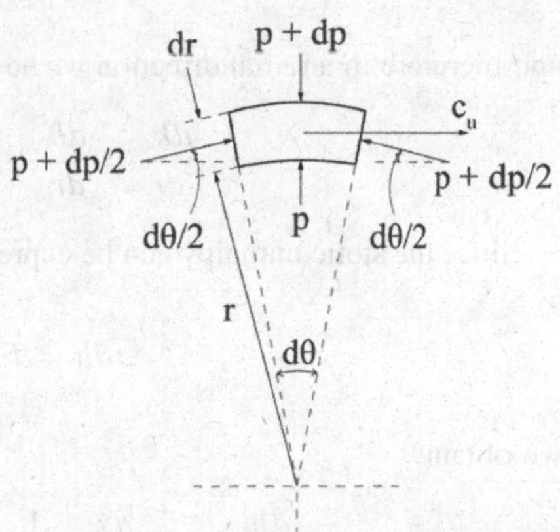

Fig. 1.22 Radial equilibrium of an element of fluid

Since for small values of $d\theta$, it is possible to assume:

$$sin\frac{d\theta}{2} \cong \frac{d\theta}{2}$$

ignoring terms of the second order of smallness, we obtain:

$$dF_p = r \cdot dp \cdot d\theta$$

This force is balanced by the centrifugal force:

$$dF_\omega = dm \cdot \frac{c_u^2}{r + \frac{dr}{2}} = \rho \cdot dV \cdot \frac{c_u^2}{r + \frac{dr}{2}}$$
$$= \rho \cdot \left(r + \frac{dr}{2}\right) \cdot d\theta \cdot dr \cdot \frac{c_u^2}{r + \frac{dr}{2}} = \rho \cdot c_u^2 \cdot d\theta \cdot dr$$

Balancing these two forces, *the equation of radial equilibrium* is obtained:

$$\frac{1}{\rho} \cdot \frac{dp}{dr} = \frac{c_u^2}{r} \tag{1.77}$$

The Eq. (1.77) demonstrates that a positive pressure gradient is generated radially by the centrifugal forces. Consequently, if in the fluid flow there is a tangential velocity component, the pressure increases from the hub to the casing.

The radial equilibrium allows also to derive the energy equation in the radial direction.

The total enthalpy (Sect. 1.8.1) is (considering $c_z = 0$):

$$h_t = h + \frac{c^2}{2} = h + \frac{c_u^2}{2} + \frac{c_m^2}{2}$$

and therefore in a radial direction we have:

$$\frac{dh_t}{dr} = \frac{dh}{dr} + c_u\frac{dc_u}{dr} + c_m\frac{dc_m}{dr}$$

Since the static enthalpy can be expressed, Eq. (1.4), through:

$$dh = T \cdot ds + \frac{dp}{\rho}$$

we obtain:

$$\frac{dh_t}{dr} = T \cdot \frac{ds}{dr} + \frac{1}{\rho} \cdot \frac{dp}{dr} + c_u\frac{dc_u}{dr} + c_m\frac{dc_m}{dr}$$

Therefore, by using the radial equilibrium Eq. (1.77), we can write the energy equation in the radial direction as:

$$\frac{dh_t}{dr} = T \cdot \frac{ds}{dr} + \frac{c_u^2}{r} + c_u \frac{dc_u}{dr} + c_m \frac{dc_m}{dr} = T \cdot \frac{ds}{dr} + \frac{c_u}{r} \cdot \frac{d}{dr}(r \cdot c_u) + c_m \frac{dc_m}{dr}$$

Generally, the total enthalpy can be assumed constant in the radial direction:

$$\frac{dh_t}{dr} = 0$$

as a consequence, work transfer is equal at all radii. Furthermore, if the fluid flow is far from the boundary layers of the wall, it is also reasonable to consider the losses constant in the radial direction. Consequently, we can assume that:

$$\frac{ds}{dr} = 0$$

By using these assumptions, the energy equation in the radial direction becomes:

$$\frac{c_u}{r} \cdot \frac{d}{dr}(r \cdot c_u) + c_m \frac{dc_m}{dr} = 0 \tag{1.78}$$

We can solve this Eq. (1.78) by assuming the meridional velocity component constant along the radius:

$$c_{m(r)} = const \quad \Rightarrow \quad \frac{dc_m}{dr} = 0$$

Under these assumptions, we obtain:

$$r \cdot c_{u(r)} = const \tag{1.79}$$

The Eq. (1.79) represents the *free vortex law*. This law establishes the angular momentum conservation in the fluid flow. By adopting this law, the velocity triangles, defined in the previous paragraphs at the mean diameter, can be calculated at any radius between the blade hub and tip (see Chaps. 3–7). This means that the blades of a turbomachine, especially when the blade height becomes significant, are twisted.

In some cases, the adoption of the free vortex law entails some drawbacks, such as for example an excessive twisting along the blade span, an excessive variation in the degree of reaction, etc. In these cases, other solutions of the vortex energy equation can be adopted (for example the adoption of the *forced vortex law*). In any case, in the following, we will generally refer to the free vortex law.

In conclusion, in this section we have studied the fluid flow outside the blade channel of turbomachinery. In fact, on the basis of what has been just illustrated, the free vortex law (1.79) can be applied in the spaces between the blade rows of axial turbomachinery (that is, in stations just upstream and downstream of each blade row), and in the vaneless spaces of radial machines (for example just upstream of the centrifugal compressor impeller and just downstream of the radial inflow turbine impeller). In these spaces, in fact, the fluid flow is as shown in Fig. 1.22. For other vaneless spaces, such as the space between the stator and rotor of a radial turbine, the vaneless diffuser of a compressor or a radial pump, the volute of the radial machines, where there is a radial velocity component, neglecting the friction, the angular momentum is constant; by combining the Eq. (1.79) with the continuity Eq. (1.57), the fluid flow can be completely calculated.

1.8 Analysis of Turbomachinery Stages

Now, stages of turbines and compressors, handling real compressible fluids, will be analyzed. We will consider adiabatic processes and negligible potential energy variations between stage inlet and outlet.

About velocity and velocity triangles, both in this Chapter and in the following ones, we will refer to sections just upstream and just downstream of the stage and its parts (stator and rotor). It is evident that just downstream and just upstream of these sections the fluid velocity undergoes alterations due to blade thickness (*blade blockage*). These alterations (contraction of flow area, accelerations, shocks) will be taken into account in the loss correlations (Chaps. 3–7) by considering blade thickness.

1.8.1 Total Quantities and Rothalpy

Preliminarily, let us introduce the total quantities and the rothalpy which are extremely effective in turbomachine stage analysis.

Total enthalpy is defined by:

$$dh_t = dh + cdc$$

It is the sum of the static enthalpy and the kinetic energy of the fluid flow. So, it is the fluid enthalpy when the flow is brought to rest isentropically:

$$h_t = h + \frac{c^2}{2} \tag{1.80}$$

In each thermodynamic state of the fluid, characterized by entropy, s, and total enthalpy, h_t, the other total quantities (real fluid) can be calculated using the state equations:

$$T_t = f_T(s, h_t)$$
$$p_t = f_p(s, h_t)$$
$$\rho_t = f_\rho(s, h_t)$$

Total pressure is defined by:

$$dp_t = dp + \rho c dc \tag{1.81}$$

The above equation ensures that, during isentropic flow (ds = 0) and without work transfer (dW = 0), we have the conservation both of total enthalpy ($dh_t = 0$) and total pressure ($dp_t = 0$). Indeed, from Eq. (1.4):

$$dh = T ds + v dp$$

by adding cdc to both sides and using (1.81), we obtain:

$$dh + cdc = T ds + v dp + cdc \quad \Rightarrow \quad dh_t = T ds + v dp_t$$

and therefore:

$$\text{if } \; ds = 0 \; \text{ and } \; dh_t = 0 \; \Rightarrow \; dp_t = 0$$

The following definition of total pressure is also found in the technical literature:

$$p_t^* = p + \frac{1}{2}\rho c^2$$

However, this definition ensures the above condition only for incompressible fluid flows (v = cost). In fact, for compressible fluid flows we have:

$$dp_t^* = dp + \rho c dc + \frac{1}{2}c^2 d\rho$$

By applying Eq. (1.4) we obtain:

$$dh_t = T ds + v dp_t^* - \frac{1}{2}c^2 \frac{d\rho}{\rho}$$

and therefore:

$$\text{if } \; ds = 0 \; \text{ and } \; dh_t = 0 \; \Rightarrow \; dp_t^* \neq 0$$

Similarly, the total relative quantities can be introduced; they are particularly effective in the study of rotor channels:

$$h_{tr} = h + \frac{w^2}{2} \tag{1.82}$$

$$T_{tr} = f_T(s, h_{tr})$$
$$p_{tr} = f_p(s, h_{tr})$$
$$\rho_{tr} = f_\rho(s, h_{tr})$$

Total relative pressure is defined by:

$$dp_{tr} = dp + \rho w dw \tag{1.83}$$

Another important quantity, very useful in the analysis of rotor channels (Sects. 1.8.2 and 1.8.3), is the *rothalpy* (Dixon and Hall 2014):

$$I = h + \frac{w^2}{2} - \frac{u^2}{2} \tag{1.84}$$

According to the total relative enthalpy definition (1.82), the rothalpy can also be expressed as:

$$I = h_{tr} - \frac{u^2}{2} \tag{1.85}$$

1.8.2 Turbine Stage

Figure 1.23 sketches a stage expansion in the h–s diagram. Sections 0, 1 and 2 indicate respectively the fluid conditions just upstream of the stator, just downstream of the stator (that is just upstream of the rotor inlet) and just downstream of the rotor.

By applying the energy Eq. (1.1) between sections 0-2:

$$cdc + dh = dW$$

we obtain the expression of isentropic (or ideal) work (ideal process) and adiabatic (or actual) work (real process):

$$W_{is} = \frac{c_0^2}{2} + h_0 - \frac{c_{2is}^2}{2} - h_{2is} = h_{0t} - h_{2t,is} = \Delta h_{t,is} \tag{1.86}$$

$$W = \frac{c_0^2}{2} + h_0 - \frac{c_2^2}{2} - h_2 = h_{0t} - h_{2t} = \Delta h_{t,ad} \tag{1.87}$$

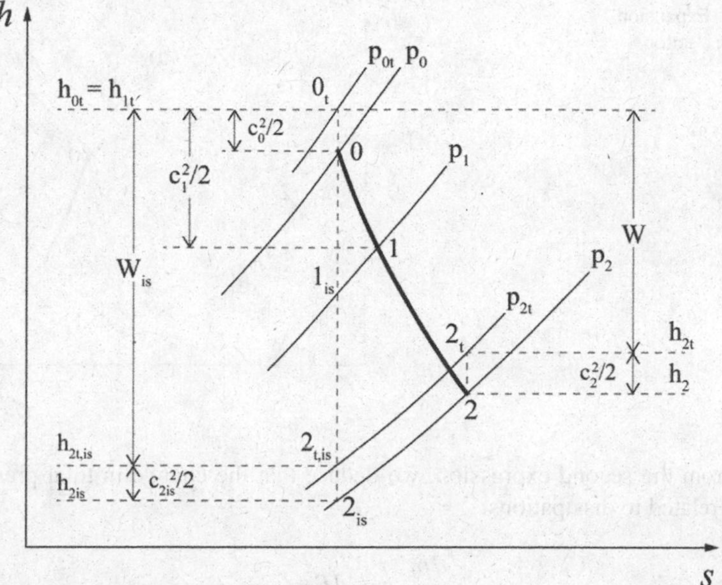

Fig. 1.23 Expansion process in a turbine stage

When, as generally happens, the absolute velocity of the fluid at stage inlet and outlet can be considered approximately equal:

$$c_0 \cong c_{2is} \cong c_2$$

isentropic (1.86) and adiabatic (1.87) works become:

$$W_{is} = h_0 - h_{2is} = \Delta h_{is}$$
$$W = h_0 - h_2 = \Delta h_{ad}$$

Stator (Sects. 0–1)

As there is no work transfer, the energy equation in the thermodynamic form (1.1) involves:

$$cdc + dh = 0$$

and in the mechanical form (1.5):

$$cdc + vdp = -dE_{diss}$$

From the first expression, we deduce the total enthalpy conservation in the stator:

$$dh_t = 0 \tag{1.88}$$

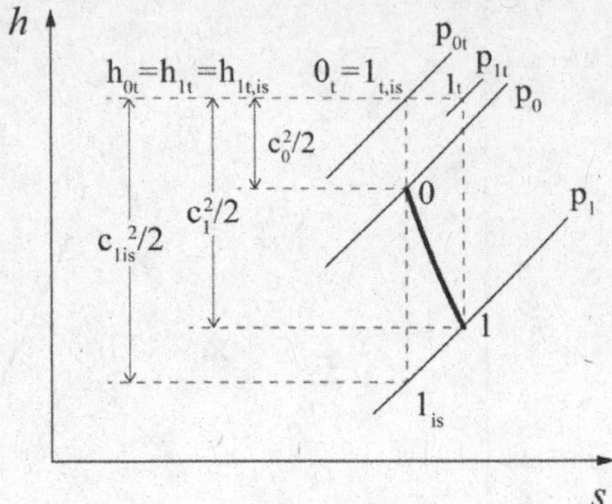

Fig. 1.24 Expansion process in a stator

while, from the second expression, we deduce that the change in total pressure is directly related to dissipations:

$$\frac{dp_t}{\rho} = -dE_{diss} \tag{1.89}$$

By using the total enthalpy conservation (1.88) and being the entropy generation due to irreversibility (adiabatic process), Eq. (1.3), we obtain:

$$\frac{dp_t}{\rho} = -Tds_{irr}$$

Figure 1.24 sketches an expansion in a stator nozzle in the h-s diagram.

Total enthalpy conservation (1.88) allows to express the fluid velocity at nozzle exit:

$$h_{0t} = h_{1t} = h_{1t,is}$$

$$c_{1is} = \sqrt{2 \cdot \Delta h_{S,is} + c_0^2}$$

$$c_1 = \sqrt{2 \cdot \Delta h_{S,ad} + c_0^2}$$

In order to take into account the effects of irreversibility (frictions in the fluid flow and between fluid and channel walls), a *velocity reduction coefficient* is usually used and it is defined as:

$$\upsilon_N = \frac{c_1}{c_{1is}}$$

The *velocity reduction coefficient* depends on the shape and roughness of the channel walls. For convergent nozzles, this coefficient assumes very high values (0.96–0.99), while for convergent-divergent nozzles the values are lower (about 0.94–0.96) because of the channel length and the high (supersonic) velocity in the divergent part.

An isentropic expansion efficiency can, however, be introduced for nozzles following the general definition given in Sect. 1.4. Since the nozzle increases the fluid flow kinetic energy, its efficiency can be defined as the ratio between the kinetic energy actually conferred to the fluid (adiabatic expansion) to the kinetic energy conferred by a reversible process (isentropic expansion):

$$\eta_S = \frac{c_1^2}{c_{1is}^2} = \upsilon_N^2$$

On the basis of this definition we have (Fig. 1.24):

$$\eta_S = \frac{h_{0t} - h_1}{h_{0t} - h_{1,is}} \tag{1.90}$$

On the other hand, if the efficiency of the nozzle is defined as the ratio between adiabatic and isentropic enthalpy drop (equal to the ratio between actual and ideal kinetic energy increase), we have:

$$\eta_S' = \frac{\Delta h_{S,ad}}{\Delta h_{S,is}} = \frac{h_0 - h_1}{h_0 - h_{1is}} = \frac{c_1^2 - c_0^2}{c_{1is}^2 - c_0^2}$$

This value coincides with the previous one only when c_0 is negligible in comparison with c_1.

Rotor (Sects. 1–2)

The energy equation in the thermodynamic form (1.1) involves:

$$cdc + dh = dW$$

and in mechanical form (1.5):

$$cdc + vdp = dW - dE_{diss}$$

By combining these last two equations with the Euler's Eq. (1.69):

$$dW = cdc + udu - wdw$$

we get:

$$wdw + dh = udu$$

$$wdw + vdp = udu - dE_{diss}$$

The energy equation in thermodynamic form states the rothalpy (Eq. 1.84) conservation in the rotor:

$$dI = dh + wdw - udu = 0 \tag{1.91}$$

and therefore:

$$I_1 = h_1 + \frac{w_1^2}{2} - \frac{u_1^2}{2} = I_2 = h_2 + \frac{w_2^2}{2} - \frac{u_2^2}{2}$$

$$h_{1tr} - \frac{u_1^2}{2} = h_{2tr} - \frac{u_2^2}{2}$$

For an axial stage (udu = 0), we have also the total relative enthalpy (Eq. 1.82) conservation:

$$dh_{tr} = dh + wdw = 0 \tag{1.92}$$

and therefore:

$$h_{1tr} = h_1 + \frac{w_1^2}{2} = h_{2tr} = h_2 + \frac{w_2^2}{2}$$

By using Eq. (1.83), the energy equation in mechanical form then provides:

$$\frac{dp_{tr}}{\rho} = udu - E_{diss} \tag{1.93}$$

Therefore in the rotor of axial stages (udu = 0) the total relative pressure variation is directly related to dissipations, while in the rotor of radial stages it is related also to the variation of the blade speed between the rotor inlet and outlet.

Figures 1.25 and 1.26 show the h-s diagram of an expansion in an axial stage (Fig. 1.25) and in a radial one (Fig. 1.26).

Degree of reaction

The stator and rotor blade shapes depend on the distribution of the enthalpy drop between the stator and the rotor (Figs. 1.23, 1.24, 1.25, 1.26).

In this regard, it is extremely effective to introduce the *degree of reaction*, defined as:

$$R = \frac{\Delta h_R}{\Delta h_{t,ad}} = \frac{\Delta h_R}{\Delta h_{t,R}} = \frac{\Delta h_R}{W} \tag{1.94}$$

By applying the energy Eq. (1.1) in the rotor, we obtain:

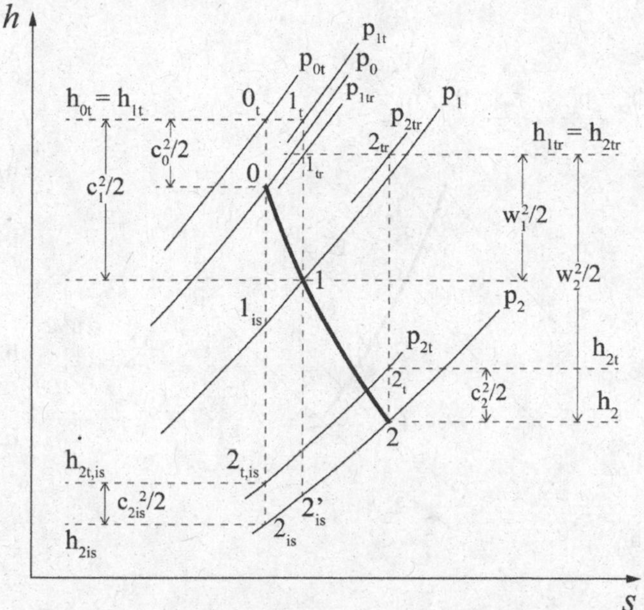

Fig. 1.25 Expansion process in an axial stage

$$\Delta h_R = W - \frac{c_1^2 - c_2^2}{2} = W - \Delta E_{c,R}$$

and therefore:

$$R = 1 - \frac{\Delta E_{c,R}}{W} \tag{1.95}$$

In driving turbomachine stages, the stator always produces an enthalpy drop and this involves $R < 1$. When all the enthalpy drop occurs in the stator, we have $R = 0$ and the stage is called *impulse stage*; in all the other cases ($R > 0$) the stage is called *reaction stage*.

For an impulse stage, by using the energy equation in thermodynamic form (1.1) applied to the rotor, we know that work is exchanged in the rotor only by variation of kinetic energy without any contribution of pressure energy. Indeed, in reversible processes, no fluid expansion occurs in an axial machine rotor ($u_1 = u_2$) so the relative velocity of the fluid is unchanged across the rotor ($w_1 = w_2$) as well as its pressure. Some examples of this type of stage are the Pelton turbine and some axial stages

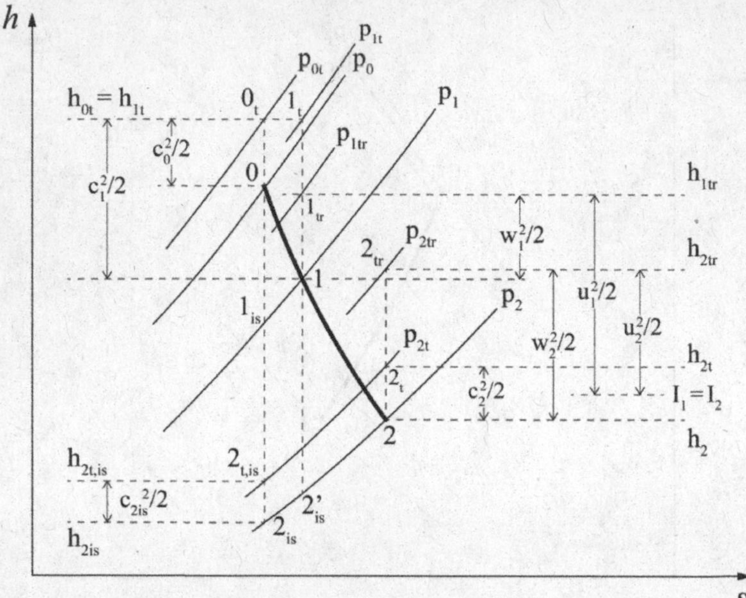

Fig. 1.26 Expansion process in a radial stage

of steam turbines (the first ones) where it is very important to be able to perform a partial admission of the fluid.

The degree of reaction can be expressed in terms of fluid velocities; by using the energy equation in thermodynamic form (1.1) and the Euler's Eq. (1.69), we can write the enthalpy drop in the rotor as:

$$\Delta h_R = \frac{u_1^2 - u_2^2}{2} + \frac{w_2^2 - w_1^2}{2}$$

$$W = \frac{c_1^2 - c_2^2}{2} + \frac{u_1^2 - u_2^2}{2} + \frac{w_2^2 - w_1^2}{2}$$

and therefore the degree of reaction (1.94) becomes:

$$R = \frac{u_1^2 - u_2^2 + w_2^2 - w_1^2}{c_1^2 - c_2^2 + u_1^2 - u_2^2 + w_2^2 - w_1^2} \tag{1.96}$$

or (1.95):

$$R = 1 - \frac{c_1^2 - c_2^2}{c_1^2 - c_2^2 + u_1^2 - u_2^2 + w_2^2 - w_1^2} \tag{1.97}$$

For axial stages, the expressions (1.96) and (1.97) are then simplified into:

$$R = \frac{w_2^2 - w_1^2}{c_1^2 - c_2^2 + w_2^2 - w_1^2} \tag{1.98}$$

$$R = 1 - \frac{c_1^2 - c_2^2}{c_1^2 - c_2^2 + w_2^2 - w_1^2} \tag{1.99}$$

With reference to axial stages, the following figures (adapted from Caputo 1994) show examples of stages with different degrees of reaction: impulse stage R = 0 (Fig. 1.27) and reaction stage R = 0.5 (Fig. 1.28).

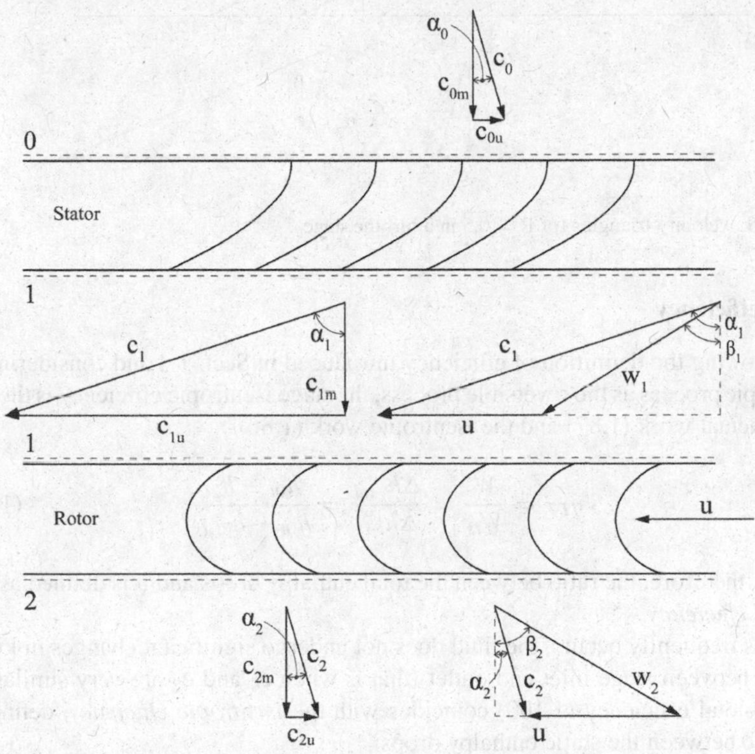

Fig. 1.27 Velocity triangles for R = 0 in a turbine stage

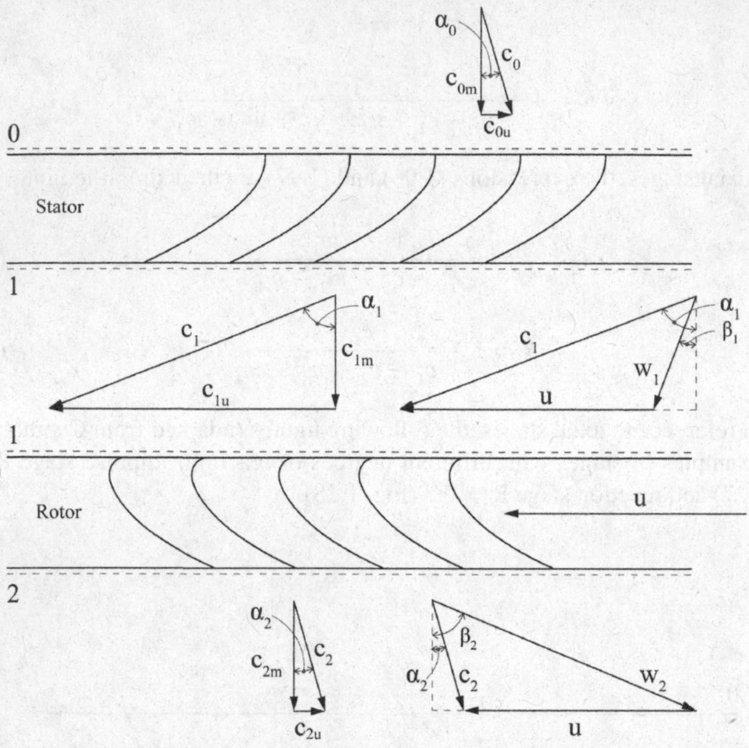

Fig. 1.28 Velocity triangles for R = 0.5 in a turbine stage

Stage efficiency

Following the definition of efficiency introduced in Sect. 1.4 and considering the isentropic process as the reversible process, the stage isentropic efficiency is the ratio of the actual work (1.87) and the isentropic work (1.86):

$$\eta_{TT} = \frac{W}{W_{is}} = \frac{\Delta h_{t,ad}}{\Delta h_{t,is}} = \frac{h_{0t} - h_{2t}}{h_{0t} - h_{2t,is}} \tag{1.100}$$

It is, therefore, the ratio between the total enthalpy drops and it is defined as *total to total efficiency*.

If, as frequently occurs, the fluid does not undergo significant changes in kinetic energy between stage inlet and outlet (that is when c_0 and c_2 are very similar) the total-to-total efficiency (1.100) coincides with the *isentropic efficiency*, defined as the ratio between the static enthalpy drops:

$$\eta_{is} = \frac{\Delta h_{ad}}{\Delta h_{is}} = \frac{h_0 - h_2}{h_0 - h_{2,is}} \tag{1.101}$$

In order to highlight also losses associated with the kinetic energy at the stage exit ($c_2^2/2$), we can introduce another efficiency, called *total-to-static efficiency*:

$$\eta_{TS} = \frac{W}{W\max} = \frac{h_{0t} - h_{2t}}{h_{0t} - h_{2,is}} = \frac{h_{0t} - h_{2t}}{\frac{c_0^2}{2} + \Delta h_{is}} \tag{1.102}$$

It is the ratio between actual work (1.86) and the total energy available in the stage during an isentropic expansion, which is the ideal work obtainable without any dissipations in the channels and without any losses due to kinetic energy at the stage exit.

1.8.3 Compressor Stage

Figure 1.29 sketches a stage compression in the h-s diagram. Sections 1, 2 and 3 indicate respectively the fluid conditions just upstream of the rotor, just downstream of the rotor (that is just upstream of the stator inlet) and just downstream of the stator.

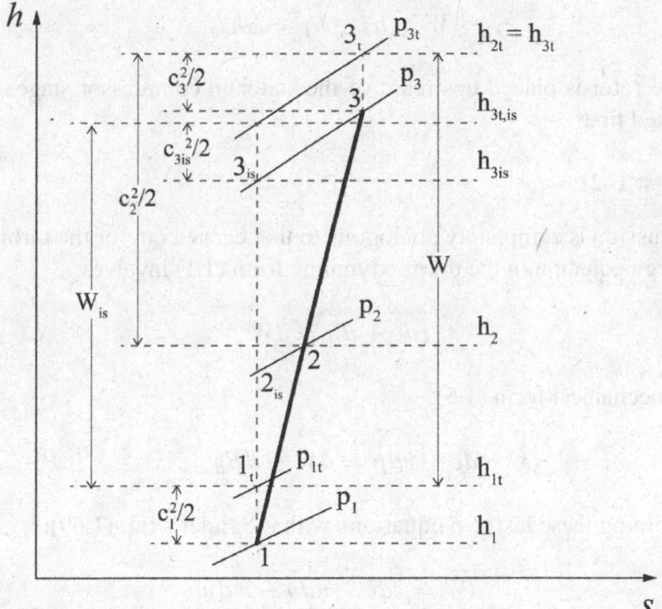

Fig. 1.29 Compression process in a compressor stage

This notation has been adopted to maintain, for both turbine and compressor stages, Sections 1 and 2 as rotor inlet and outlet sections.

By applying the energy Eq. (1.1) between Sects. 1.1–1.3:

$$cdc + dh = dW$$

we obtain the expression of isentropic (or ideal) work (ideal process) and adiabatic (or real) work (real process):

$$W_{is} = \frac{c_{3is}^2}{2} + h_{3is} - \frac{c_1^2}{2} - h_1 = h_{3t,is} - h_{1t} = \Delta h_{t,is} \tag{1.103}$$

$$W = \frac{c_3^2}{2} + h_3 - \frac{c_1^2}{2} - h_1 = h_{3t} - h_{1t} = \Delta h_{t,ad} \tag{1.104}$$

When, as generally occurs, the absolute velocity of the fluid at stage inlet and outlet can be considered approximately equal:

$$c_1 \cong c_{3is} \cong c_3$$

isentropic (1.103) and adiabatic (1.104) works become:

$$W_{is} = h_{3is} - h_1 = \Delta h_{is}$$
$$W = h_3 - h_1 = \Delta h_{ad}$$

Since the rotor is placed upstream of the stator in compressor stages, the rotor will be treated first.

Rotor (Sects. 1–2)

The discussion is completely analogous to that carried out for the turbine stages. The energy equation in the thermodynamic form (1.1) involves:

$$cdc + dh = dW$$

and in the mechanical form (1.5):

$$cdc + vdp = dW - dE_{diss}$$

By combining these last two equations with the Euler's Eq. (1.69):

$$dW = cdc + udu - wdw$$

we get:

$$wdw + dh = udu$$

$$wdw + vdp = udu - dE_{diss}$$

The energy equation in thermodynamic form states the rothalpy (Eq. 1.84) conservation in the rotor:

$$dI = dh + wdw - udu = 0 \tag{1.105}$$

and therefore:

$$I_1 = h_1 + \frac{w_1^2}{2} - \frac{u_1^2}{2} = I_2 = h_2 + \frac{w_2^2}{2} - \frac{u_2^2}{2}$$
$$h_{1tr} - \frac{u_1^2}{2} = h_{2tr} - \frac{u_2^2}{2}$$

For an axial stage (udu = 0), we have also the total relative enthalpy (Eq. 1.82) conservation:

$$dh_{tr} = dh + wdw = 0 \tag{1.106}$$

and therefore:

$$h_{1tr} = h_1 + \frac{w_1^2}{2} = h_{2tr} = h_2 + \frac{w_2^2}{2}$$

By using (Eq. 1.83), the energy equation in the mechanical form then provides:

$$\frac{dp_{tr}}{\rho} = udu - E_{diss} \tag{1.107}$$

Therefore in the rotor of axial stages (udu = 0) the total relative pressure variation is directly related to dissipations, while in the rotor of radial stages it is related also to the variation of the blade speed between the rotor inlet and outlet.

Figures 1.30 and 1.31 show the h–s diagram of a compression in an axial stage (Fig. 1.30) and in a radial one (Fig. 1.31).

Stator (Sects. 2–3)

As there is no work transfer, the energy equation in the thermodynamic form (1.1) involves:

$$cdc + dh = 0$$

and in the mechanical form (1.5):

$$cdc + vdp = -dE_{diss}$$

From the first expression we deduce the total enthalpy conservation in the stator:

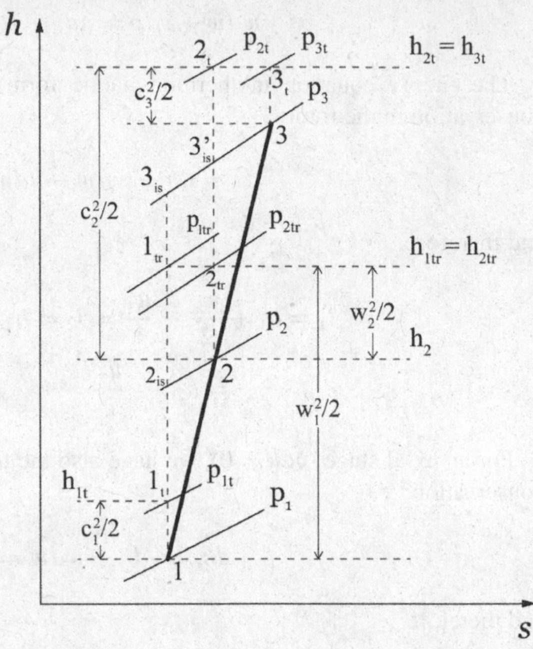

Fig. 1.30 Compression process in an axial stage

$$dh_t = 0 \tag{1.108}$$

while from the second expression we deduce that the change in total pressure is directly related to dissipations:

$$\frac{dp_t}{\dot{\rho}} = -dE_{diss} \tag{1.109}$$

By using the total enthalpy conservation (1.108) and being the entropy generation due to irreversibility (adiabatic process), Eq. (1.3), we obtain:

$$\frac{dp_t}{\rho} = -TdS_{irr}$$

Figure 1.32 sketches a compression in a stator diffuser in the h-s diagram.

Total enthalpy conservation (1.108) allows to express the fluid velocity at diffuser exit:

$$h_{3t} = h_{2t} = h_{3't,is}$$

$$c_{3'is} = \sqrt{c_2^2 - 2 \cdot \Delta h_{S,is}}$$

$$c_3 = \sqrt{c_2^2 - 2 \cdot \Delta h_{S,ad}}$$

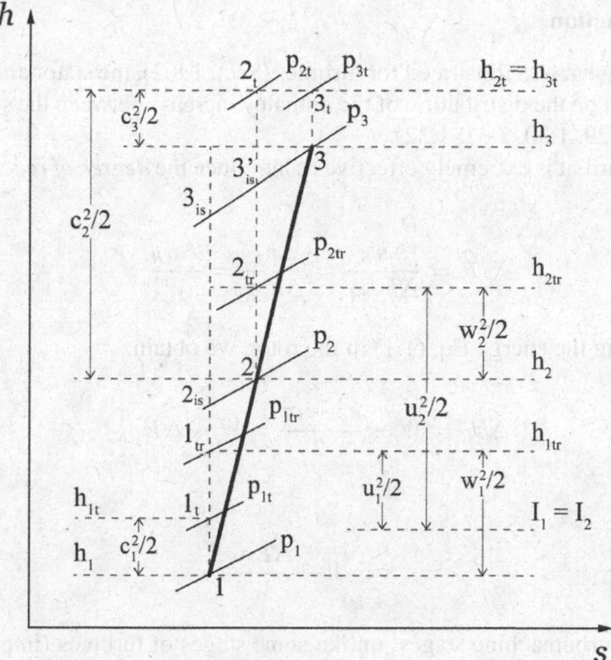

Fig. 1.31 Compression process in a radial stage

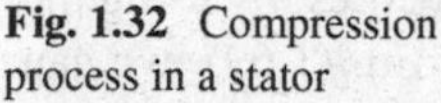

Fig. 1.32 Compression process in a stator

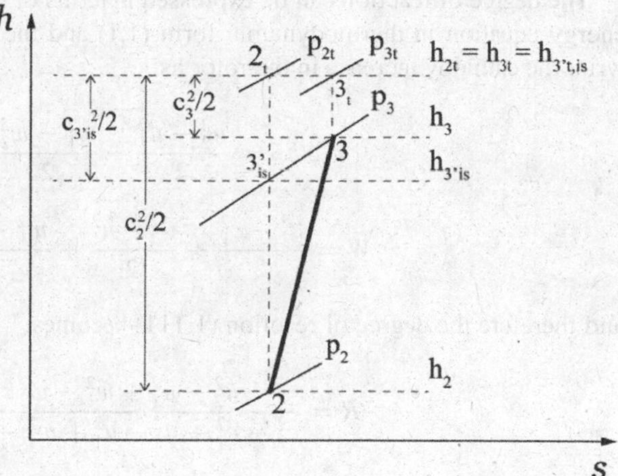

For the diffuser, efficiency, defined as the ratio between isentropic and adiabatic enthalpy increase, is used (Fig. 1.32):

$$\eta_S = \frac{h_{3'is} - h_2}{h_3 - h_2} = \frac{c_2^2 - c_{3'is}^2}{c_2^2 - c_3^2} \tag{1.110}$$

Degree of reaction

Similar to what was illustrated for turbines (Sect. 1.8.2), the stator and rotor blade shapes depend on the distribution of the enthalpy increase between the stator and the rotor (Figs. 1.29, 1.30, 1.31, 1.32).

In this regard, it is extremely effective to introduce the *degree of reaction*, defined as:

$$R = \frac{\Delta h_R}{\Delta h_{t,ad}} = \frac{\Delta h_R}{\Delta h_{t,R}} = \frac{\Delta h_R}{W} \tag{1.111}$$

By applying the energy Eq. (1.1) in the rotor, we obtain:

$$\Delta h_R = W - \frac{c_2^2 - c_1^2}{2} = W - \Delta E_{c,R}$$

and therefore:

$$R = 1 - \frac{\Delta E_{c,R}}{W} \tag{1.112}$$

In driven turbomachine stages, unlike some stages of turbines (impulse stages), the rotor always confers an enthalpy increase upon the fluid, therefore R > 0. In some configurations of axial compressors, the degree of reaction can even be slightly higher than 1.

The degree of reaction can be expressed in terms of fluid velocities; by using the energy equation in thermodynamic form (1.1) and the Euler's Eq. (1.69), we can write the enthalpy increase in the rotor as:

$$\Delta h_R = \frac{u_2^2 - u_1^2}{2} + \frac{w_1^2 - w_2^2}{2}$$

$$W = \frac{c_2^2 - c_1^2}{2} + \frac{u_2^2 - u_1^2}{2} + \frac{w_1^2 - w_2^2}{2}$$

and therefore the degree of reaction (1.111) becomes:

$$R = \frac{u_2^2 - u_1^2 + w_1^2 - w_2^2}{c_2^2 - c_1^2 + u_2^2 - u_1^2 + w_1^2 - w_2^2} \tag{1.113}$$

or (1.112):

$$R = 1 - \frac{c_2^2 - c_1^2}{c_2^2 - c_1^2 + u_2^2 - u_1^2 + w_1^2 - w_2^2} \tag{1.114}$$

For axial stages, the expressions (1.113) and (1.114) are then simplified into:

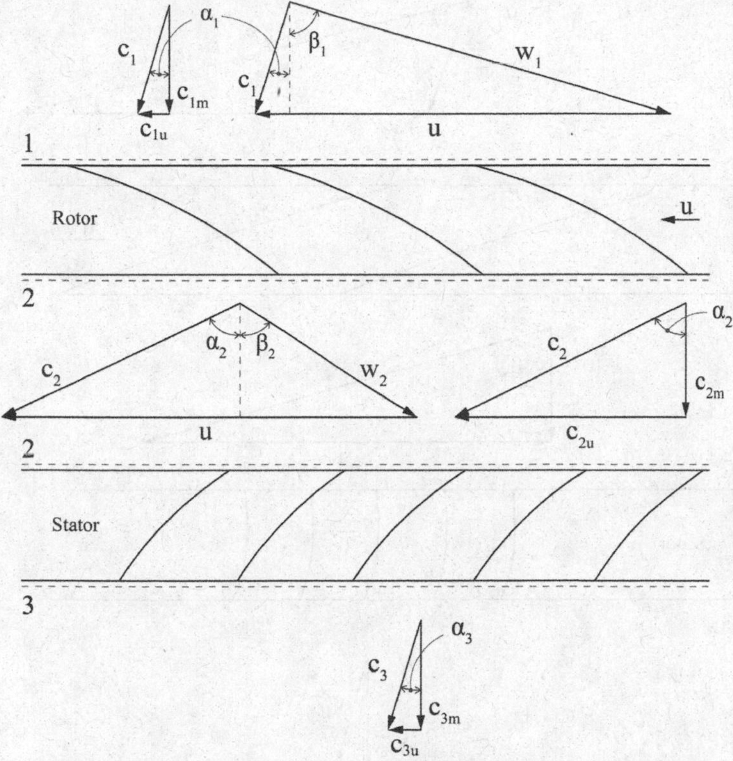

Fig. 1.33 Velocity triangles for R < 1 in a compressor stage

$$R = \frac{w_1^2 - w_2^2}{c_2^2 - c_1^2 + w_1^2 - w_2^2} \tag{1.115}$$

$$R = 1 - \frac{c_2^2 - c_1^2}{c_2^2 - c_1^2 + w_1^2 - w_2^2} \tag{1.116}$$

With reference to axial stages, the following figures (adapted from Caputo 1994) show examples of stages with different degrees of reaction: reaction stage R < 1 (Fig. 1.33) and reaction stage R > 1 (Fig. 1.34).

Stage efficiency

Similarly to turbines, the *total-to-total efficiency* is defined as the ratio between the reversible (isentropic) work (1.103) and the adiabatic work (1.104):

$$\eta_{TT} = \frac{W_{is}}{W} = \frac{\Delta h_{t,is}}{\Delta h_{t,ad}} = \frac{h_{3t,is} - h_{1t}}{h_{3t} - h_{1t}} \tag{1.117}$$

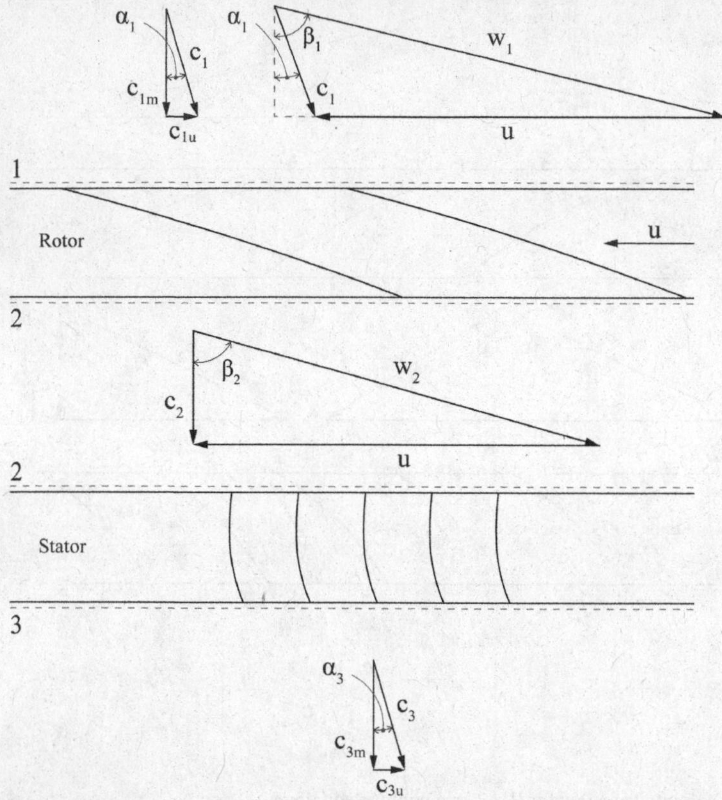

Fig. 1.34 Velocity triangles for R > 1 in a compressor stage

If, as frequently occurs, the fluid does not undergo significant changes in kinetic energy between stage inlet and outlet (that is when c_1 and c_3 are very similar) the total-to-total efficiency (1.117) coincides with the *isentropic efficiency*, defined as the ratio between the static enthalpy increases:

$$\eta_{is} = \frac{\Delta h_{is}}{\Delta h_{ad}} = \frac{h_{3is} - h_1}{h_3 - h_1} \tag{1.118}$$

In order to highlight losses associated with the kinetic energy at the stage exit ($c_3^2/2$), we can introduce another efficiency, called *total-to-static efficiency* and defined as:

$$\eta_{TS} = \frac{W_{\min}}{W} = \frac{h_{3is} - h_{1t}}{h_{3t} - h_{1t}} \tag{1.119}$$

1.9 Losses in Turbomachinery Stages

Losses can be subdivided into two main categories:

- *internal losses*: these losses occur in the fluid flow across the stator, rotor and whole turbomachinery stage and they affect the thermodynamic state of the fluid during compression/expansion processes;
- *external losses*: they occur outside the stator and rotor and do not alter the thermodynamic state of the fluid but are responsible for dissipating mechanical energy. They are given by mechanical friction on bearings, mass losses (leakage flow, recirculation flow that does not alter the state of the fluid, and so on) and losses due to friction on rotating parts of the turbomachinery immersed in the fluid (especially in radial machines where the impeller backplate is immersed in the fluid).

The stage efficiencies defined in the previous sections (polytropic and isentropic efficiency) take into account internal losses; external losses are included in the overall efficiency and in the calculation of machine net power (Sect. 1.9.6). Postponing, therefore, to this paragraph the considerations on external losses, we will now focus on internal losses as they represent the absolutely predominant share of the overall turbomachinery losses.

1.9.1 Loss Sources in Turbomachinery

As illustrated in Sect. 1.1, energy dissipation (and therefore internal losses) are related to entropy generation by irreversibilities, Eq. (1.3):

$$dE_{diss} = T \cdot ds_{irr}$$

Starting from this consideration, Denton (1993) developed an in-depth and complete analysis of loss mechanisms in turbomachinery; he therefore pinpointed the sources of irreversibilities that generate losses. In the following, reference will be made to this fundamental article to illustrate the loss mechanisms in turbomachines.

In fluid-dynamic processes entropy generation is due to:

- viscous friction in boundary layers and in region of boundary layer separation, in secondary flows and in mixing processes, including wakes, vortices and trailing edge flows;
- heat transfers across finite temperature differences;
- non-equilibrium processes that occur in extremely rapid expansions or in shock waves and in the interactions of these shock waves with the boundary layers.

Leaving aside the irreversibilities related to heat exchange, which become significant only for cooled turbine stages (first stages of gas turbines with fluid temperatures

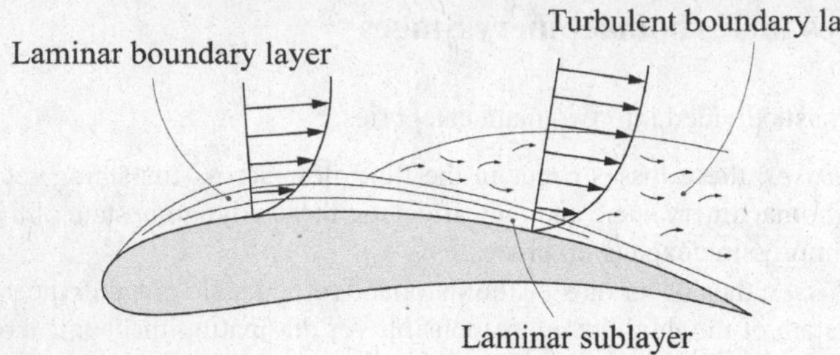

Fig. 1.35 The boundary layer on a blade profile

above 1000 °C), we therefore consider an adiabatic stage (Sect. 1.4): the most significant dissipations are correlated to the boundary layers and to mixing processes, as well as to shock waves for stages operating in a transonic or supersonic regime.

Boundary layers

The boundary layers are due to the fluid viscosity and they are placed on the stator and rotor channel walls and therefore along the blade profiles and on the tip and hub end wall surfaces.

Figure 1.35 illustrates the boundary layer along a blade profile. There is a laminar region (low Reynolds number) closest to the profile, a transition region and a turbulent region. Even in the turbulent boundary layer, however, there is a region near the profile which maintains laminar flow (laminar sublayer). Outside the boundary layer, the fluid flow is turbulent and the effects of viscosity can be neglected (unless there are regions affected by wakes) at least for fluids commonly used in turbomachinery which are characterized by low viscosities.

The boundary layer shape strongly depends on fluid flow: in turbine channels the fluid acceleration tends to reduce its thickness and therefore dissipative effects decrease. In compressor channels, the fluid deceleration, and therefore the unfavorable pressure gradient, tends to thicken the boundary layer and, if the main fluid flow is not able to drag the boundary layer, a boundary layer separation occurs, with velocity profile reverse flows (Fig. 1.36) and, therefore, creation of high energy dissipation regions.

What was illustrated for the boundary layers on blade profiles also applies to the boundary layers on tip and hub end wall surfaces.

In order to express turbomachinery losses, we must introduce some parameters of the boundary layer:

- boundary layer thickness, δ: it is the distance from the wall where the velocity is equal to 99% of fluid velocity outside the boundary layer (inviscid flow);
- displacement thickness, δ^*: it is the fictitious thickness to balance the mass flow reduction caused by the boundary layer;

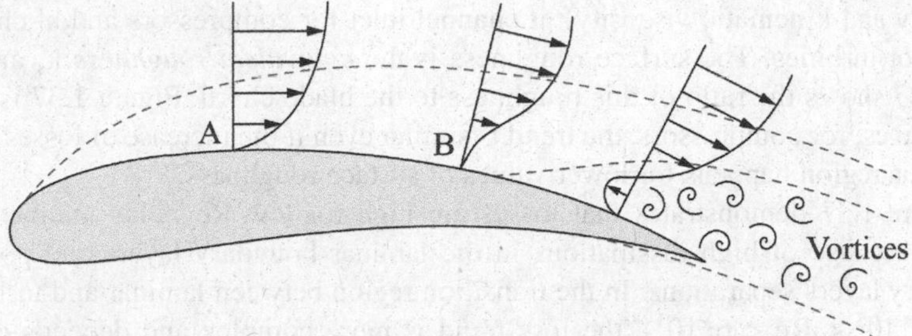

Fig. 1.36 Separation of the boundary layer on a blade profile

- momentum thickness, θ: it is the thickness corresponding to the momentum loss caused by the boundary layer;
- boundary layer shape factor, H: it is the ratio δ^*/θ and has specific values for laminar and turbulent boundary layer.

Dissipations connected with the boundary layer depend significantly on the Reynolds number and surface roughness. Figure 1.37 (Denton 1993) shows these dissipations (more precisely the profile losses, see Sect. 1.9.2) as a function of the Reynolds number and surface roughness; more precisely in this figure dissipations are expressed as the fraction of dissipations evaluated at Re $= 10^6$. The Reynolds number is calculated by using the blade chord (Chaps. 3–7) and the fluid conditions

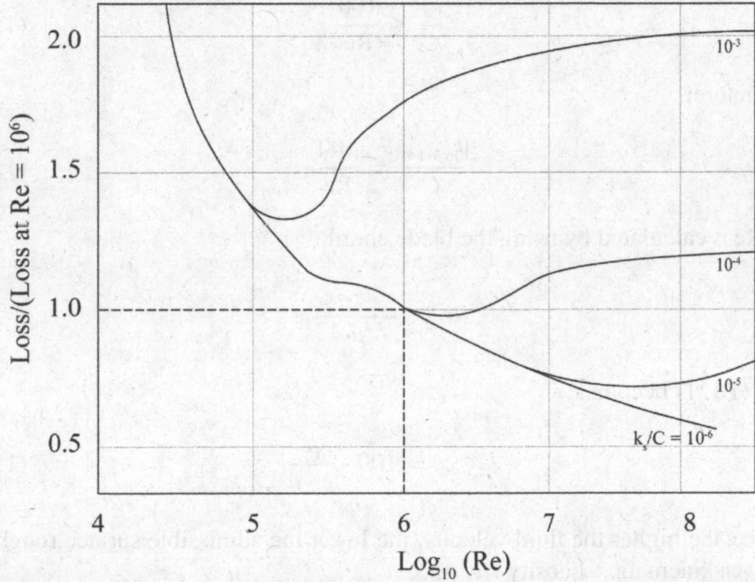

Fig. 1.37 Variation of profile loss with Reynolds number and surface roughness (adapted from Denton (1993)

(velocity and kinematic viscosity) at channel inlet for compressors and at channel outlet for turbines. The surface roughness is the *equivalent roughness* k_s and the Fig. 1.37 shows the ratio of this roughness to the blade chord. Figure 1.37 is valid for turbines; for compressors, the trend is similar even if the increase of losses in the transition region happens for lower values of surface roughness.

Figure 1.37 demonstrates that losses are high for low Reynolds numbers ($Re < 10^5$) because of high dissipations in the laminar boundary layers and possible boundary layers separations. In the transition region between laminar and turbulent flow ($2*10^5 < Re < 6*10^5$), the loss trend is more complex and depends on the combination of two factors: loss reduction with Re increase and loss increase due to upstream displacement of the transition point. In turbulent flow ($Re > 6*10^5$) losses are reduced with a trend proportional to $Re^{-0.16}$ (referred to smooth surfaces). However, in turbulent flow, losses are greatly influenced by surface roughness so that in turbomachines, which operate with high Reynolds numbers, the surface finish becomes a very important parameter in order to contain losses. From Fig. 1.37 it is clear, in fact, that for $Re > 1*10^6$ losses decrease only if surface roughness has an ever-greater reduction. In this regard, the concept of admissible surface roughness $k_{s,adm}$ (Schlichting 1978) can be recalled: this is the minimum roughness below which roughness does not affect losses and, therefore, walls can be considered smooth. This condition occurs when the surface roughness is contained within the laminar sublayer of the boundary layer.

With reference to flat plates, Schlichting demonstrated that the previous situation happens when:

$$\frac{k_s}{C} \le \frac{100}{\mathrm{Re}} \tag{1.120}$$

and therefore:

$$\frac{k_{s,adm}}{C} = \frac{100}{\mathrm{Re}} \tag{1.121}$$

As Re is calculated by using the blade chord:

$$\mathrm{Re} = \frac{\rho \cdot c \cdot C}{\mu}$$

the Eq. (1.121) becomes:

$$k_{s,adm} = 100 \cdot \frac{\mu}{\rho \cdot c} \tag{1.122}$$

Hence, the higher the fluid velocity, the lower the admissible surface roughness for a given kinematic viscosity (μ/ρ).

The expression (1.121) can also be used to evaluate the maximum Reynolds number beyond which roughness has an influence on losses. In fact, defining the critical Reynolds number as:

$$\text{Re}_{cr} = 100 \cdot \frac{C}{k_s} \tag{1.123}$$

this value identifies two ranges: when $\text{Re} < \text{Re}_{cr}$ roughness has no effect on losses; when $\text{Re} > \text{Re}_{cr}$ roughness has great effects on losses.

In the turbomachinery field, several authors suggested that the factor of 100 actually varies between 90 and 120, but the value 100 is generally acceptable. In turbomachines, the kinematic and thermodynamic quantities (c, ρ, μ) used to calculate $k_{s,adm}$ are referred to channel inlet for compressors and to channel outlet for turbines; moreover, the fluid velocity is the absolute one for the stator and the relative one for the rotor.

The calculation of $k_{s,\,adm}$ is very important as it allows to establish the finishing level of the surfaces in order to contain dissipations.

Taking into account that turbomachinery surface roughness generally falls within the range:

$$5\mu m < k_s < 50\mu m$$

the evaluation of $k_{s,adm}$ allows deciding if a surface treatment, aimed at respecting this admissible roughness value, is possible and economically convenient as well as to give up very high finishing levels where not required.

Mixing processes

Mixing processes are the second most important source of loss; they are, however, largely and intimately connected with the boundary layers.

For example, let us consider blade profiles. The boundary layer is present on both blade profiles (suction and pressure surfaces) but its thickness is different (δ_p and δ_s, Fig. 1.38).

Downstream of the trailing edge, the two boundary layers mix and wakes are generated (Fig. 1.39) together with other dissipative processes.

The blade trailing edge is an example of interaction among the different sources of loss: here separation of the boundary layer (Fig. 1.40) can happen together with non-equilibrium processes such as instantaneous expansions and shock waves (Fig. 1.41 shows a supersonic flow at trailing edge).

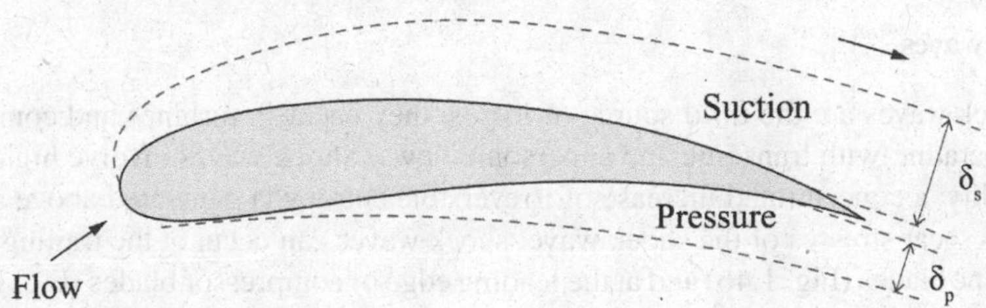

Fig. 1.38 The boundary layer on suction and pressure surfaces

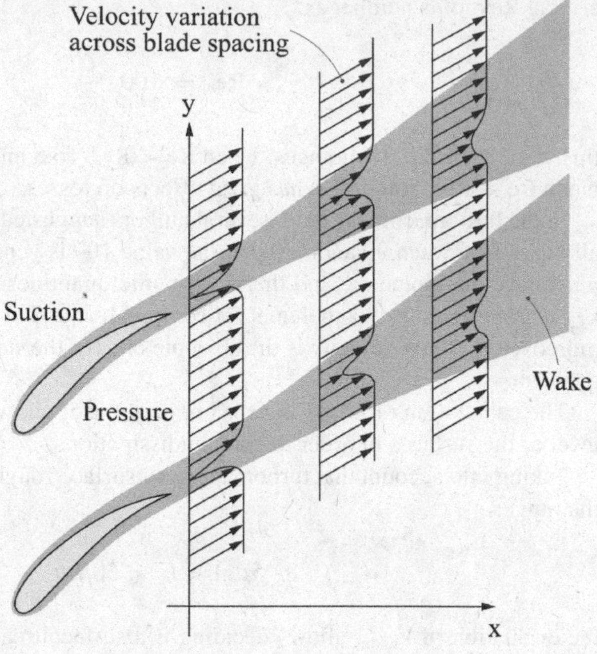

Fig. 1.39 Wakes formation at blade trailing edge

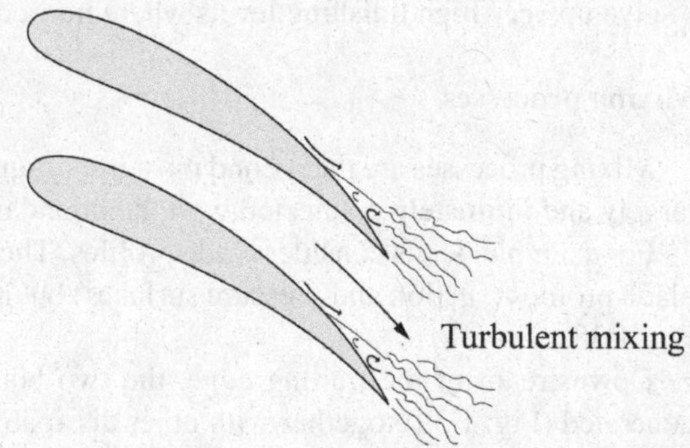

Fig. 1.40 Trailing edge with a separated boundary layer

Mixing processes then take place in the boundary layers of the hub and tip end wall, in swirling flow perpendicular to the main flow (Fig. 1.42) as well as in vortices generated at blade tip (Figs. 1.43, 1.44, 1.45) between pressure and suction surfaces.

Shock waves

Shock waves are the third source of losses; they occur in turbines and compressors operating with transonic and supersonic flows. Shock waves involve high irreversibility; a concentrated increase of irreversible entropy is generated above all by viscous shear stresses of the shock wave; shock waves can occur at the trailing edge of turbine blades (Fig. 1.46) and at the leading edge of compressor blades (Fig. 1.47).

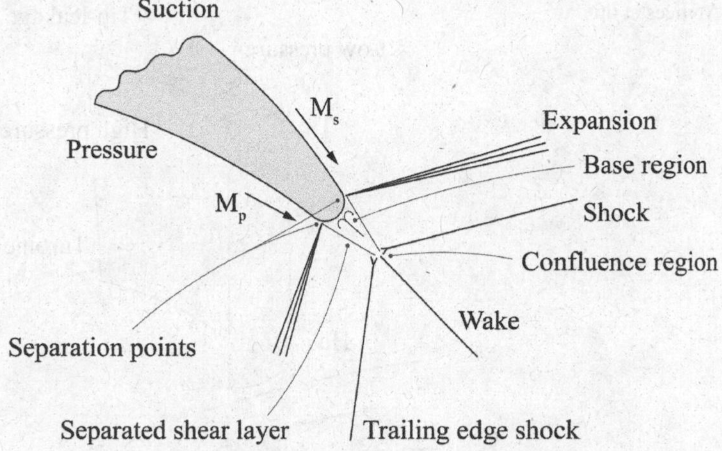

Fig. 1.41 Trailing edge with supersonic fluid flow

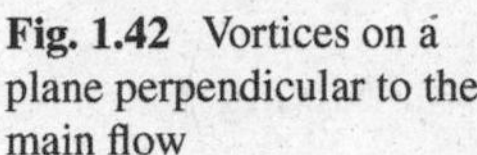
Fig. 1.42 Vortices on a plane perpendicular to the main flow

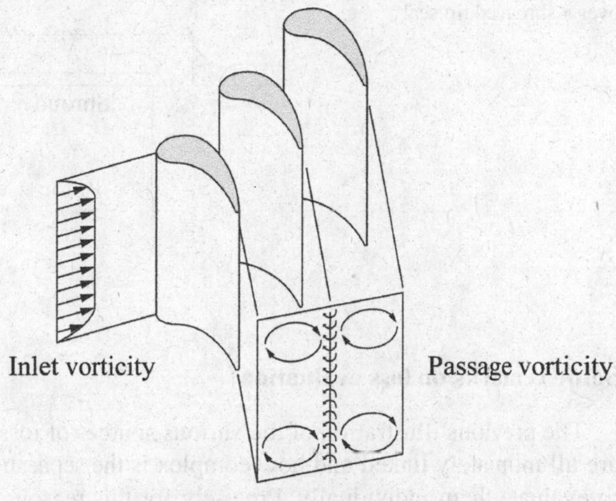

Shock waves also interact strongly with mixing processes and losses associated with the boundary layers. Shock waves, in fact, cause an alteration of the transition region of the boundary layer and can also lead to its complete separation.

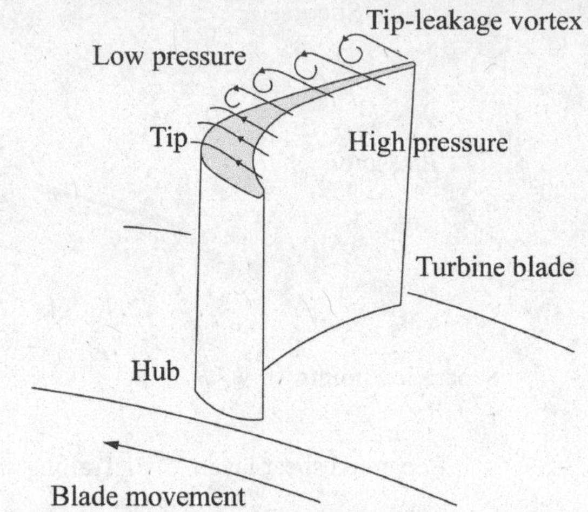

Fig. 1.43 Vortices at tip clearance

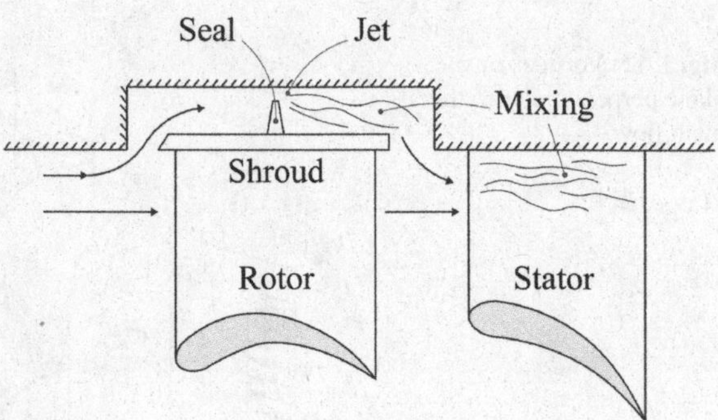

Fig. 1.44 Mixing processes over a shrouded tip seal

Some remarks on loss evaluation

The previous illustration of the various sources of losses clearly shows how they are all intimately linked and how complex is the separation of their effects in order to evaluate them individually. Precisely for this reason, in the technical literature, empirical correlations are provided for loss calculation; they are valid during the preliminary design of turbomachinery because they provide a loss assessment in the various physical area of the machine: blade profile, annulus (hub and tip end wall), blade trailing edge, blade tip and so on. In the following therefore we will illustrate the loss sources divided into physical zones of axial and radial stages; Chaps. 3–7 instead will provide specific loss correlations to be used for each turbomachine.

We remark the excellence of the seminal study of Denton (1993), partly illustrated above; he masterfully identified sources of loss and after him, other authors (see Chapter 4, for example) could develop interesting analyses aimed at evaluating the maximum efficiency of turbomachinery; this efficiency takes into account only the losses that can be reduced but in any case not eliminated (viscous effects in the boundary layers and mixing processes). These analyses will be discussed in the following Chapters when guidelines for the optimal choice of input parameters in the turbomachinery preliminary design procedures will be illustrated.

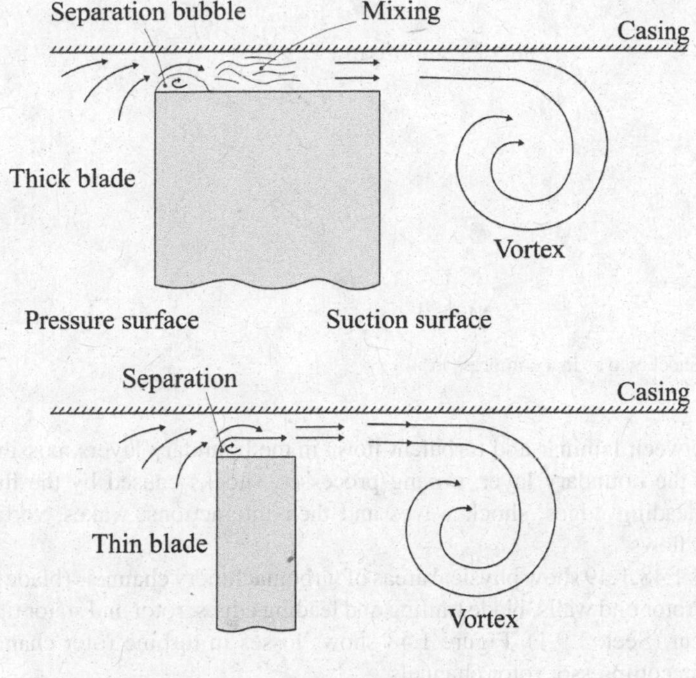

Fig. 1.45 Mixing processes over tip gap of an unshrouded blade

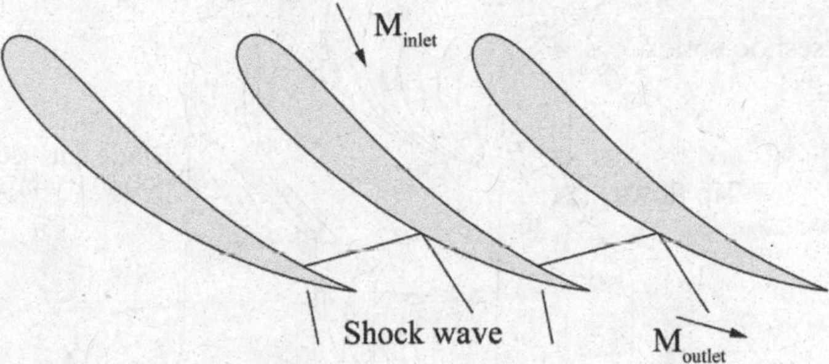

Fig. 1.46 Shock waves in a turbine

1.9.2 Loss Models in Axial Stages

In axial turbomachinery, the fluid flow is three-dimensional, viscous and unsteady; therefore this flow is very complex to analyze. Moreover, as already illustrated in Sect. 1.9.1, other phenomena must also be taken into account, such as: transition

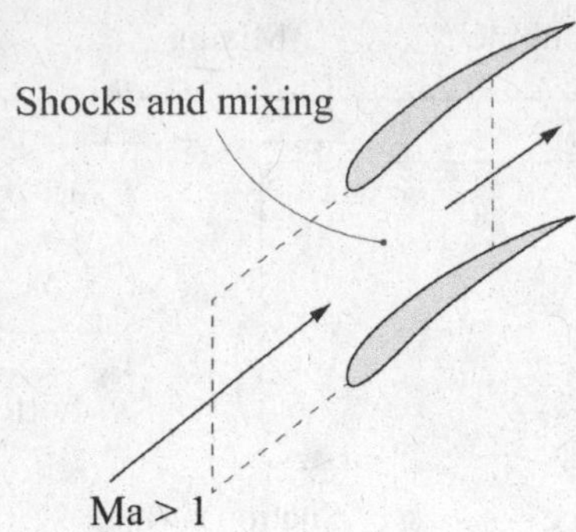

Fig. 1.47 Shock waves in a compressor

region (between laminar and turbulent flow) in the boundary layers, possible separations of the boundary layer, mixing processes, shocks caused by the fluid flow on blade leading edges, shock waves and their interactions, wakes, vortices and secondary flows.

Figures 1.48, 1.49 show physical areas of turbomachinery channels (blade profiles, stator and rotor end walls, blade trailing and leading edges, rotor and stator tip) where losses occur (Sect. 1.9.1). Figure 1.48 shows losses in turbine rotor channels and Fig. 1.49 in compressor rotor channels.

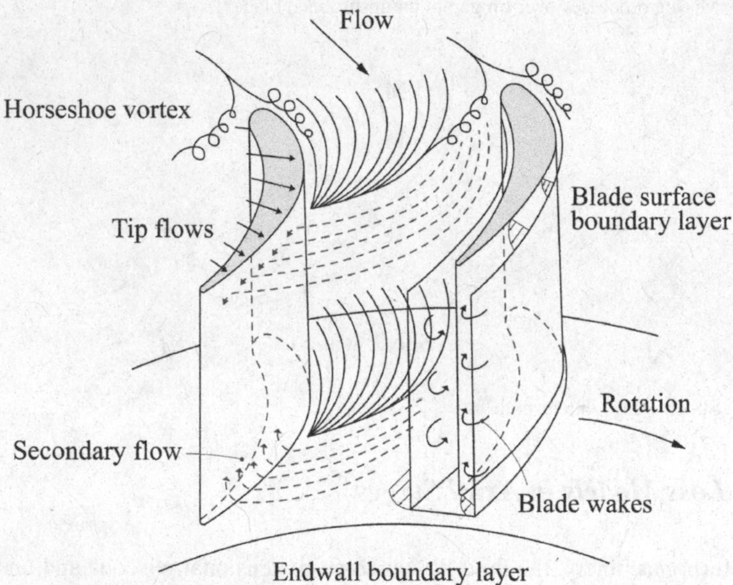

Fig. 1.48 Losses in turbine rotor channels

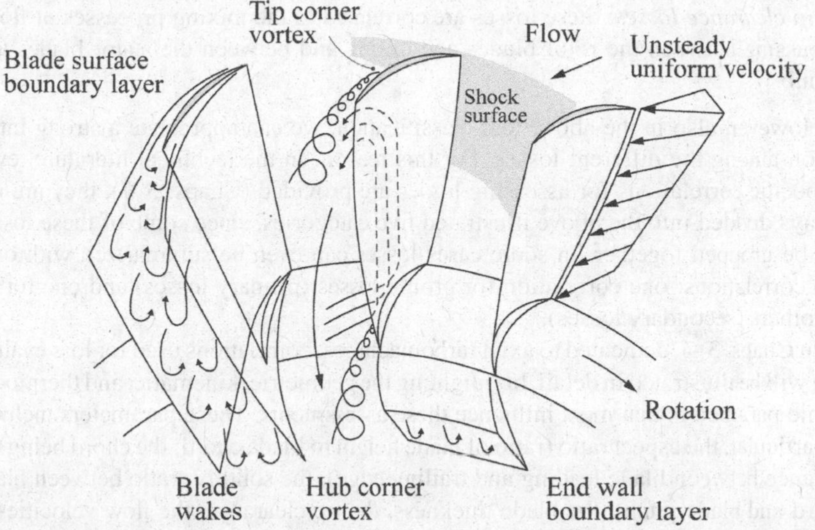

Fig. 1.49 Losses in compressor rotor channels

With reference to turbomachinery physical areas, fluid dynamic losses can be generally classified into five categories:

- *profile losses*: these losses are produced by energy dissipation in the boundary layer along the contact surface between fluid flow and blade surfaces and are due to viscous friction. The profile losses also include incidence losses, particularly important in off-design conditions, correlated to the fluid flow on the blade leading edge. For turbines, losses due to shock waves at blade trailing edge are sometimes included in the profile losses; while for transonic and supersonic compressors, losses due to shock waves near the blade leading edge are evaluated separately;
- *trailing edge losses*: these losses are mainly due to the formation of wakes and their mixing processes downstream of the blade trailing edge. In many cases these losses are included in the profile losses;
- *end wall losses*: these losses are produced by energy dissipation in the boundary layer along the contact surfaces between fluid flow and hub and tip end wall surfaces and are due to viscous friction;
- *secondary losses*: these losses are due to three-dimensional velocity components within channels; these velocity components are generated by imbalances between pressure forces and centrifugal forces and they lie on a plane orthogonal to the main flow direction. Since these flows are closely correlated to the boundary layers of the end wall surfaces, several authors group these losses together with end wall losses. In any case, it is very complex to separate these losses from all the others as they are also closely related to processes occurring at blade leading and trailing edges;

- *tip clearance losses*: these losses are correlated to the mixing processes of flows passing between the rotor blades and casing and between the stator blades and hub.

However, also in the above loss classification, we can appreciate a strong interaction among the different losses. For this reason, in the technical literature, even if specific correlations for assessing losses are provided (Chaps. 3–7), they are not always divided into the above illustrated five categories, since some of these losses can be grouped together. In some cases losses can even be summarized with only two correlations: one correlation for profile losses (primary losses) and one for all the others (secondary losses).

In Chaps. 3–4, dedicated to axial turbomachines, correlations used for loss evaluation will be illustrated in detail, highlighting the geometric, kinematic, and thermodynamic parameters that most influence these assessments. These parameters include, in particular, the aspect ratio (ratio of blade height to blade chord; the chord being the distance between blade leading and trailing edge), the solidity (ratio between blade chord and blade pitch), the blade thickness, the tip clearance, the flow velocities at the channel inlet and outlet, the flow deflection in channels, the surface roughness, Reynolds and Mach numbers.

1.9.3 Loss Models in Radial Stages

Loss sources (Sect. 1.9.1) are the same as for axial stages but their impact on turbomachine performance is different due precisely to the fluid radial flow.

The blade speed u, in fact, varies considerably across the rotor. For this reason, the fluid relative velocity varies lesser in comparison with an axial stage (assuming the same enthalpy change). Consequently, a centrifugal compressor can generate the same enthalpy increase as an axial compressor with less fluid diffusion in the rotor and a radial turbine can generate the same enthalpy drop with lesser acceleration of fluid flow in comparison with axial turbomachines.

The lower change in relative velocity through the rotor is particularly advantageous for compressors since the diffusion of the relative velocity causes the growth and the possible separation of the boundary layer. Therefore, a centrifugal compressor can obtain higher compression ratios for the same rotor diffusion factor than an axial compressor. This is because most of the increase in pressure is generated by the centrifugal force field and the remaining part by deceleration of the relative velocity. In a radial turbine, on the other hand, the reduced change in relative velocity (in comparison with axial turbines) means that higher expansion ratios can be obtained before choking and without losses associated with the transonic flows.

The reduced change in relative velocity also entails a lower average relative velocity in the rotor than that of an axial stage (always assuming the same blade tip speed and the same enthalpy change).

Based on these considerations, we would expect greater efficiency of radial turbomachines than axial ones. Actually, if it is true that a radial stage can obtain higher pressure ratios than that of an axial stage, it is also true that radial machines have lower efficiency than axial ones. It is substantially due to the greater complexity of the fluid flow: it changes direction becoming an axial flow from a radial one (turbines) and vice versa (compressors). This 90° curvature causes more secondary flows than most axial machines, also due to a significant variation in the blade height with the radius.

On the basis of these observations, even if the loss sources are substantially the same as in axial stages, this particular radial flow suggests a different aggregation and/or separation of the different loss formulations in comparison with axial turbomachines.

Let us consider, for example, a radial turbine. Figure 1.50 shows the different physical areas of the machine where these losses occur.

The flow through the rotor, as already illustrated, is even more complex than that of axial stages due to the 90° rotation between rotor inlet (radial direction) and outlet (axial direction). In this case, unlike axial stages, it is not possible to separate contributions correlated to profile losses, end wall losses and secondary flows. All these loss sources are therefore merged into the passage loss: it represents losses generated when the fluid crosses the rotor. Passage loss, therefore, includes:

- frictional losses generated by fluid viscosity through the rotor;

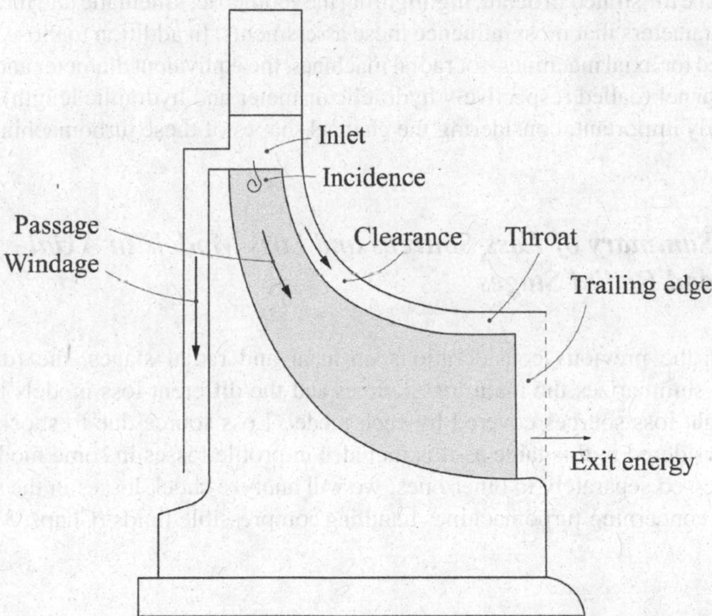

Fig. 1.50 Rotor losses in a radial turbine

- secondary losses generated by three-dimensional flow through the rotor;
- blockage losses and kinetic energy dissipation correlated to the growth of the boundary layer and losses due to possible flow separation.

Since this loss includes all the dissipations occurring between rotor inlet and throat section (Fig. 1.50), the other dissipations, between the throat section and rotor outlet, are assessed separately through trailing edge losses. Tip clearance losses and kinetic energy losses (last stage exit), as well as windage losses, are also expressed separately.

In radial compressors, unlike radial turbines, passage losses (including trailing edge losses) are divided into:

- frictional losses through the rotor generated by fluid viscosity (skin friction);
- blade loading losses which take into account secondary flows generated in the rotor by the pressure gradient between pressure and suction surfaces of each blade. These losses depend on the diffusion factor which measures the flow deceleration in the rotor;
- hub to shroud losses: while blade loading loss is related to secondary flows generated by the pressure gradient between blade surfaces, these losses are related to secondary flows generated by the pressure gradient along the blade span between blade hub and tip;
- mixing losses due to flow mixing at the blade trailing edge of the rotor blades.

In Chaps. 5–7, dedicated to radial turbomachines, correlations used for loss evaluation will be illustrated in detail, highlighting the geometric, kinematic and thermodynamic parameters that most influence these assessments. In addition to those already mentioned for axial machines, for radial machines, the equivalent diameter and length of the channel (called respectively hydraulic diameter and hydraulic length) will be particularly important, considering the channel shapes of these turbomachines.

1.9.4 Summary of Loss Sources and Loss Models in Axial and Radial Stages

Based on the previous considerations on axial and radial stages, the following Fig. 1.51 summarizes the main loss sources and the different loss models in order to highlight loss sources covered by each model. Loss source due to shock waves is not considered in this table as it is included in profile losses in some models and it is expressed separately in other ones; we will analyze shock losses in the specific Chapters concerning turbomachines handling compressible fluids (Chaps. 3–6).

<table>
<tr><th rowspan="3">Loss sources</th><th colspan="4">Loss models</th></tr>
<tr><th colspan="2">Axial stages</th><th colspan="2">Radial stages</th></tr>
<tr><th>Turbines</th><th>Compressors</th><th>Turbines</th><th>Compressors</th></tr>
<tr><td>Blade Boundary Layer</td><td>Profile losses</td><td rowspan="2">Profile losses</td><td rowspan="4">Passage and trailing edge losses</td><td rowspan="3">Frictional, blade loading and mixing losses</td></tr>
<tr><td>Mixing at trailing edge</td><td>Trailing edge losses</td></tr>
<tr><td>End wall boundary layer</td><td rowspan="2">Secondary losses</td><td>End wall losses</td></tr>
<tr><td>Secondary flow mixing</td><td>Secondary losses</td><td>Hub to shroud losses</td></tr>
<tr><td>Tip clearance mixing</td><td>Tip clearance losses</td><td>Tip clearance losses</td><td>Tip clearance losses</td><td>Tip clearance losses</td></tr>
</table>

Fig. 1.51 Correspondence between loss sources and loss models

1.9.5 Loss Quantification in Loss Models

The loss models of each type of stage will be detailed in Chaps. 3–7. Here we will only illustrate how to express the losses described above.

These losses can be quantified by enthalpy increases in comparison with an isentropic process and computed at rotor and stator exit:

turbine (Figs. 1.25–1.26)

$$\Delta h_{S,irr} = h_1 - h_{1,is} \tag{1.124}$$

$$\Delta h_{R,irr} = h_2 - h_{2',is} \tag{1.125}$$

compressor (Figs. 1.30–1.31)

$$\Delta h_{R,irr} = h_2 - h_{2,is} \tag{1.126}$$

$$\Delta h_{S,irr} = h_3 - h_{3',is} \tag{1.127}$$

Losses, however, can also be effectively expressed in terms of total pressure (stator) and total relative pressure (rotor) drops. For example, referring to an axial stage (where the term *udu* does not appear) the following pressure loss coefficients can be defined:

turbine (Fig. 1.25)

$$Y_S = \frac{p_{0t} - p_{1t}}{p_{1t} - p_1} \tag{1.128}$$

$$Y_R = \frac{p_{1tr} - p_{2tr}}{p_{2tr} - p_2} \tag{1.129}$$

compressor (Fig. 1.30)

$$Y_R = \frac{p_{1tr} - p_{2tr}}{p_{1tr} - p_1} \tag{1.130}$$

$$Y_S = \frac{p_{2t} - p_{3t}}{p_{2t} - p_2} \tag{1.131}$$

In Chaps. 3–7 loss models, expressed both in enthalpy and total pressure terms, will be illustrated in details.

A final clarification concerns *external losses*, previously introduced: since these losses occur outside the stator and rotor channels, they do not entail fluid total pressure losses but only enthalpy increases computed at constant pressure at the stage exit; these enthalpy increases modify shaft work and do not alter work exchanged between the fluid and the blades (Euler's work), as illustrated in the following Sect. 1.9.6.

Since the pressure loss coefficients, Y_R and Y_S, are not affected by the external losses, in order to uniform the isentropic efficiency calculation by using these pressure loss coefficients or the enthalpy increases due to irreversibility, Δh_{irr}, it is advisable not to include the external losses in the isentropic efficiency, also considering that these losses do not alter fluid thermodynamic states. For these reasons external losses will be counted separately (Sect. 1.9.6).

1.9.6 Overall Efficiency and Shaft Power of a Stage

To evaluate mechanical power at the stage shaft, it is necessary to take into account the internal and external losses as well as the mechanical losses (power dissipated in bearings, gears, lubrication, fluid leakages and so on).

To calculate this power, let us first consider power exchanged between fluid flow and blades (associated with the Euler's work). By indicating fluid mass flow rate with m, the power for a turbine is:

$$P_W = m \cdot W = m \cdot \Delta h_{t,ad} = m \cdot \Delta h_{t,is} \cdot \eta_{is} = P_{is} \cdot \eta_{is} \tag{1.132}$$

External and mechanical losses reduce this power, so shaft power becomes:

$$P_m = P_W - P_{ext,loss} - P_{m,loss} = m \cdot \left(W - W_{ext,loss} - W_{m,loss}\right) \tag{1.133}$$

where $P_{ext,\,loss}$ and $P_{m,\,loss}$ are powers dissipated due to external and mechanical losses.

An overall efficiency can be defined as:

$$\eta_{ov} = \frac{P_m}{P_{is}} = \frac{W}{W_{is}} \cdot \frac{W - W_{ext,loss}}{W} \cdot \frac{W - W_{ext,loss} - W_{m,loss}}{W - W_{ext,loss}} = \eta_{is} \cdot \eta_{ext,loss} \cdot \eta_m \tag{1.134}$$

By defining an overall isentropic efficiency:

$$\eta_{is,ov} = \eta_{is} \cdot \eta_{ext,loss} \tag{1.135}$$

the stage overall efficiency becomes:

$$\eta_{ov} = \eta_{is,ov} \cdot \eta_m \tag{1.136}$$

and therefore the shaft power is:

$$P_m = \eta_{ov} \cdot P_{is} = \eta_{ov} \cdot m \cdot \Delta h_{t,is} \tag{1.137}$$

Extension to a compressor is straightforward:

$$P_W = m \cdot W = m \cdot \Delta h_{t,ad} = m \cdot \frac{\Delta h_{t,is}}{\eta_{is}} = \frac{P_{is}}{\eta_{is}} \tag{1.138}$$

Shaft power must compensate for external and mechanical losses, so:

$$P_m = P_W + P_{ext,loss} + P_{m,loss} = m \cdot \left(W + W_{ext,loss} + W_{m,loss}\right) \tag{1.139}$$

where $P_{ext,\,loss}$ and $P_{m,\,loss}$ are powers dissipated due to external and mechanical losses.

An overall efficiency can be defined as:

$$\eta_{ov} = \frac{P_{is}}{P_m} = \frac{W_{is}}{W} \cdot \frac{W}{W + W_{ext,loss}} \cdot \frac{W + W_{ext,loss}}{W + W_{ext,loss} + W_{m,loss}} = \eta_{is} \cdot \eta_{ext,loss} \cdot \eta_m \tag{1.140}$$

By defining an overall isentropic efficiency:

$$\eta_{is,ov} = \eta_{is} \cdot \eta_{ext,loss} \tag{1.141}$$

the stage overall efficiency becomes:

$$\eta_{ov} = \eta_{is,ov} \cdot \eta_m \tag{1.142}$$

and therefore the shaft power is:

$$P_m = \frac{P_{is}}{\eta_{ov}} = \frac{m \cdot \Delta h_{t,is}}{\eta_{ov}} \tag{1.143}$$

For pumps, the discussion is the same as for compressors, considering hydraulic efficiency instead of isentropic efficiency and reversible work, as defined in Sect. 1.5, instead of isentropic enthalpy change.

References

Caputo C (1994) Turbomachinery (in Italian). Masson, Milano

Casey M.: Accounting for losses and definitions of efficiency in turbomachinery stages. Proc IMechE Part A: J. Power and Energy (2007)

Denton J.D.: Loss Mechanisms in Turbomachines. Journal of Turbomachinery (1993)

Dixon S.L., Hall C.A.: Fluid mechanics and thermodynamics of turbomachinery, 7th edn., Elsevier (2014)

Gambini M, Vellini M (2007) Lecture notes on Fluid Machines (in Italian). Texmat, Rome

Schlichting H (1978) Boundary Layer Theory, 7th edn. McGraw-Hill, New York

Chapter 2
Turbomachinery Selection

Abstract The turbomachinery selection process, as well as the consequent preliminary design, is based on the application of the *similitude theory*, which allows to "capitalize" all previous experience in the turbomachinery sector by transferring the results obtained for an existing machine, considered as a model, to another machine, to be designed, which is "similar" to the reference model. First of all (Sect. 2.1), we briefly recall the similitude theory through the dimensional analysis and the Π theorem (Buckingham's theorem), and then (Sect. 2.2), we illustrate the effective and fundamental *Baljé's method*, through which we can proceed not only with the choice of the turbomachine stage configuration (axial, radial, mixed flow) but we can also define the number of stages, the speed of rotation and the characteristic diameter of the turbomachine (Sect. 2.3). The numerical application of the proposed procedure will instead be developed in Chapter 8 and will concern the selection of turbomachines for advanced conversion cycles.

2.1 Dimensional Analysis

Generally, the description and analysis of a physical phenomenon involve knowledge of a large number of physical (dimensional) parameters.

The dimensional analysis allows reducing the number of parameters describing a phenomenon by using "dimensionless" parameters obtained by appropriately combining the dimensional parameters.

In complete analogy to the study of heat transfer, dimensional analysis and Π theorem are used for turbomachinery analysis. Briefly, this theorem states that *a dimensional equation among "n" physical parameters expressed through "m" fundamental quantities (e.g. in thermo-mechanics: mass, length, time, temperature) can be reduced to an equation among "n-m" dimensionless groups Π*.

Considering, therefore, the generic equation:

$$f(P_1, P_2, \dots P_n) = 0$$

M. Gambini and M. Vellini, *Turbomachinery*, Springer Tracts in Mechanical
Engineering, https://doi.org/10.1007/978-3-030-51299-6_2

which correlates "n" dimensional parameters P_i, expressed in "m" fundamental independent quantities (units of measurement), the same equation can be expressed as a function of "n-m" dimensionless groups Π_i, obtained by combining the parameters P_i:

$$f(\Pi_1, \Pi_2, \ldots \Pi_{n-m}) = 0$$

2.1.1 Application to Turbomachines

By using the dimensional analysis, it can be demonstrated that turbomachinery efficiency, η, is a function of four dimensionless parameters:

$$\eta = f(\Pi_1, \Pi_2, \Pi_3, \Pi_4) \tag{2.1}$$

where each parameter Π_i is correlated to:

Π_1 flow rate
Π_2 work transfer
Π_3 fluid viscosity
Π_4 fluid compressibility

Π_1 *evaluation*
Since the volumetric flow rate is proportional to:

$$V \propto c \cdot A \propto u \cdot D^2 \propto \omega \cdot D^3$$

Π_1 can be expressed through the following dimensionless parameter Φ (*flow factor*):

$$\Pi_1 \equiv \Phi = \frac{V}{\omega \cdot D^3} \tag{2.2}$$

Π_2 *evaluation*
Since the work transfer is proportional to:

$$W \propto c^2 \propto u^2 \propto \omega^2 \cdot D^2$$

Π_2 can be expressed through the following dimensionless parameter Ψ (*work factor*):

$$\Pi_2 \equiv \Psi = \frac{W}{\omega^2 \cdot D^2} \tag{2.3}$$

Π_3 *evaluation*

Π_3 is correlated to the *Reynolds number* (Re), which is the ratio between inertial and viscous forces:

$$\Pi_3 \equiv R_e = \frac{\rho \cdot u \cdot D}{\mu} \tag{2.4}$$

Π_4 *evaluation*

Π_4 is correlated to the fluid compressibility and therefore to the *Mach number* (generally blade speed is used and then we refer to the *peripheral Mach number*):

$$\Pi_4 \equiv M_u = \frac{u}{c_s} \tag{2.5}$$

Therefore, the Eq. 2.1 can be expressed as:

$$\eta = f(\Phi, \Psi, R_e, M_u) \tag{2.6}$$

Hydraulic turbomachines

In the hydraulic sector, since the fluid density is nearly constant, there are no effects of the fluid compressibility, so the previous equation becomes:

$$\eta = f(\Phi, \Psi, R_e)$$

Completely turbulent flows (Re > 10^6), generally occurring in turbomachines, allow to neglect the effects of the Reynolds number on turbomachinery efficiency (for high Re, friction losses depend only on surface roughness, as seen in Fig. 1.37); in this case, we can consider:

$$\eta = f(\Phi, \Psi) \tag{2.7}$$

If the Reynolds number is lower than that ensuring fully turbulent flow, the efficiency could be corrected in order to take into account this effect:

$$\eta' = \eta \cdot f(R_e) \quad \text{and} \quad f(R_e) < 1$$

Thermal turbomachines

The Reynolds number has the same influence as in hydraulic turbomachines.

Moreover, it can be assumed that efficiency is not affected by fluid compressibility if $M_u < 0.49$; therefore Eq. (2.7) is still valid:

$$\eta = f(\Phi, \Psi)$$

When the Reynolds number is lower than that ensuring completely turbulent flow, or when $M_u > 0.49$, the previous efficiency must be corrected in order to take into account these effects:

$$\eta' = \eta \cdot f(R_e) \cdot f(M_u)$$

where:

$$f(R_e) < 1 \text{ if } R_e < 10^6$$
$$f(M_u) < 1 \text{ if } M_u > 0.49$$

Preliminary remarks

Based on the previous considerations, let us examine two turbomachines: a machine to be designed and an existing turbomachine assumed as a model. *If these two machines are geometrically similar* (the two machines have the same ratios between geometric parameters and, therefore, one is in scale of the other) *and operate in kinematic similarity* (the two machines have the same velocity triangles), *these two turbomachines have the same factors,* Φ *and* Ψ.

It follows that, assuming completely turbulent flow (Re greater than 10^6) and neglecting the effect of compressibility (M_u less than 0.5), *the two machines have the same efficiency.*

In other words, two geometrically similar machines, having the same Φ and Ψ factors, exhibit the same efficiency, always taking the considerations on Re and M_u into account.

Since similitude is not applicable to volumetric and mechanical losses, this efficiency is the isentropic efficiency (thermal machines) or the hydraulic efficiency (hydraulic machines).

2.2 The Baljè's Method

In the previous section, the turbomachinery efficiency was expressed as:

$$\eta = f(\Phi, \Psi)$$

This efficiency can also be expressed as a function of two different dimensionless parameters (Balje 1962a, 1962b):

- the specific speed ω_s
- the specific diameter D_s

$$\eta = f(\omega_s, D_s) \tag{2.8}$$

This formulation is particularly effective for turbomachinery selection and preliminary design (Casey et al. 2010).

The specific speed: ω_s

This dimensionless parameter is obtained by combining the Φ and Ψ factors, Eqs. (2.2) and (2.3), and eliminating the diameter D from them:

$$\left.\begin{array}{ll} \Phi = \dfrac{V}{\omega \cdot D^3} & \Rightarrow \ \mathrm{D} = \left(\dfrac{V}{\omega \cdot \Phi}\right)^{1/3} \\ \Psi = \dfrac{W}{\omega^2 \cdot D^2} & \Rightarrow \ \mathrm{D} = \dfrac{1}{\omega}\left(\dfrac{W}{\Psi}\right)^{1/2} \end{array}\right\} \Rightarrow \quad \frac{\Phi^{1/2}}{\Psi^{3/4}} = \omega \cdot \frac{V^{1/2}}{W^{3/4}}$$

Assuming:

$$\omega_s = \frac{\Phi^{1/2}}{\Psi^{3/4}}$$

we obtain:

$$\omega_s = \omega \cdot \frac{V^{1/2}}{W^{3/4}} \tag{2.9}$$

Since the turbomachinery dimension (D) has been eliminated, we can state that: *the specific speed is indicative of the shape and not of the size of a turbomachine; different-sized turbomachines, having the same specific speed, are in the same shape (for example: axial, radial, mixed flow).*

The specific diameter: D_s

The specific diameter takes into account the size of a turbomachine. This dimensionless parameter is obtained by combining the Φ and Ψ factors, Eqs. (2.2) and (2.3), and eliminating the rotational speed, ω, from them:

$$\left.\begin{array}{ll} \Phi = \dfrac{V}{\omega \cdot D^3} & \Rightarrow \ \omega = \dfrac{V}{\Phi \cdot D^3} \\ \Psi = \dfrac{W}{\omega^2 \cdot D^2} & \Rightarrow \ \omega = \dfrac{W^{1/2}}{D \cdot \Psi^{1/2}} \end{array}\right\} \Rightarrow \quad \frac{\Psi^{1/4}}{\Phi^{1/2}} = D \cdot \frac{W^{1/4}}{V^{1/2}}$$

Assuming:

$$D_s = \frac{\Psi^{1/4}}{\Phi^{1/2}}$$

we obtain:

$$D_s = D \cdot \frac{W^{1/4}}{V^{1/2}} \tag{2.10}$$

Based on the previous considerations, the turbomachine efficiency is a function of the specific speed and the specific diameter:

$$\eta = f(\omega_s, D_s)$$

The Φ and Ψ factors can be immediately expressed through ω_s and D_s. As a matter of fact:

$$\omega_s = \frac{\Phi^{1/2}}{\Psi^{3/4}} \qquad D_s = \frac{\Psi^{1/4}}{\Phi^{1/2}}$$

and, therefore:

$$\Phi = \frac{1}{\omega_s \cdot D_s^3} \qquad \Psi = \frac{1}{\omega_s^2 \cdot D_s^2}$$

For each specific speed, we can identify the specific diameter optimizing the turbomachine efficiency:

$$\eta_{opt} = f(\omega_s, D_{s,opt})$$

and, therefore, we can deduce a univocal correlation between the specific speed and the optimum specific diameter:

$$D_{s,opt} = f(\omega_s)$$

Cordier (1953) identified a statistical correlation between the specific speed and the optimum specific diameter; this correlation, called Cordier's line, is shown in Fig. 2.1. This line initially was referred to pumps; then, it was extended to all driven turbomachines.

Sometimes technical and/or economic reasons advise against adopting the optimum specific diameter. Furthermore, the Cordier line does not provide information on the efficiency attained for each specific speed and on other design parameters (Epple et al. 2011).

Baljé (1962a, 1962b) expanded Cordier's idea by reporting in statistical diagrams the turbomachine efficiency as a function of the specific speed and specific diameter; these diagrams were obtained by examining a large number of both driving and driven

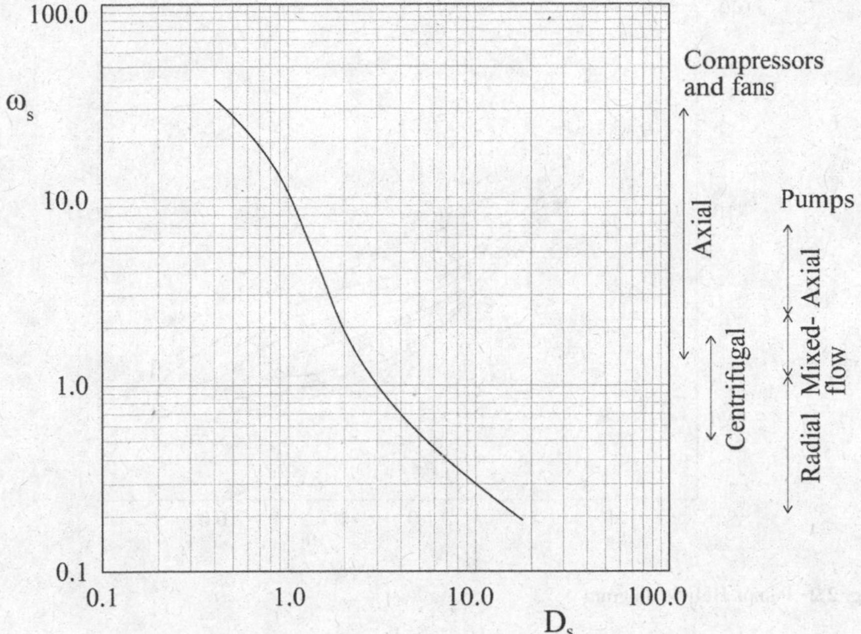

Fig. 2.1 Cordier line (adapted from Dixon and Hall 2014)

turbomachines. These diagrams are called the Baljè diagrams and show graphically Eq. (2.8).

The Baljè diagrams in the following Figs. 2.2, 2.3, 2.4, 2.5, 2.6, 2.7, 2.8, 2.9 (adapted from Cornetti and Millo 2015a, 2015b, 2015c) are referred to the stages of both hydraulic and thermal turbomachines.

Figures 2.2, 2.3 are for different configurations of pump stages: centrifugal, mixed flow, and axial stages.

Work, W, in the expressions of ω_s and D_s is the reversible work, exchanged between stage and fluid flow, and, therefore, it is (Sect. 1.5):

$$W = g \cdot \Delta H$$

The diameter D in the expression of D_s is the rotor external diameter (outlet diameter for radial impellers and tip diameter for axial rotors); the efficiency is the hydraulic efficiency of the stage.

Figure 2.2 shows the iso-efficiency lines on the ω_s–D_s diagram. In this figure, the optimal D_s line (that is, Cordier line) is also sketched.

As the specific speed increases, we find centrifugal stages ($\omega_s = 0.2 \div 1.5$), mixed flow stages ($\omega_s = 1.0 \div 3.5$) and finally axial stages ($\omega_s = 3.0 \div 6.0$). In addition, Fig. 2.3 shows the variation of the efficiency, referred to the optimal D_s line, with the specific speed. Figure 2.3 reveals that efficiencies close to the maximum values require specific speed ranges narrower than the previous ones: $\omega_s = 0.5 \div 1.3$ for

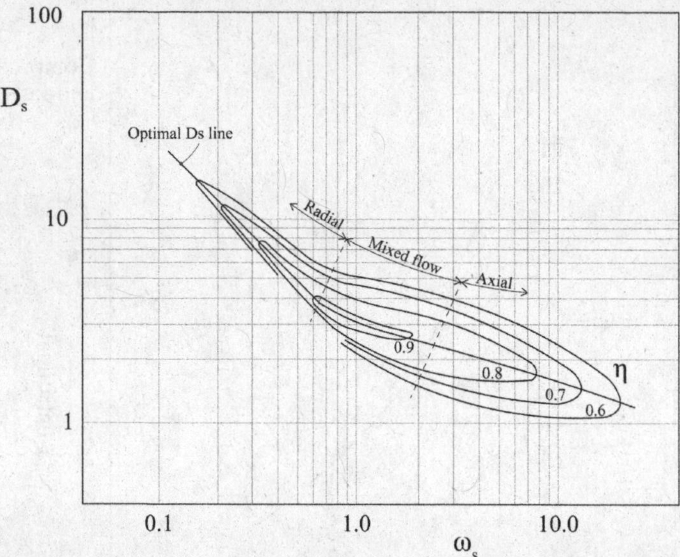

Fig. 2.2 Pumps Baljè's diagram

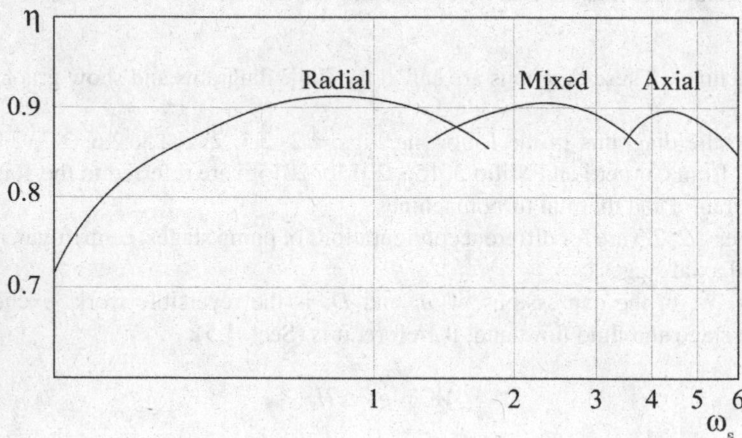

Fig. 2.3 Variation of pump efficiency with specific speed

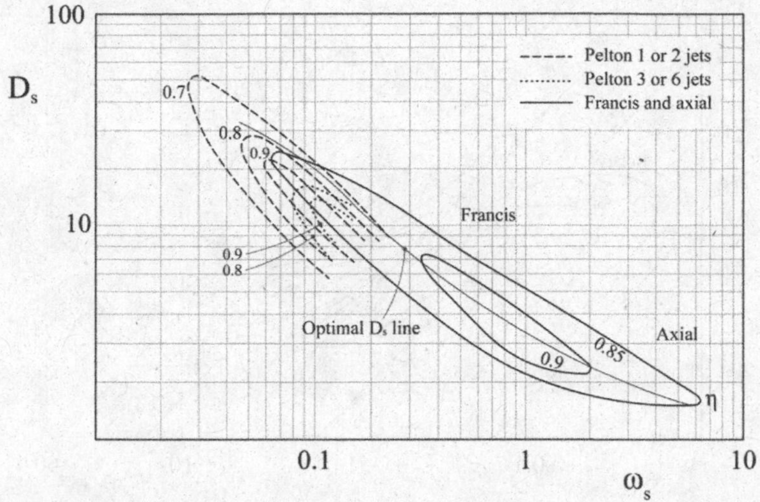

Fig. 2.4 Hydraulic turbines Baljè's diagram

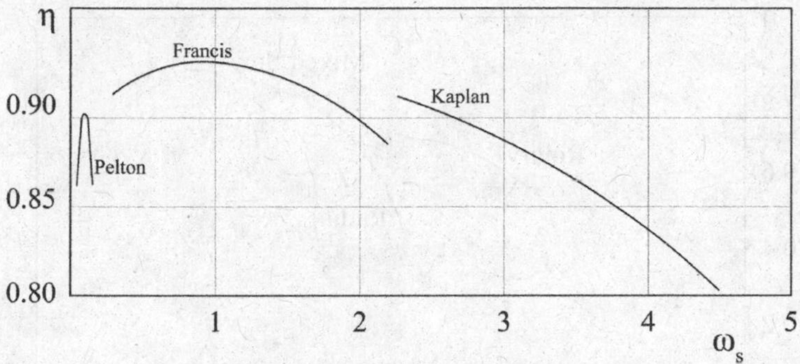

Fig. 2.5 Variation of hydraulic turbine efficiency with specific speed

radial stages, $\omega_s = 1.9 \div 3.0$ for mixed flow stages, and $\omega_s = 4.0 \div 5.0$ for axial stages.

Figures 2.4, 2.5 are for different configurations of hydraulic turbines (single-stage): Pelton, Francis (radial and mixed flow turbines), and Kaplan (axial turbines).

Work, W, in the expressions of ω_s and D_s is the reversible work, exchanged between stage and fluid flow, and, therefore, it is:

$$W = g \cdot \Delta H$$

where ΔH is the head.

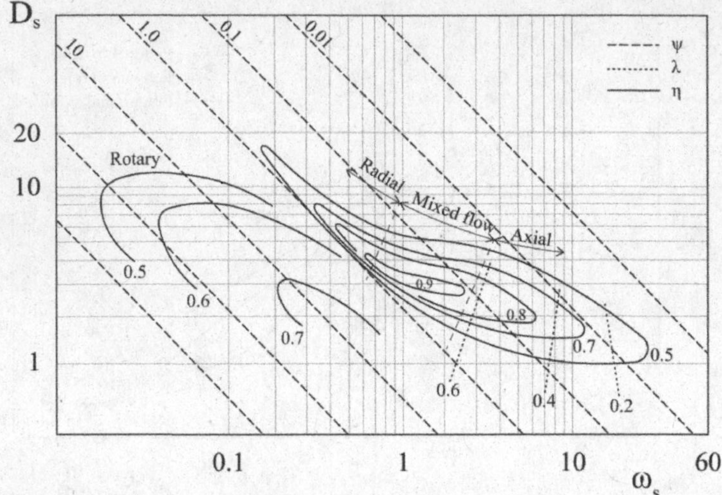

Fig. 2.6 Compressors Baljè's diagram

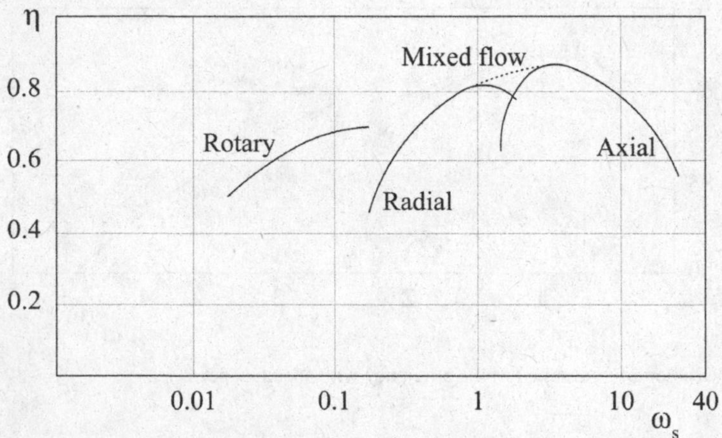

Fig. 2.7 Variation of compressor efficiency with specific speed

The diameter D in the expression of D_s is the rotor external diameter (inlet diameter for radial turbines and tip diameter for axial turbines); the efficiency is the hydraulic efficiency of the stage.

Figure 2.4 shows the iso-efficiency lines on the ω_s–D_s diagram together with the optimal D_s line.

As the specific speed increases, we find Pelton turbines (from single jet configuration, $\omega_s = 0.02 \div 0.1$, to six-jet configuration, $\omega_s \cong 0.3$), Francis turbines ($\omega_s = 0.2 \div 2.5$), and finally Kaplan turbines ($\omega_s = 2.0 \div 5.0$). In addition, Fig. 2.5 shows the variation of the efficiency, referred to the optimal D_s line, with the specific

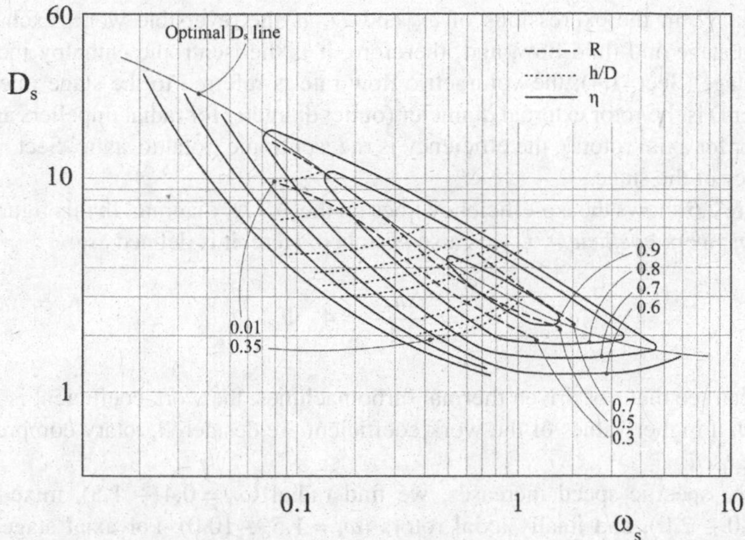

Fig. 2.8 Axial turbines Baljè's diagram

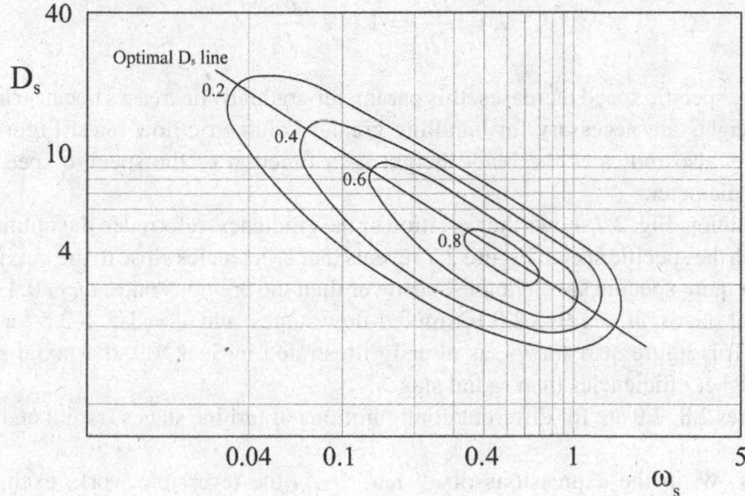

Fig. 2.9 Radial turbines Baljè's diagram

speed. In this figure, it is more evident that efficiencies close to the maximum values require specific speed ranges narrower than the previous ones: $\omega_s = 0.04 \div 0.1$ for Pelton turbines, $\omega_s = 0.5 \div 2.0$ for Francis turbines, and $\omega_s = 2.5 \div 3.5$ for Kaplan turbines.

Figures 2.6, 2.7 are for different configurations of compressor stages (radial, mixed flow and axial stages).

Work, W, in the expressions of ω_s and D_s is the reversible work, exchanged between stage and fluid flow, and, therefore, it is the isentropic enthalpy increase in the stage (Sect. 1.4); the volumetric flow rate is referred to the stage inlet; the diameter D is the rotor external diameter (outlet diameter for radial impellers and tip diameter for axial rotors); the efficiency is the isentropic (total to static, Sect. 1.8.3) efficiency of the stage.

Figure 2.6 shows the iso-efficiency lines on the ω_s–D_s diagram. In this figure, the *isentropic work coefficient,* ψ_{is}, is also sketched. Here, it is defined as:

$$\psi_{is} = \frac{\Delta h_{is}}{u_2^2} = 4 \cdot \Psi_{is}$$

We can see that, for driven thermal turbomachines, the work coefficient is lower than 1.0; if higher values of the work coefficient are desidered, rotary compressors must be used.

As the specific speed increases, we find radial (ω_s= 0.4 ÷ 1.5), mixed-flow (ω_s= 1.0 ÷ 2.0), and finally axial rotors (ω_s= 1.5 ÷ 10.0). For axial stages, the parameter λ is also shown in Fig. 2.6; this parameter is correlated to the blade height (h_B), and is defined as the ratio between the hub (D_h) and tip (D_t) diameters:

$$\lambda = \frac{D_h}{D_t} = 1 - 2 \cdot \frac{h_B}{D_t}$$

As the specific speed increases, this parameter gradually decreases because higher blade heights are necessary for handling greater volumetric flow rates. Figure 2.6, therefore, also indicates the blade height as a function of the specific speed and specific diameter.

In addition, Fig. 2.7 shows the variation of the efficiency, referred to the optimal D_s line, with the specific speed. Figure 2.7 reveals that efficiencies close to the maximum values require specific speed ranges narrower than the previous ones: ω_s= 0.4 ÷ 1.0 for radial stages, ω_s= 1.0 ÷ 2.0 for mixed-flow stages, and ω_s= 1.5 ÷ 2.5 for axial stages. This figure also shows, as already illustrated in Sect. 1.9, that axial stages attain higher efficiencies than radial stages.

Figures 2.8, 2.9 are for different configurations of turbine stages (radial and axial stages).

Work, W, in the expressions of ω_s and D_s is the reversible work, exchanged between stage and fluid flow, and, therefore, it is the isentropic enthalpy drop in the stage (Sect. 1.4); the volumetric flow rate is referred to the stage outlet; the diameter D is the rotor external diameter (inlet diameter for radial turbines and tip diameter for axial turbines); the efficiency is the isentropic (total to static, Sect. 1.8.2) efficiency of the stage.

Figure 2.8 shows the iso-efficiency lines on the ω_s–D_s diagram for axial stage turbine. In this figure the degree of reaction (R) and the ratio of the blade height to the rotor tip diameter (h_B/D_t) are also shown. As the specific speed increases, blade heights increase in relation to tip diameters, while the degree of reaction, closer to the Cordier line and therefore to the optimum efficiency, is around 0.5.

Analogously to compressors, axial turbine stages (Fig. 2.8) attain higher efficiencies than those of radial stages (Fig. 2.9).

Radial turbine stages have a specific speed in the $\omega_s = 0.2 \div 1.0$ range and axial turbines in the $\omega_s = 0.4 \div 3.0$ range. However, achieving efficiencies close to the maximum values requires specific speed ranges narrower than the previous ones: $\omega_s = 0.4 \div 0.8$ for radial stages and $\omega_s = 0.6 \div 1.2$ for axial stages.

2.3 Selection Process

The similitude theory, and in particular the Baljè's method, allows identifying the type of turbomachine stages (axial, radial, mixed flow) for a specific application.

Indeed, Baljè diagrams, as illustrated in Sect. 2.2, provide the specific speed range for each type of turbomachine stage.

These ranges are summarized in Table 2.1 as well as in Fig. 2.10.

For each type of turbomachine stage, the specific speed range in Table 2.1 and Fig. 2.10 becomes narrower if we want to attain high-efficiency stages. Table 2.2 reports the specific speed ranges ensuring stage efficiencies close to the maximum values.

The turbomachinery selection implies the preliminary evaluation of ω_s.

The specific application provides the following information:

- type of fluid
- mass flow rate (kg/s)
- inlet temperature (° C)
- inlet pressure (bar)
- outlet pressure (bar)

Table 2.1 Turbomachinery specific speed ranges

Turbomachines	Type	Specific speed
Pumps	Radial	0.2÷1.5
	Mixed flow	1.0÷3.5
	Axial	3.0÷6.0
Hydraulic turbines	Pelton	0.02÷0.3
	Francis	0.2÷2.5
	Kaplan	2.0÷5.0
Compressors	Radial	0.4÷1.5
	Mixed flow	1.0÷2.0
	Axial	1.5÷10
Turbines	Radial	0.2÷1.0
	Axial	0.4÷3.0

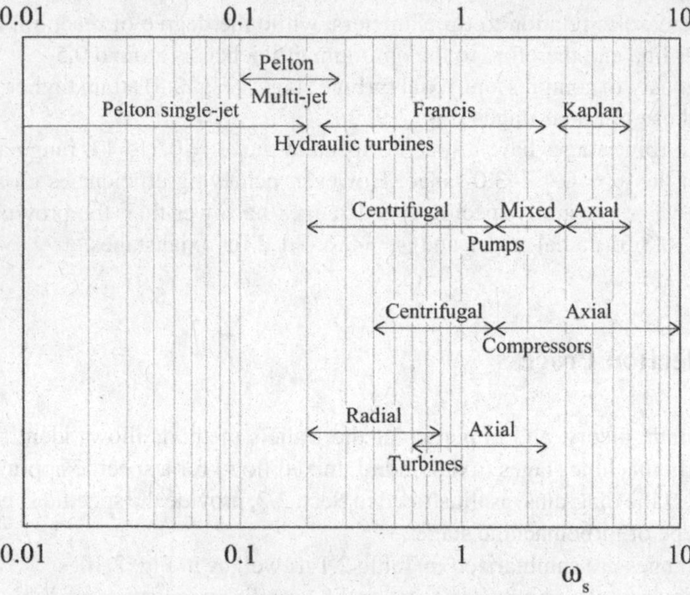

Fig. 2.10 Turbomachines specific speed (adapted from Lewis 1996)

Table 2.2 High-efficiency turbomachinery specific speed ranges

Turbomachines	Type	Specific speed
Pumps	Radial	0.5÷1.3
	Mixed flow	1.9÷3.0
	Axial	4.0÷5.0
Hydraulic turbines	Pelton	0.04÷0.1
	Francis	0.5÷2.0
	Kaplan	2.5÷3.5
Compressors	Radial	0.4÷1.0
	Mixed flow	1.0÷2.0
	Axial	1.5÷2.5
Turbines	Radial	0.4÷0.8
	Axial	0.6÷1.2

By using these parameters, the turbomachinery reversible work calculation is straightforward: it is the isentropic enthalpy change in thermal machines, Δh_{is}, or the head in hydraulic machines, ΔH. However, the specific speed calculation is not so immediate as it is also necessary to know the speed of rotation (n or ω) and, with the exception of the hydraulic turbines (always single-stage turbomachines), the number of stages z, as ω_s is the specific speed of a single stage.

Consider, for example, turbomachines handling compressible fluids; the specific speed is:

$$\omega_s = \omega \cdot \frac{V^{1/2}}{W_{rev}^{3/4}} = \frac{2 \cdot \pi \cdot n}{60} \cdot \frac{V^{1/2}}{\Delta h_{is,stage}^{3/4}} \tag{2.11}$$

where V is the volumetric flow rate at the compressor stage inlet or turbine stage outlet.

For hydraulic turbomachines (pumps, since turbines are always single-stage turbomachines), the specific speed is expressed as:

$$\omega_s = \omega \cdot \frac{V^{1/2}}{W_{rev}^{3/4}} = \frac{2 \cdot \pi \cdot n}{60} \cdot \frac{V^{1/2}}{\left(g \cdot \Delta H_{stage}\right)^{3/4}} \tag{2.12}$$

For these reasons, the reversible work must be distributed among the stages composing the turbomachine. This operation can be based on different criteria: it is possible to establish the pressure ratio of each stage (for example, the same pressure ratio in all stages), or the specific speed (for example, the same specific speed in all stages), or the stage reversible work (for example, the same reversible work in all stages).

By choosing the latter criterion and assuming to divide equally the reversible work among the z stages:

$$\Delta h_{is,stage} = \frac{\Delta h_{is}}{z} \quad ; \quad \Delta \mathrm{H}_{stage} = \frac{\Delta H}{z}$$

the specific speed of the stage can be expressed as:

$$\omega_s = \frac{2 \cdot \pi \cdot n}{60} \cdot \frac{V^{1/2}}{\left(\frac{\Delta h_{is}}{z}\right)^{3/4}} \quad ; \quad \omega_s = \frac{2 \cdot \pi \cdot n}{60} \cdot \frac{V^{1/2}}{\left(\frac{g \cdot \Delta H}{z}\right)^{3/4}}$$

and, therefore:

$$\omega_s = f(n, z)$$

Under these assumptions, pumps will have the same specific speed in each stage (volumetric flow rate is constant). Thermal turbomachines will have instead different ones in each stage: the stage-specific speed increases from the first to the last stage for turbines and decreases for compressors, since the fluid volumetric flow rate varies between the turbomachinery inlet and outlet.

For pumps, diagrams such as the one in Fig. 2.11 allow choosing immediately n and z to ensure a suitable specific speed, compatibly with the machine size (z), and with possible constraints on the speed of rotation (e.g. cavitation limits, Chap. 7).

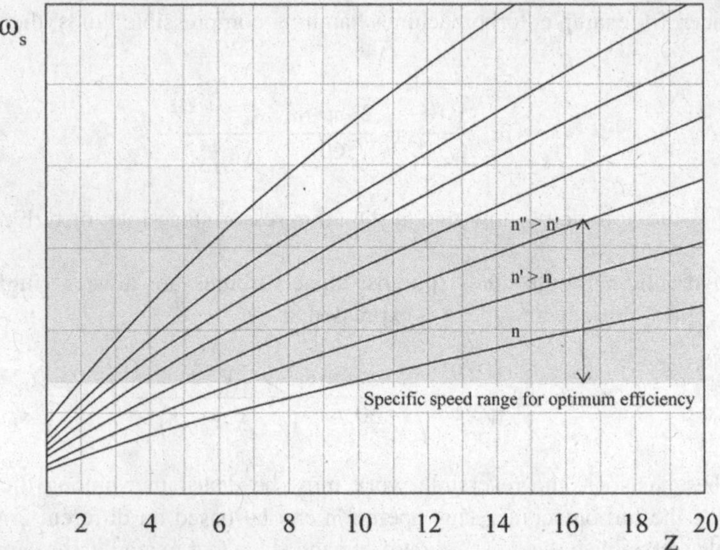

Fig. 2.11 Basic diagram for n-z selection (pumps)

For thermal turbomachines, on the other hand, a preliminary assessment of the volumetric flow rate at the stage inlet (compressors) or outlet (turbines) is necessary. Since pressures and temperatures in upstream/downstream stages can be calculated by using $\Delta h_{is,stage}$ (Figs. 2.12–2.13, respectively for turbines and compressors), volumetric flow rates are consequently known. In this way, diagrams such as those shown in Figs. 2.14, 2.15, respectively for turbines and compressors, can be sketched: on these diagrams, the specific speed of the first and last stage are shown. These diagrams allow us to choose the speed of rotation, n, and the number of stages, z, to ensure a suitable specific speed, compatibly with the machine size (z), and

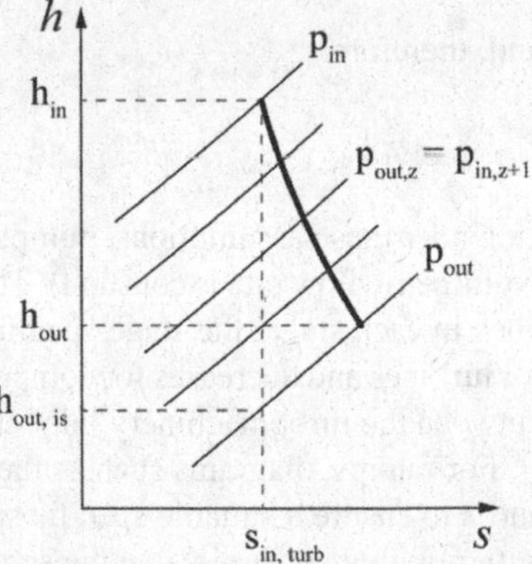

Fig. 2.12 Expansion subdivision among stages

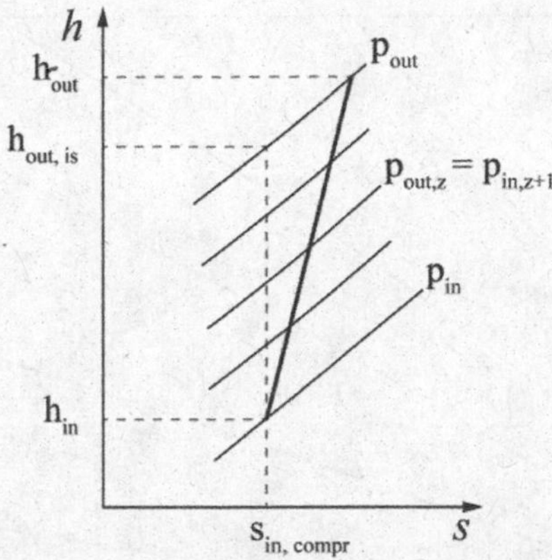

Fig. 2.13 Compression subdivision among stages

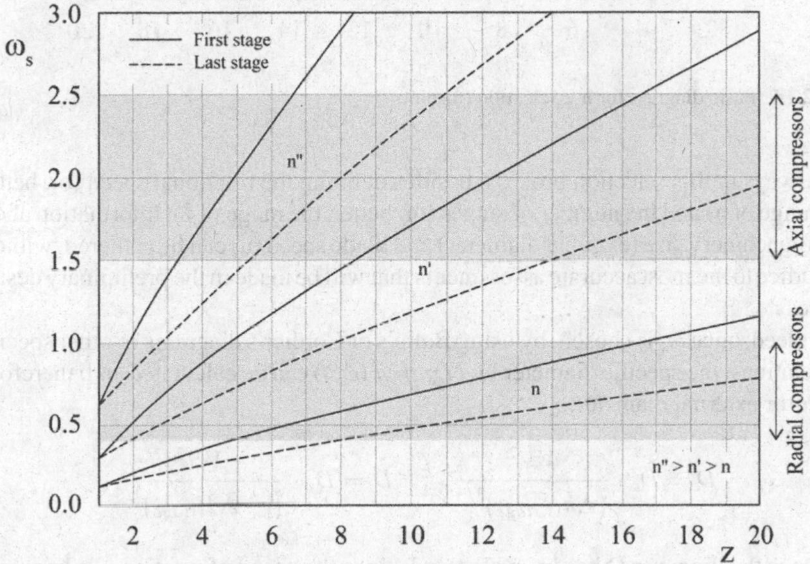

Fig. 2.14 Basic diagram for n-z selection (compressor)

with possible constraints on the speed of rotation (e.g. blade speed, u, limits due to mechanical stresses).

In this regard, however, a clarification is needed. These diagrams (Figs. 2.11, 2.14, and 2.15) do not provide, in general, a single value of n and z, ensuring an appropriate specific speed range. It is, therefore, more correct to affirm that this turbomachinery selection process provides some ranges of n ($n_{min} < n < n_{max}$) and z ($z_{min} < z < z_{max}$) constituting the input for the following preliminary design process (Chaps. 3-7), where a specific value of n and z will be identified to attain the best performance (in terms of efficiency and/or overall dimensions of the machine).

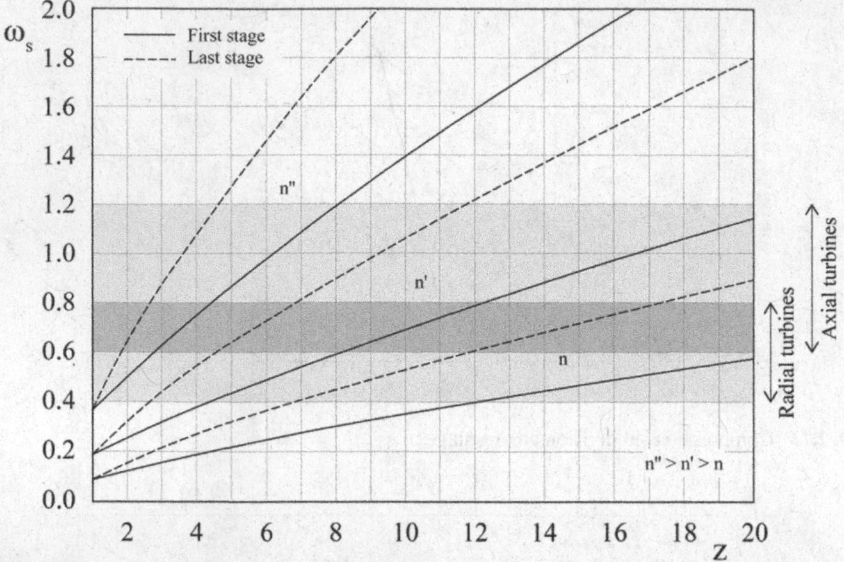

Fig. 2.15 Basic diagram for n-z selection (turbine)

Always in this selection process, besides defining the rotational speed (or, better, the range of n) and the number of stages (or, better, the range of z), information about turbomachinery size (external diameter) and blade speed (u) can be gathered, without prejudice to the most accurate assessments that will be made in the preliminary design process.

Indeed, once ω_s is chosen, by using Baljé's or Cordier's diagrams or other specific correlations, the specific diameter D_s ($D_s = f(\omega_s)$) can be calculated and, therefore, the rotor external diameter:

$$D = D_s \cdot \frac{V^{1/2}}{\left(\Delta h_{is,stage}\right)^{1/4}} \quad ; \quad D = D_s \cdot \frac{V^{1/2}}{\left(g \cdot \Delta H_{stage}\right)^{1/4}}$$

Once the diameter D has been assessed, since the speed of rotation n is known, it is immediate to calculate the blade speed, u, corresponding to the diameter D:

$$u = \frac{\pi \cdot n}{60} \cdot D$$

The values of the blade speed, u, are fundamental in the turbomachinery selection process because there are specific limits to be respected both for thermal and hydraulic turbomachines (Chaps. 3-8).

Finally, in order to calculate the specific diameter, besides the well-established Baljé's or Cordier's diagrams, some analytical correlations are also available, such as:

$$D_s = 2.781 \cdot \omega_s^{-0.988} \quad \text{when} \quad \omega_s < 0.84$$
$$D_s = 3.147 \cdot \omega_s^{-0.281} \quad \text{when} \quad \omega_s \geq 0.84 \tag{2.13}$$

for pumps and compressors, and:

$$D_s = 2.56 \cdot \omega_s^{-0.812} \tag{2.14}$$

for turbines.

It is important to highlight that these correlations can only be used to have a preliminary indication of the machine size; in the following Chaps. (3-7), indeed, we will proceed to a more accurate and in-depth assessment of the geometry of all the stages constituting the turbomachine.

In conclusion, the output of the selection process, illustrated in this chapter, consists of:

- type of stages (radial, mixed flow, axial stages)
- ranges of the speed of rotation ($n_{\text{min}} < n < n_{\text{max}}$) and of the number of stages ($z_{\text{min}} < z < z_{\text{max}}$) which ensure a suitable specific speed of each stage.

This output will be the input of the procedures developed in the following Chaps. 3-7.

References

Balje OE (1962) A study on design criteria and matching of turbomachines: part A–similarity relations and design criteria of turbines. Trans ASME

Balje OE (1962) A study on design criteria and matching of turbomachines: part B–compressor and pump performance and matching of turbocomponents. Trans ASME

Casey M, Zwyssig C, Robinson C (2010) The Cordier line for mixed flow compressors. In: Proceedings of ASME Turbo Expo 2010, GT2010, Glasgow, UK

Cordier O (1953) Ähnlichkeitsbedingungen für Strömungsmaschinen, BWK Bd. 6, Nr. 10

Cornetti G, Millo F (2015a) Hydraulic machines (in Italian). Il Capitello, Torino

Cornetti G, Millo F (2015b) Gas machines (in Italian). Il Capitello, Torino

Cornetti G, Millo F (2015c) Thermal science and steam machines (in Italian). Il Capitello, Torino

Dixon SL, Hall CA (2014) Fluid mechanics and thermodynamics of turbomachinery, 7th edn. Elsevier

Epple Ph, Durst F, Delgado A (2011) Theoretical derivation of the cordier diagram for turbomachines. Proc IMechE Part C: J Mech Eng Sci

Lewis RI (1996) Turbomachinery performance analysis. Elsevier

Chapter 3
Preliminary Design of Axial Flow Turbines

Abstract With reference to an axial turbine stage, a procedure for the calculation of kinematic parameters at mean diameter (Sect. 3.1), thermodynamic parameters (Sect. 3.2), geometric parameters (Sect. 3.3), parameters in the radial direction (Sect. 3.4) and stage losses (Sect. 3.5) is provided. Then, Sect. 3.6 discusses the input parameters of this procedure and suggests their numerical values to be used in the calculations. Finally, Sect. 3.7 illustrates the procedure for extending calculations to multistage turbines. The numerical application of the proposed procedure, aimed at the preliminary design of multistage axial turbines, is instead developed in Chap. 8.

3.1 Kinematic Parameters

In this section, we will deduce the kinematic correlations of an axial stage; they are correlations among kinematic parameters of a stage (degree of reaction R, absolute and relative flow angles, α and β) and dimensionless performance parameters (flow coefficient, φ, and work coefficient, ψ).

Variations of the previously introduced (Chap. 2) dimensionless factors Φ and Ψ are used in practice.

In particular, the ratio of the meridional component of the absolute velocity to the blade speed is the *flow coefficient*. Consequently, for axial stages where $u_1=u_2$, assuming c_m=const, the flow coefficient is:

$$\varphi = \frac{c_m}{u}$$

Note that there is a proportionality between this flow coefficient and the flow factor, previously introduced (Chap. 2):

$$\Phi = \frac{V}{\omega \cdot D^3} \propto \frac{c_m \cdot D^2}{u \cdot D^2} = \frac{c_m}{u} = \varphi$$

M. Gambini and M. Vellini, *Turbomachinery*, Springer Tracts in Mechanical Engineering, https://doi.org/10.1007/978-3-030-51299-6_3

Analogously, the ratio of actual work exchanged between fluid flow and rotor to blade speed squared is the *work coefficient*:

$$\psi = \frac{W}{u^2} = \frac{\Delta h_{t,ad}}{u^2}$$

which is proportional to the work factor, Ψ, previously introduced (Chap. 2).

In an axial stage, the fluid flow is studied (Sect. 1.7) on a cylindrical meridional section of blades developed on a plane (Fig. 1.14, for convenience, still reproduced in Fig. 3.1a).

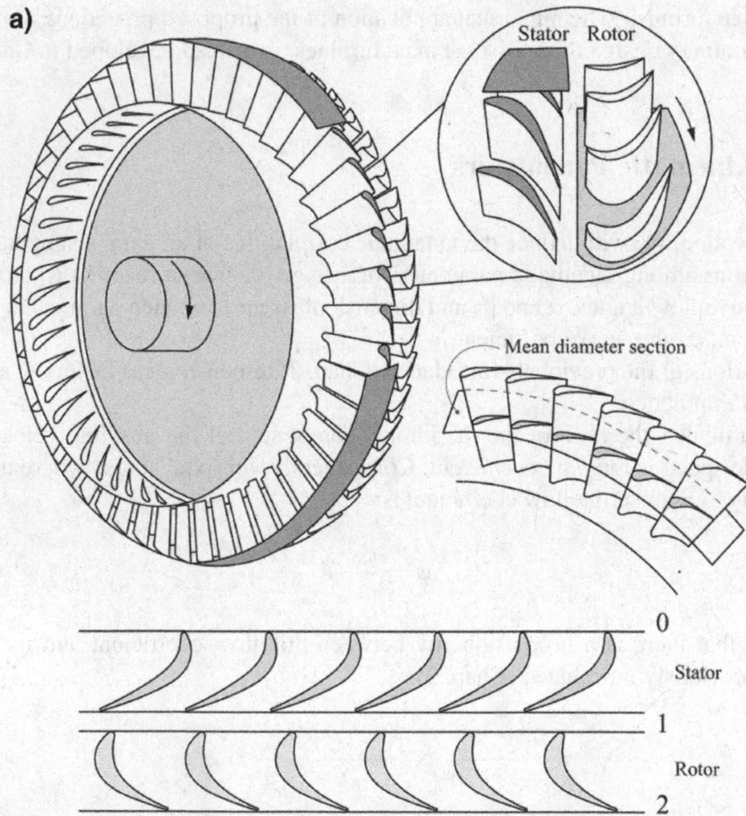

Fig. 3.1 **a** Axial turbine stage: stator and rotor representation. **b** Velocity triangles in an axial turbine stage

Fig. 3.1 (continued)

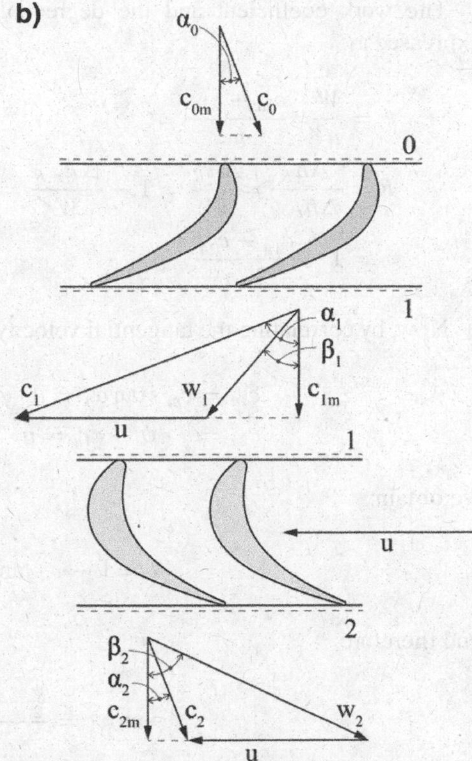

With reference to this representation, Fig. 3.1b depicts the velocity triangles of an axial turbine stage. These triangles refer to the mean line, that is, to the mean diameter of the stage (Sect. 1.7.1).

For an axial stage $u_1 = u_2 = u$ (Sect. 1.7.3). In addition, we can assume that the meridional component of the absolute velocity is constant, as well as the mean diameter:

$$c_m = const$$
$$D_M = const$$

Next, assuming the angles shown in Fig. 3.1 as positive (i.e. c_{1u} and c_{2u} are in opposite directions), the energy equation applied to a turbine stage (Sect. 1.8.2) and Euler's equation (Sect. 1.6.1) provide:

$$W = \Delta h_{t,ad} = h_{0t} - h_{2t} = u \cdot \Delta c_u = u \cdot (c_{1u} + c_{2u}) = u \cdot \Delta w_u = u \cdot (w_{2u} + w_{1u})$$

The work coefficient and the degree of reaction (Sect. 1.8.2) are therefore expressed as:

$$\psi = \frac{W}{u^2} = \frac{c_{1u} + c_{2u}}{u}$$

$$R = \frac{\Delta h_R}{\Delta h_{t,ad}} = \frac{\Delta h_R}{W} = 1 - \frac{\Delta E_{c,R}}{W} = 1 - \frac{c_1^2 - c_2^2}{2 \cdot W} = 1 - \frac{c_{1u}^2 - c_{2u}^2}{2 \cdot W}$$
$$= 1 - \frac{c_{1u} - c_{2u}}{2 \cdot u}$$

Now, by correlating the tangential velocity components to the angle α_1:

$$c_{1u} = c_m \cdot \tan \alpha_1 = u \cdot \varphi \cdot \tan \alpha_1$$
$$c_{2u} = u \cdot \psi - c_{1u} = u \cdot \psi - u \cdot \varphi \cdot \tan \alpha_1$$

we obtain:

$$R = 1 - \varphi \cdot \tan \alpha_1 + \frac{\psi}{2}$$

and therefore:

$$\tan \alpha_1 = \frac{\frac{\psi}{2} - R + 1}{\varphi}$$

Following a similar procedure, we can also obtain the angle α_2:

$$c_{2u} = c_m \cdot \tan \alpha_2 = u \cdot \varphi \cdot \tan \alpha_2$$
$$c_{1u} = u \cdot \psi - c_{2u} = u \cdot \psi - u \cdot \varphi \cdot \tan \alpha_2$$
$$R = 1 + \varphi \cdot \tan \alpha_2 - \frac{\psi}{2}$$

and therefore:

$$\tan \alpha_2 = \frac{\frac{\psi}{2} + R - 1}{\varphi}$$

The above equation shows that if the angle α_2 was chosen as input to minimize the kinetic energy exiting the stage (that is, $\alpha_2 = 0$), the degree of reaction, R, would have been set by the work coefficient ψ:

$$R_{(\alpha_2=0)} = 1 - \frac{\psi}{2}$$

For angles of relative velocities we obtain:

$$\tan\beta_1 = \frac{w_{1u}}{c_m} = \frac{c_{1u}-u}{c_m} = \frac{1}{\varphi}\cdot\frac{c_{1u}-u}{u} = \frac{1}{\varphi}\cdot(\varphi\cdot\tan\alpha_1 - 1) = \tan\alpha_1 - \frac{1}{\varphi}$$

$$\tan\beta_2 = \frac{w_{2u}}{c_m} = \frac{c_{2u}+u}{c_m} = \frac{1}{\varphi}\cdot\frac{c_{2u}+u}{u} = \frac{1}{\varphi}\cdot(\varphi\cdot\tan\alpha_2 + 1) = \tan\alpha_2 + \frac{1}{\varphi}$$

and therefore:

$$\tan\beta_1 = \frac{\frac{\psi}{2} - R}{\varphi}$$

$$\tan\beta_2 = \frac{\frac{\psi}{2} + R}{\varphi}$$

Through the above expressions, we can obtain the stator and rotor deflection angles; they represent the flow direction changes in the stator and in the rotor.

For a stator, considering $\alpha_0 = \alpha_2$, we have:

$$\delta_S = \alpha_1 + \alpha_0 = \alpha_1 + \alpha_2$$

and so:

$$\tan\delta_S = \tan(\alpha_1 + \alpha_2) = \frac{\tan\alpha_1 + \tan\alpha_2}{1 - \tan\alpha_1\cdot\tan\alpha_2}$$

and:

$$\tan\delta_S = \frac{\frac{\psi}{\varphi}}{1 + \frac{1}{\varphi^2}\cdot\left[(R-1)^2 - \left(\frac{\psi}{2}\right)^2\right]}$$

Similarly, the rotor deflection angle is:

$$\delta_R = \beta_2 + \beta_1$$

and so:

$$\tan\delta_R = \tan(\beta_2 + \beta_1) = \frac{\tan\beta_2 + \tan\beta_1}{1 - \tan\beta_1\cdot\tan\beta_2}$$

$$\tan\delta_R = \frac{\frac{\psi}{\varphi}}{1 + \frac{1}{\varphi^2}\cdot\left[R^2 - \left(\frac{\psi}{2}\right)^2\right]}$$

We can observe that if R = 0.5, the stator deflection angle is equal to the rotor deflection angle.

Remarks on input parameters

From the kinematic correlations reported above, we infer that once the three independent parameters φ, ψ and R are set (Sect. 3.6), all flow angles can be calculated. Once the blade speed, u, is also established (in order to determine this speed, it is necessary to know, in addition to the work coefficient ψ, also the stage isentropic efficiency (or the polytropic efficiency, as illustrated in Sect. 3.6), the kinematics at the mean diameter of the rotor can be completely calculated:

$$\begin{aligned} c_m &= u \cdot \varphi \\ c_{1u} &= u \cdot \varphi \cdot \tan \alpha_1 \\ c_{2u} &= u \cdot \varphi \cdot \tan \alpha_2 \\ w_{1u} &= u \cdot \varphi \cdot \tan \beta_1 \\ w_{2u} &= u \cdot \varphi \cdot \tan \beta_2 \\ c_1 &= u \cdot \varphi \cdot \sqrt{1 + \tan^2 \alpha_1} \\ c_2 &= u \cdot \varphi \cdot \sqrt{1 + \tan^2 \alpha_2} \\ w_1 &= u \cdot \varphi \cdot \sqrt{1 + \tan^2 \beta_1} \\ w_2 &= u \cdot \varphi \cdot \sqrt{1 + \tan^2 \beta_2} \end{aligned}$$

In order to complete the kinematics of the stage, the velocity c_0, the absolute velocity at the stage inlet, must also be determined.

Generally (*repeating stage,* also called *normal stage*) we assume:

$$c_0 = c_2 \quad ; \quad \alpha_0 = \alpha_2$$

This assumption also allows ensuring, for multistage turbines, that the absolute velocity at the stator inlet is equal to the absolute velocity at the upstream stage outlet.

If $c_0 \neq c_2$, the blade speed, u, can still be calculated as a function of the work coefficient, but now:

$$\Delta h_{t,ad} \neq \Delta h_{ad}$$

so we must remember that:

$$\Delta h_{t,ad} = \Delta h_{ad} + \frac{c_0^2 - c_2^2}{2} \cdot$$

3.2 Thermodynamic Parameters

The pressure p_0 and temperature T_0 at the inlet of the stator and the pressure p_2 at the outlet of the rotor are assigned based on what is illustrated in Chap. 2 (for multistage turbines, these values are calculated by the procedure described in Sect. 3.7).

Now, all the thermodynamic parameters at the stator inlet and at the rotor inlet/outlet can be calculated as follows.

If the working fluid cannot be assimilated to a perfect gas, for the following calculations, we need appropriate state functions, such as those of the NIST libraries. In the following, we will refer to a real working fluid.

Knowing p_0, T_0, we can calculate:

$$h_0 = h_{p,T}(p_0, T_0)$$
$$s_0 = s_{p,T}(p_0, T_0)$$
$$\rho_0 = \rho_{p,T}(p_0, T_0)$$

Once c_0 is established (Sect. 3.1), the total quantities (Sect. 1.8.1) can be evaluated (Fig. 3.2):

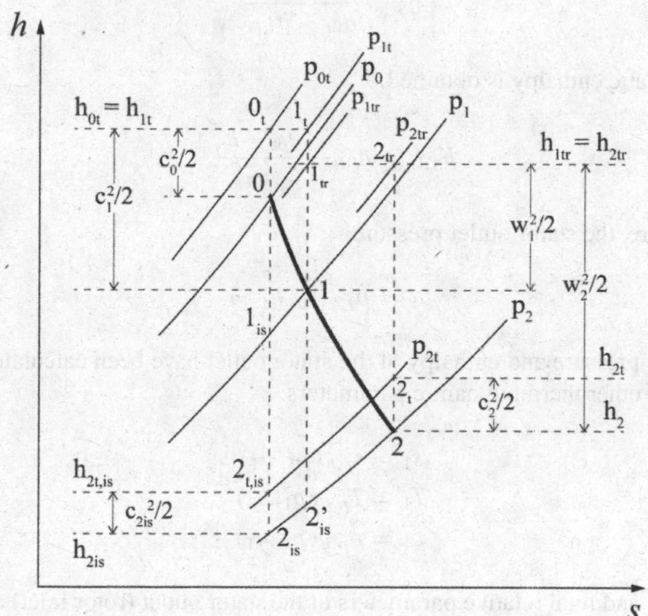

Fig. 3.2 Expansion process in an axial turbine stage

$$h_{0t} = h_0 + \frac{c_0^2}{2}$$
$$p_{0t} = p_{h,s}(h_{0t}, s_0)$$
$$T_{0t} = T_{h,s}(h_{0t}, s_0)$$
$$\rho_{0t} = \rho_{h,s}(h_{0t}, s_0)$$

To calculate the thermodynamic parameters at the stator outlet, we must know the stator efficiency and remember the total enthalpy conservation in the nozzle (Sect. 1.8.2):

$$h_{1t} = h_{0t}$$

In fact, the static enthalpy is calculated through the previous equation:

$$h_1 = h_{1t} - \frac{c_1^2}{2}$$

while using the expression of the stator efficiency (Sect. 1.8.2):

$$\eta_S = \frac{h_{0t} - h_1}{h_{0t} - h_{1,is}}$$

isentropic static enthalpy is obtained:

$$h_{1,is} = h_{0t} - \frac{h_{0t} - h_1}{\eta_S}$$

and, therefore, the stator outlet pressure:

$$p_1 = p_{s,h}(s_0, h_{1,is})$$

Once the pressure and enthalpy at the stator outlet have been calculated, we can evaluate the other thermodynamic parameters:

$$s_1 = s_{p,h}(p_1, h_1)$$
$$T_1 = T_{p,h}(p_1, h_1)$$
$$\rho_1 = \rho_{p,h}(p_1, h_1)$$

The total and total relative parameters at the stator outlet (rotor inlet) are then:

$$p_{1t} = p_{h,s}(h_{1t}, s_1)$$
$$T_{1t} = T_{h,s}(h_{1t}, s_1)$$
$$\rho_{1t} = \rho_{h,s}(h_{1t}, s_1)$$

$$h_{1tr} = h_1 + \frac{w_1^2}{2}$$
$$p_{1tr} = p_{h,s}(h_{1tr}, s_1)$$
$$T_{1tr} = T_{h,s}(h_{1tr}, s_1)$$

Losses in the stator are related to the entropy increase $(s_1 - s_0)$ due to irreversibilities and entail an increase in the final enthalpy $(h_1 - h_{1,is})$ and a reduction in the total pressure expressed by the loss coefficient (Sect. 1.9.5):

$$Y_S = \frac{p_{0t} - p_{1t}}{p_{1t} - p_1}$$

At the rotor outlet, the pressure is known; therefore the enthalpy, referred to the isentropic expansion, can be calculated as:

$$h_{2,is} = h_{p,s}(p_2, s_0)$$

The final enthalpy in the adiabatic expansion can be determined by applying the conservation of the total relative enthalpy in the rotor (Sect. 1.8.2):

$$h_{2tr} = h_{1tr}$$

$$h_2 = h_{2tr} - \frac{w_2^2}{2}$$

or by applying the stage isentropic efficiency (referred to static enthalpy drops, Sect. 1.8.2):

$$\eta_{is} = \frac{h_0 - h_2}{h_0 - h_{2,is}}$$
$$h_2 = h_0 - \eta_{is} \cdot (h_0 - h_{2,is})$$

The calculation of the thermodynamic parameters can, therefore, be completed:

$$s_2 = s_{p,h}(p_2, h_2)$$
$$T_2 = T_{p,h}(p_2, h_2)$$
$$\rho_2 = \rho_{p,h}(p_2, h_2)$$
$$h_{2t} = h_2 + \frac{c_2^2}{2}$$
$$p_{2t} = p_{h,s}(h_{2t}, s_2)$$
$$T_{2t} = T_{h,s}(h_{2t}, s_2)$$
$$\rho_{2t} = \rho_{h,s}(h_{2t}, s_2)$$

$$p_{2tr} = p_{h,s}(h_{2tr}, s_2)$$
$$T_{2tr} = T_{h,s}(h_{2tr}, s_2)$$

Losses in the rotor are related to the entropy increase $(s_2 - s_1)$ due to irreversibilities and entail an increase in the final enthalpy $(h_2 - h_{2',is})$ and a reduction in the total relative pressure expressed by the loss coefficient (Sect. 1.9.5):

$$Y_R = \frac{p_{1tr} - p_{2tr}}{p_{2tr} - p_2}$$

From the knowledge of the kinematics and thermodynamics of the stage, the absolute (stator inlet/outlet) and relative (rotor inlet/outlet) Mach numbers can be calculated. From the properties of the working fluid, first, the speeds of sound are calculated:

$$c_{s,0} = c_{s,p,T}(p_0, T_0)$$
$$c_{s,1} = c_{s,p,T}(p_1, T_1)$$
$$c_{s2} = c_{s,p,T}(p_2, T_2)$$

and, therefore, the Mach numbers:

$$Ma_0 = \frac{c_0}{c_{s0}}$$
$$Ma_{1,S} = \frac{c_1}{c_{s1}}$$
$$Ma_{1,R} = \frac{w_1}{c_{s1}}$$
$$Ma_2 = \frac{w_2}{c_{s2}}$$

By using the pressure and temperature conditions of the working fluid, we can calculate the fluid viscosities (necessary to assess the Reynolds numbers, Sect. 3.3):

$$\mu_0 = \mu_{p,T}(p_0, T_0)$$
$$\mu_1 = \mu_{p,T}(p_1, T_1)$$
$$\mu_2 = \mu_{p,T}(p_2, T_2)$$

Remarks on input parameters

Based on what is illustrated in this Sect. 3.2, in addition to the parameters already identified in Sect. 3.1 (φ, ψ, R e η_{is}), the stator efficiency must be assumed in order to calculate the thermodynamics of the stage.

3.3 Geometric Parameters

Now we will deduce the geometric correlations of an axial stage; they are correlations among geometric (blade height, tip and hub diameters, chord and blade pitch, and so on), thermodynamic (density, volumetric flow rate) and kinematic parameters (meridional velocities, absolute velocity angles α, relative velocity angles β, stator and rotor deflection angles) of the stage.

With reference to Fig. 3.3, it is assumed that between the stage inlet and outlet:

- the mean diameter is constant:

$$D_M = const$$

consequently, blade speed at mean diameter is constant:

$$u_M = const$$

- the meridional component of the velocity is constant:

$$c_m = const$$

The three parameters listed above depend on the kinematics of the stage (Sect. 3.1) and on the rotational speed (n, rpm) of the machine (Sect. 3.6):

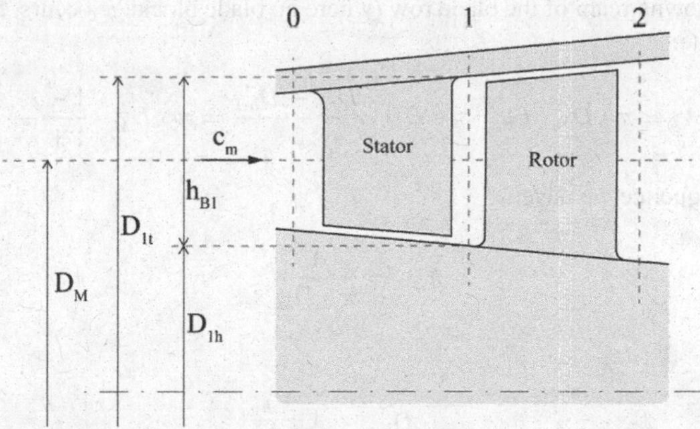

Fig. 3.3 Stage geometry (meridional section)

$$D_M = \frac{60}{\pi} \cdot \frac{u_M}{n}$$

Volumetric flow rates in the stage (0—stator inlet, 1—stator outlet/rotor inlet, 2—rotor outlet) are a function of the assigned mass flow rate and fluid densities calculated in Sect. 3.2:

$$V_i = \frac{m}{\rho_i} \qquad i = 0, 1, 2$$

By using these volumetric flow rates, we can calculate the flow areas:

$$A_i = \frac{V_i}{c_m}$$

They are correlated with the geometric parameters of a turbomachine's meridional section (tip diameter, D_t, hub diameter, D_h, and blade height, h_B) represented in Fig. 3.3.

By introducing the following geometric ratios:

$$\lambda_i = \frac{D_{h,i}}{D_{t,i}} \quad \text{and} \quad \frac{h_{B,i}}{D_M} = \frac{D_{t,i} - D_{h,i}}{2 \cdot D_M}$$

the previous geometric parameters can be expressed as a function of the mean diameter of the stage. In fact:

$$D_M = \frac{D_{t,i} + D_{h,i}}{2} = \mathrm{D}_{t,i} \cdot \frac{1 + \lambda_i}{2} = \mathrm{D}_{h,i} \cdot \frac{1 + \lambda_i}{2 \cdot \lambda_i}$$

and considering the expression of the flow area in the sections just upstream/downstream of the blade row (where no blade blockage occurs, Sect. 1.8), we can write:

$$A_i = \pi \cdot D_M \cdot h_{B,i} = \pi \cdot D_M \cdot \frac{D_{t,i} - D_{h,i}}{2} = \pi \cdot D_M^2 \cdot \frac{1 - \lambda_i}{1 + \lambda_i}$$

As a consequence we have:

$$h_{B,i} = \frac{A_i}{\pi \cdot D_M}$$

and:

$$\lambda_i = \frac{D_{h,i}}{D_{t,i}} = \frac{1 - \frac{h_{B,i}}{D_M}}{1 + \frac{h_{B,i}}{D_M}}$$

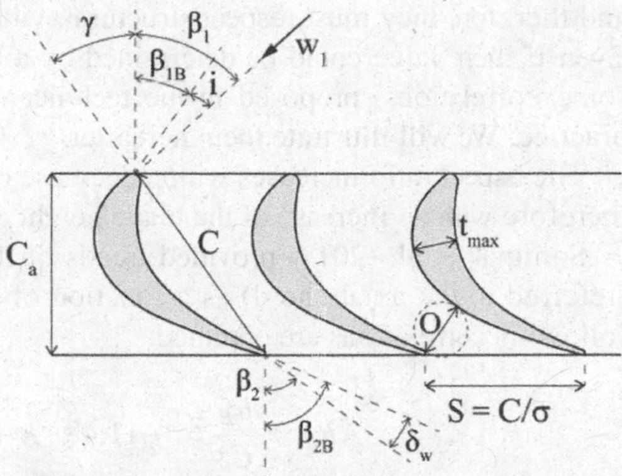

Fig. 3.4 Blade row geometry

The *hub to tip diameter ratio*, also called *boss ratio*, λ, decreases as the blade height increases. Since, as the blade height increases, the variation of the parameters R, φ and ψ goes up between the hub and the tip of the blade, it is necessary to verify the value of λ at the hub diameter (Sect. 3.4). In fact, here, the work and flow coefficients assume the maximum values while the degree of reaction the minimum value; therefore λ must be higher than a threshold value which ensures the congruence of the parameters R, φ and ψ, chosen at the mean diameter.

After evaluating these first geometric parameters (h_B, D_h, D_t), it is necessary to calculate other geometric parameters correlated with the blade rows (Fig. 3.4), and in particular: chord (C), blade stagger angle (γ), axial chord (C_a), blade angles (α_{1B}, α_{2B}, β_{1B}, β_{2B}), blade pitch (S) and number of blades (N_B).

In the following, we will refer explicitly to the rotor; the extension to the stator is obtained by replacing the angles β with the angles α and the relative velocity w with the absolute velocity c.

It is necessary to introduce two other parameters (Fig. 3.4) which will be very important in the calculation of stage losses (Sect. 3.5): the *aspect ratio*, AR, expressed as the ratio of blade height to chord, and the *blade solidity*, σ, expressed as the ratio of chord to blade pitch:

$$AR = \frac{h_B}{C} \qquad \sigma = \frac{C}{S}$$

In fact, if the blade height is established, these two parameters allow to determine the geometry of the channels and, especially through the chord, the blade surfaces related to profile losses (Sect. 1.9), and, through the blade pitch, the end wall surfaces related to end wall losses and secondary losses (Sect. 1.9).

Precisely, because of their importance in the calculation of stage losses, these two parameters can be chosen in such a way as to minimize the overall losses, even if it must be taken into account that these parameters also influence mechanical aspects and therefore they must respect structural, vibrational and acoustic constraints too. Even if their value could be determined by a loss optimization routine (Sect. 3.7), some correlations, proposed in the technical literature, may be very effective in practice. We will illustrate them hereafter.

The aspect ratio increases with a decrease of the *hub to tip diameter ratio* λ and therefore with an increase of the blade height and volumetric flow rate in the stage.

Sammak et al. (2013) provided trends of the rotor and stator axial aspect ratio (referred to the axial chord) as a function of λ; by interpolating these trends, the following correlations are obtained:

$$AR_a = \frac{h_B}{C_a} = -11.48 \cdot \lambda + 11.39 \qquad \text{stator}$$

$$AR_a = \frac{h_B}{C_a} = -12.44 \cdot \lambda + 12.59 \qquad \text{rotor}$$

These equations immediately allow to calculate the axial chord:

$$C_a = \frac{h_B}{AR_a}$$

To calculate the aspect ratio and therefore the chord, it is necessary to determine the chord slope (called *blade stagger angle* γ); in fact, we have:

$$AR = AR_a \cdot \cos\gamma$$

$$C = \frac{C_a}{\cos\gamma}$$

The blade stagger angle calculation will be carried out within the following two criteria for assessing the solidity.

The first criterion is based on the Zweifel coefficient, ZW, expressed as (Benner et al. 2006a, b):

$$ZW = 2 \cdot \frac{S}{C_a} \cdot (\tan\beta_2 + \tan\beta_1) \cdot \cos^2\beta_2$$

which provides:

$$\sigma = 2 \cdot \frac{\cos^2\beta_2}{ZW \cdot \cos\gamma} \cdot (\tan\beta_2 + \tan\beta_1)$$

This expression allows calculating the solidity considering that the ZW coefficient is 0.8–1.15 for the stator and 0.8–1.25 for the rotor and that the blade stagger angle can be approximated by:

$$\tan\gamma = \frac{1}{2}\cdot(\tan\beta_2 - \tan\beta_1)$$

Alternatively, the method presented by Tournier and El-Genk (2010) can be used to more accurately calculate the solidity, the blade stagger angle as well as the deviation angle at the blade trailing edge.

The optimum solidity is calculated by interpolating the optimum solidities of two types of blade cascades (respectively characterized by $\beta_{1B} = 0$ and by $\beta_{1B} = \beta_2$):

$$\left(\frac{1}{\sigma_{opt}}\right) = \left(\frac{1}{\sigma_{opt}}\right)_{opt}^{(\beta_{1B}=0)} + \left|\frac{\beta_{1B}}{\beta_2}\right|\cdot\left(\frac{\beta_{1B}}{\beta_2}\right)\cdot\left[\left(\frac{1}{\sigma_{opt}}\right)_{opt}^{(\beta_{1B}=\beta_2)} - \left(\frac{1}{\sigma_{opt}}\right)_{opt}^{(\beta_{1B}=0)}\right]$$

where we can use the following correlations provided by Aungier (2006):

$$\left(\frac{1}{\sigma_{opt}}\right)_{opt}^{(\beta_{1B}=0)} = 0.427 + \frac{90-\beta_2}{58} - \left(\frac{90-\beta_2}{93}\right)^2$$

$$\left(\frac{1}{\sigma_{opt}}\right)_{opt}^{(\beta_{1B}=\beta_2)} = 0.224 + \left(0.575 + \frac{\beta_2}{90}\right)\cdot\left(1 - \frac{\beta_2}{90}\right)$$

For the calculation of the optimum solidity it is necessary to know the blade angle at the rotor inlet; it depends on the flow angle and the incidence angle, i (Fig. 3.4):

$$\beta_{1B} = \beta_1 - i$$

For turbines operating at nominal conditions, we can assume $i = 0$ and therefore

$$\beta_{1B} = \beta_1$$

The blade stagger angle is then calculated by:

$$\gamma = a + b\cdot\beta_{1B} + c\cdot\beta_{1B}^2$$

where:

$$a = 0.710 + 0.288\cdot\beta_{2B} + 6.93\times10^{-3}\cdot\beta_{2B}^2$$
$$b = 0.489 - 0.0407\cdot\beta_{2B} + 4.27\times10^{-4}\cdot\beta_{2B}^2$$
$$c = -3.65\times10^{-3} + 2.66\times10^{-5}\cdot\beta_{2B} + 8.33\times10^{-8}\cdot\beta_{2B}^2$$

The blade stagger angle, therefore, depends on the blade angle at the trailing edge which in turn depends on the deviation angle at the blade trailing edge, δ_w:

$$\beta_2 = \beta_{2B} - \delta_w$$

The deviation angle can be calculated using the expression provided by Zhu and Sjolander (2005):

$$\delta_w = 17.3 \cdot \frac{(1/\sigma)^{0.05} \cdot (\beta_1 + \beta_{2B})^{0.63} \cdot \cos^2\gamma \cdot (t_{\max}/C)^{0.29}}{\left(30 + 0.01 \cdot \beta_{1B}^{2.07}\right) \cdot \tanh(\mathrm{Re}_2/200,000)}$$

where

- $t_{\max}/C$ is the ratio between the maximum blade thickness and the chord. For this ratio we can assume:

$$\frac{t_{\max}}{C} \cong 0.2$$

Alternatively, Kacker and Okapuu (1982) and Krumme (2016) provided a graphical correlation between the ratio $t_{\max}/C$ and the deflection angle δ (Sect. 3.1). This correlation may be interpolated by:

$$\frac{t_{\max}}{C} = 0.0014 \cdot \delta + 0.093$$

- Re_2 is the Reynolds number at the blade trailing edge:

$$\mathrm{Re}_2 = \frac{\rho_2 \cdot w_2 \cdot C_a}{\mu_2 \cdot \cos\gamma}$$

The γ and δ_w calculations are therefore iterative. At the end of this computation, the blade pitch (S) and the number of blades (N_B) can be assessed:

$$S = \frac{C}{\sigma}$$

$$N_B = \mathrm{int}\left(\frac{\pi \cdot D_M}{S}\right)$$

The *blade distance* at the trailing edge, O (Fig. 3.4), can be assessed by:

$$O \cong S \cdot \cos\beta_{2B}$$

by assuming as a right triangle the triangle having a hypotenuse equal to S, a base equal to O and an angle between them equal to β_{2B}. This parameter O also represents the flow area width perpendicular to the fluid velocity, assuming the flow angle very close to the blade angle ($\beta_2 \cong \beta_{2B}$).

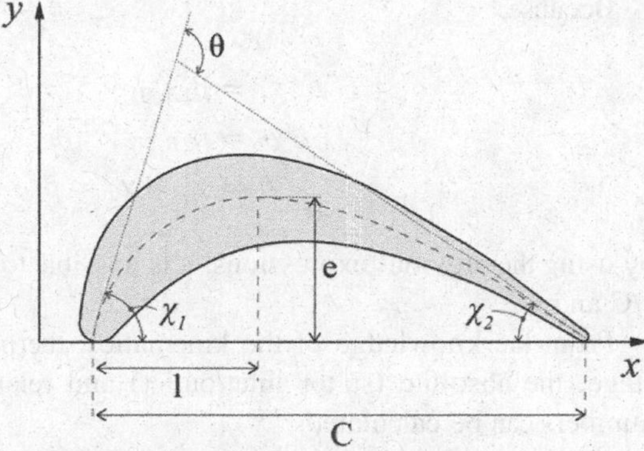

Fig. 3.5 Blade profile geometry

The ratio between the blade thickness at the trailing edge and the blade pitch can be assumed as:

$$\frac{t_{TE}}{S} \cong 0.02$$

This parameter will be necessary for calculating the trailing edge losses (Sect. 3.5).

Regarding the blade tip clearance (necessary for calculating the tip clearance losses, Sect. 3.5), we can assume that:

$$\frac{\tau}{h_B} = 0.6\% \div 1.5\%$$

but the minimum values must be not lower than $\tau = 0.3\text{ mm} \div 0.5\text{ mm}$. In the case of shrouded blades, an effective clearance value is considered as:

$$\frac{\tau_{eff}}{h_B} = \frac{\tau / n_{ts}^{0.42}}{h_B}$$

where n_{ts} is to the number of tip seals (generally equal to 2–3).

The geometric parameters of the blade profile can be completed with the calculation of the quantities shown in Fig. 3.5.

If the blade profile follows a parabolic-arc camberline, we can obtain:

$$\tan \chi_1 = \frac{e/C}{l/C - 1/4}$$

$$\tan \chi_2 = \frac{e/C}{3/4 - l/C}$$

Because:

$$\chi_1 = \beta_{1B} + \gamma$$
$$\chi_2 = \beta_{2B} - \gamma$$
$$\theta = \chi_1 + \chi_2 = \beta_{1B} + \beta_{2B}$$

by using the previous expressions, it is possible to calculate explicitly the two ratios e/C and l/C.

From the knowledge of the kinematics, thermodynamics and geometry of the stage, the absolute (stator inlet/outlet) and relative (rotor inlet/outlet) Reynolds numbers can be calculated:

$$\mathrm{Re}_0 = \frac{\rho_0 \cdot c_0 \cdot C_S}{\mu_0}$$
$$\mathrm{Re}_{1,S} = \frac{\rho_1 \cdot c_1 \cdot C_S}{\mu_1}$$
$$\mathrm{Re}_{1,R} = \frac{\rho_1 \cdot w_1 \cdot C_R}{\mu_1}$$
$$\mathrm{Re}_2 = \frac{\rho_2 \cdot w_2 \cdot C_R}{\mu_2}$$

Once the Reynolds numbers have been calculated, considerations on surface roughness can be made according to Sect. 1.9.1.

In particular, the *admissible surface roughness*, $k_{s,adm}$, can be assessed; it is the limit roughness below which there are no effects on losses allowing the walls to be considered smooth. This condition occurs when the surface roughness is contained within the laminar sub-layer of the boundary layer.

This roughness (Sect. 1.9.1) can be evaluated, for the stator and the rotor, by:

$$\frac{k_{s,adm,S}}{C_S} = \frac{100}{\mathrm{Re}_{1,S}}$$
$$\frac{k_{s,adm,R}}{C_R} = \frac{100}{\mathrm{Re}_2}$$

The calculation of $k_{s,adm}$ assumes considerable importance as it allows to establish the finishing level of the channel surfaces in order to contain dissipations.

Taking into account that turbomachinery surface roughness generally falls within the range:

$$5\ \mu\mathrm{m} < k_s < 50\ \mu\mathrm{m}$$

the knowledge of $k_{s,adm}$ allows to evaluate a possible, and economically convenient, surface treatment aimed at respecting this admissible roughness value as well as not resorting to very high finish levels where not required.

In other words, if the admissible surface roughness, $k_{s,adm}$, is lower than the minimum roughness that is technologically and economically achievable, $k_{s,min}$, we assume:

$$k_s = k_{s,\min}$$

otherwise:

$$k_{s,min} \leq k_s \leq k_{s,adm}$$

In particular, in the latter case, a roughness lower than the admissible one can be assumed in order to take into account that the roughness tends to increase with the turbomachine operation.

Remarks on input parameters

The procedure illustrated in this Sect. 3.3 does not presuppose any further input parameters, with respect to those already identified in Sects. 3.1 and 3.2, taking into account the assumptions on the aspect ratio, AR, the solidity, σ, the ratios t_{TE}/S, $t_{\max}/C$, τ/h_B, k_s/C, and that the mean diameter, D_M, depends on the blade speed, u_M, and on the speed of rotation, n (Sect. 3.6).

3.4 Kinematic Parameters in the Radial Direction

After evaluating the blade heights at the stator and rotor inlet/outlet, all the kinematic parameters previously calculated at the mean diameter can be assessed along these blade heights, and in particular at the hub and tip diameters.

In this regard, the free-vortex flow is adopted (Sect. 1.7.4); it ensures that work transfer is equal at all radii:

$$\frac{\partial h_t}{\partial r} = 0 \quad ; \quad \frac{\partial c_m}{\partial r} = 0$$

from these equations we obtain:

$$c_{u(r)} \cdot r = const$$
$$c_{m(r)} = const$$

The velocity triangles along the blade height, and consequently the Mach and Reynolds numbers, can be deduced by calculating the flow and work coefficients and the degrees of reaction along the blade height. With reference to the blade hub and tip, we have:

$$\varphi_t = \varphi_M \cdot \frac{D_M}{D_t} \quad ; \quad \varphi_h = \varphi_M \cdot \frac{D_M}{D_h}$$

$$\psi_t = \psi_M \cdot \left(\frac{D_M}{D_t}\right)^2 \quad ; \quad \psi_h = \psi_M \cdot \left(\frac{D_M}{D_h}\right)^2$$

$$R_t = 1 - (1 - R_M) \cdot \frac{D_M^2}{D_t^2} \quad ; \quad R_h = 1 - (1 - R_M) \cdot \frac{D_M^2}{D_h^2}$$

By using these parameters and the correlations deduced in Sect. 3.1, the absolute and relative flow angles along the blade height can be calculated, and in particular at the blade hub and tip:

$$\tan\alpha_{1,t} = \frac{\frac{\psi_t}{2} - R_t + 1}{\varphi_t} \qquad \tan\alpha_{1,h} = \frac{\frac{\psi_h}{2} - R_h + 1}{\varphi_h}$$

$$\tan\alpha_{2,t} = \frac{\frac{\psi_t}{2} + R_t - 1}{\varphi_t} \qquad \tan\alpha_{2,h} = \frac{\frac{\psi_h}{2} + R_h - 1}{\varphi_h}$$

$$\tan\beta_{1,t} = \frac{\frac{\psi_t}{2} - R_t}{\varphi_t} \qquad \tan\beta_{1,h} = \frac{\frac{\psi_h}{2} - R_h}{\varphi_h}$$

$$\tan\beta_{2,t} = \frac{\frac{\psi_t}{2} + R_t}{\varphi_t} \qquad \tan\beta_{1,h} = \frac{\frac{\psi_h}{2} - R_h}{\varphi_h}$$

Consequently, the kinematics and geometry of the blade profiles, previously calculated at the mean diameter (Sects. 3.1 and 3.3), can also be completely evaluated in the radial direction.

Since the maximum values of the flow and work coefficients and the minimum degree of reaction occur at the blade hub, a positive degree of reaction at the hub imposes a minimum value on the hub to tip ratio (Tournier and El-Genk 2010).

Taking up the expression of the degree of reaction at the hub and imposing:

$$R_{h,\min} = 1 - (1 - R_M) \cdot \frac{D_M^2}{D_{h,\min}^2} = 0$$

introducing:

$$\lambda_{\min} = \frac{D_{h,\min}}{D_t}$$

and therefore:

$$D_M = \frac{D_t}{2} \cdot (1 + \lambda_{\min})$$

we obtain:

$$(1 - R_M) \cdot \left[\frac{1}{2} \cdot \left(\frac{1 + \lambda_{\min}}{\lambda_{\min}}\right)\right]^2 = 1$$

and as a consequence:

$$\lambda = \frac{D_h}{D_t} \geq \lambda_{\min} = \left(\frac{2}{\sqrt{1 - R_M}} - 1\right)^{-1}$$

If this condition is not verified, the hub to tip diameter ratio, λ, must be increased by choosing a different mean-diameter degree of reaction and/or a different flow coefficient in order to reduce the blade height.

Remarks on input parameters

This procedure does not presuppose the introduction of further input parameters, compared to those already identified in the previous sections.

3.5 Stage Losses and Efficiency

The loss sources and the loss models in the different types of turbomachinery are extensively described in Sect. 1.9. In this section, we will provide specific correlations for loss models able to quantify the losses introduced in Sect. 1.9.

Losses in the stator and rotor of an axial stage (Sect. 1.9) depend on the kinematic (Sects. 3.1 and 3.4), thermodynamic (Sect. 3.2) and geometric (Sect. 3.3) parameters of the stage.

Two fundamental loss models are available in the technical literature:

- C&C model, developed by Craig and Cox (1971)
- AMDCKO model, developed by Ainley and Mathieson (1951), later revised by Dunham and Came (1970) and subsequently by Kacker and Okapuu (1982).

Here we will refer to the AMDCKO model, with the modifications and additions provided by Zhu and Sjolander (2005), Benner et al. (2006a, b), Aungier (2006); in particular, reference will be made to the complete and in-depth analysis carried out by Tournier and El-Genk (2010).

This loss model provides results in agreement with CFD simulations (Sect. 3.6.1) for usual aspect ratio values (AR > 1). Beschorner et al. (2013) have then proposed modifications to this loss model to extend its validity to ultra-low aspect ratios (AR up to 0.2).

The pressure loss coefficients are respectively for the stator and the rotor (Sect. 1.9.5):

$$Y_S = \frac{p_{0t} - p_{1t}}{p_{1t} - p_1}$$

$$Y_R = \frac{p_{1tr} - p_{2tr}}{p_{2tr} - p_2}$$

In the following, we will refer explicitly to the rotor; the extension to the stator is obtained by replacing the relative angles, β, with the absolute angles, α, the relative velocities, w, with the absolute velocities, c, and taking into account what will be expressly illustrated for the stator.

The pressure loss coefficient is composed of (Sect. 1.9.4):

$$Y = Y_p + Y_s + Y_{TE} + Y_{TC}$$

where:

- Y_p represents the profile losses;
- Y_s represents the secondary losses;
- Y_{TE} represents the trailing edge losses;
- Y_{TC} represents the tip clearance losses.

3.5.1 *Profile and Secondary Losses*

Benner et al. (2006a, b) proposed the following expression for the sum of the profile and secondary losses:

$$Y_p + Y_s = \left(1 - \frac{Z_{TE}}{h_B}\right) \cdot Y'_p + Y'_s$$

where Z_{TE} is the spanwise penetration depth (m) of the separation line between the primary and the secondary loss regions along blade height (see below).

Y'_p evaluation

Benner et al. (2006a, b) proposed:

$$Y'_p = 0.914 \cdot \left(K_{in} \cdot Y'_{p,AM} \cdot K_p + Y_{shock}\right) \cdot K_{Re}$$

where the K coefficients are:

- K_{in} = 2/3 according to Kacker and Okapuu (1982), Zhu and Sjolander (2005) proposed K_{in} = 0.825. For reaction stages, however, the first value (2/3) is mostly used.
- K_p was proposed by Kacker and Okapuu:

$$K_p = 1 - K_2 \cdot (1 - K_1)$$

where:

$$K_1 = 1 \qquad \text{when} \quad Ma_2 \le 0.2$$
$$K_1 = 1 - 1.25 \cdot (Ma_2 - 0.2) \qquad \text{when} \quad Ma_2 > 0.2$$
$$K_2 = \left(\frac{Ma_{1,R}}{Ma_2}\right)^2$$

- K_{Re} was proposed by Zhu and Sjolander to take into account the Reynolds number:

$$K_{Re} = \left(\frac{2 \times 10^5}{Re_2}\right)^{0.575} \qquad \text{when} \quad Re_2 < 2 \times 10^5$$
$$K_{Re} = 1 \qquad \text{when} \quad Re_2 \ge 2 \times 10^5$$

Since this formulation does not take into account the influence of surface roughness, it is preferable to follow the procedure proposed by Aungier (2006). A critical Reynolds number is introduced; above this value the effects of the surface roughness k_s (Sect. 1.9.1) become relevant:

$$Re_{cr} = 100 \cdot \frac{C}{k_s}$$

By using this parameter, Aungier proposed the following correction for the Reynolds number and surface roughness:

$$K_{Re} = \left(\frac{1 \times 10^5}{Re_2}\right)^{0.5} \qquad \text{when} \quad Re_2 < 1 \times 10^5$$
$$K_{Re} = 1 \qquad \text{when} \, 1 \times 10^5 \le Re_2 \le 5 \times 10^5$$
$$K_{Re} = 1 + \left\{\left[\frac{\log_{10}(5 \times 10^5)}{\log_{10}(Re_{cr})}\right]^{2.58} - 1\right\} \cdot \left(1 - \frac{5 \times 10^5}{Re_2}\right) \qquad \text{when} \quad Re_2 > 5 \times 10^5$$

Y_{shock} was proposed by Kacker and Okapuu:

$$Y_{shock} = \frac{\rho_1 \cdot w_1^2}{\rho_2 \cdot w_2^2} \cdot \frac{D_h}{D_t} \cdot \frac{3}{4} \cdot \left(Ma_{1,h} - 0.4\right)^{1.75} \qquad \text{when} \quad Ma_{1,h} > 0.4$$
$$Y_{shock} = 0 \qquad \text{when} \quad Ma_{1,h} \le 0.4$$

where for a rotor:

$$\frac{Ma_{1,h}}{Ma_1} = 5.716 \cdot \left(\frac{D_h}{D_t}\right)^2 - 10.85 \cdot \left(\frac{D_h}{D_t}\right) + 6.152 \quad \text{when} \quad \left(\frac{D_h}{D_t}\right) \le 0.95$$

$$\frac{Ma_{1,h}}{Ma_1} = 1 \quad \text{when} \quad \left(\frac{D_h}{D_t}\right) > 0.95$$

and for a stator:

$$\frac{Ma_{1,h}}{Ma_1} = 4.072 \cdot \left(\frac{D_h}{D_t}\right)^2 - 6.644 \cdot \left(\frac{D_h}{D_t}\right) + 3.705 \quad \text{when} \quad \left(\frac{D_h}{D_t}\right) \le 0.8$$

$$\frac{Ma_{1,h}}{Ma_1} = 1 \quad \text{when} \quad \left(\frac{D_h}{D_t}\right) > 0.8$$

The profile loss coefficient, $Y'_{p,AM}$, was proposed by Ainley and Mathieson (1951) and it is an interpolation between the results of two sets of cascade tests (respectively $\beta_{1B} = 0$ and $\beta_{1B} = \beta_2$):

$$Y'_{p,AM} = \left\{ Y_{p,AM}^{(\beta_{1B}=0)} + \left|\frac{\beta_{1B}}{\beta_2}\right| \cdot \left(\frac{\beta_{1B}}{\beta_2}\right) \cdot \left[Y_{p,AM}^{(\beta_{1B}=\beta_2)} - Y_{p,AM}^{(\beta_{1B}=0)} \right] \right\} \cdot \left(\frac{t_{\max}/C}{0.2}\right)^{K_m \cdot \beta_{1B}/\beta_2}$$

Regarding each term of the previous expression, we note that:

- the coefficients $Y_{p,AM}^{(\beta_{1B}=0)}$ and $Y_{p,AM}^{(\beta_{1B}=\beta_2)}$, proposed by Ainley and Mathienson, depend on the blade solidity ($\sigma = C/S$) and on the relative flow angle, β_2. The first coefficient, $Y_{p,AM}^{(\beta_{1B}=0)}$, can be expressed by:

$$Y_{p,AM}^{(\beta_{1B}=0)} = a + b \cdot \sigma + \frac{1}{\sigma} \cdot \left(c + \frac{d}{\sigma}\right)$$

where, when $\beta_2 \le 63.2°$, the coefficients a, b, c and d are:

$$\begin{aligned} a &= -5.58 \times 10^{-5} \cdot \beta_2^2 + 1.03 \times 10^{-2} \cdot \beta_2 - 0.275 \\ b &= 1.553 \times 10^{-5} \cdot \beta_2^2 - 2.32 \times 10^{-3} \cdot \beta_2 + 8.02 \times 10^{-2} \\ c &= -8.54 \times 10^{-3} \cdot \beta_2 + 0.238 \\ d &= 4.83 \times 10^{-5} \cdot \beta_2^2 + 2.83 \times 10^{-4} \cdot \beta_2 - 2.93 \times 10^{-2} \end{aligned}$$

and when $\beta_2 > 63.2°$, they are:

$$\begin{aligned} a &= 2.44 \times 10^{-4} \cdot \beta_2^2 - 4.33 \times 10^{-2} \cdot \beta_2 + 1.92 \\ b &= -4.02 \times 10^{-5} \cdot \beta_2^2 + 6.94 \times 10^{-3} \cdot \beta_2 - 0.282 \\ c &= -2.23 \times 10^{-4} \cdot \beta_2^2 + 4.96 \times 10^{-2} \cdot \beta_2 - 2.548 \\ d &= 5.39 \times 10^{-5} \cdot \beta_2^2 - 1.57 \times 10^{-2} \cdot \beta_2 + 0.958 \end{aligned}$$

The second coefficient, $Y_{p,AM}^{(\beta_{1B}=\beta_2)}$, can be expressed by:

$$Y_{p,AM}^{(\beta_{1B}=\beta_2)} = Y_{p,\min} + A \cdot \left|\left(\frac{1}{\sigma}\right) - \left(\frac{1}{\sigma}\right)_{\min}\right|^n$$

The solidity value that ensures minimum losses is:

$$\left(\frac{1}{\sigma}\right)_{\min} = -5.14 \times 10^{-4} \cdot \beta_2^2 + 5.48 \times 10^{-2} \cdot \beta_2 - 0.798 \qquad \text{when} \quad \beta_2 > 60°$$

$$\left(\frac{1}{\sigma}\right)_{\min} = -8.63 \times 10^{-6} \cdot \beta_2^3 + 9.68 \times 10^{-4} \cdot \beta_2^2 - 3.76 \times 10^{-2} \cdot \beta_2 + 1.272 \qquad \text{when} \quad \beta_2 \leq 60°$$

and, therefore, the minimum value of the loss coefficient is given by:

$$Y_{p,\min} = 0.280 \cdot \left[1.0 - \left(\frac{1}{\sigma}\right)_{\min}\right]$$

Finally, as regards the coefficient A and exponent n, the following correlations can be used:

- if $\left(\frac{1}{\sigma}\right) \geq \left(\frac{1}{\sigma}\right)_{\min}$:

$$n = 1.524 \times 10^{-4} \cdot \beta_2^2 - 0.031 \cdot \beta_2 + 2.992$$
$$A = 5.407 \times 10^{-3} \cdot \beta_2 - 0.19642 \qquad \text{when} \quad \beta_2 > 60°$$
$$A = -2.91 \times 10^{-3} \cdot \beta_2 + 0.30260 \qquad \text{when} \quad \beta_2 \leq 60°$$

- if $\left(\frac{1}{\sigma}\right) < \left(\frac{1}{\sigma}\right)_{\min}$:

$$n = 1.174 \times 10^{-2} \cdot \beta_2^2 - 1.5731 \cdot \beta_2 + 54.85 \qquad \text{when} \quad \beta_2 > 60°$$
$$n = 3.271 \times 10^{-3} \cdot \beta_2^2 - 0.3010 \cdot \beta_2 + 9.023 \qquad \text{when} \quad \beta_2 \leq 60°$$
$$A = 9.240 \times 10^{-3} \cdot \beta_2^2 - 1.2067 \cdot \beta_2 + 40.04 \qquad \text{when} \quad \beta_2 > 60°$$
$$A = 2.701 \times 10^{-3} \cdot \beta_2^2 - 0.2456 \cdot \beta_2 + 5.909 \qquad \text{when} \quad \beta_2 \leq 60°$$

- since these evaluations are referred to blade cascades having a maximum blade thickness to chord ratio (t_{max}/C) equal to 0.2, in the expression of $Y'_{p,AM}$ there is the term, $\left(\frac{t_{max}/C}{0.2}\right)^{K_m \cdot \beta_{1B}/\beta_2}$, to take into account different values of that ratio. The exponent K_m of this term was provided by Zhu and Sjolander (2005):

$$K_m = 1 \quad \text{when} \quad t_{max}/C \leq 0.2$$
$$K_m = -1 \quad \text{when} \quad t_{max}/C > 0.2$$

Z_{TE} evaluation

Benner et al. (2006a, b) provided the following correlation:

$$\frac{Z_{TE}}{h_B} = \frac{0.1 \cdot |F_t|^{0.79}}{\sqrt{\cos\beta_1/\cos\beta_2} \cdot (h_B/C)^{0.55}} + 32.7 \cdot \left(\frac{\delta^*}{h_B}\right)$$

The tangential loading parameter, F_t, is given by:

$$F_t = 2 \cdot \frac{S}{C \cdot \cos\gamma} \cdot \cos^2\beta_M \cdot (\tan\beta_1 + \tan\beta_2)$$

where the mean velocity vector angle, β_M, is given by:

$$\tan\beta_M = \frac{1}{2} \cdot (\tan\beta_1 - \tan\beta_2)$$

The displacement thickness of the boundary layer (Sect. 1.9.1), δ^*, was finally provided by Schlichting (1978):

$$\delta^* = \frac{\delta}{8} = \frac{0.0463 \cdot x}{(\rho_1 \cdot w_1 \cdot x/\mu_1)^{0.2}}$$

where the reference length, x, is taken as half the blade axial chord:

$$x = \frac{C_a}{2}$$

Y'_s evaluation

The secondary loss coefficient was proposed by Benner et al. (2006a, b) and it is given by:

$$Y'_s = F_{AR} \cdot \frac{0.038 + 0.41 \cdot \tanh(1.2 \cdot \delta^*/h_B)}{\sqrt{\cos\gamma} \cdot (\cos\beta_1/\cos\beta_2) \cdot (C \cdot \cos\beta_2/C_a)^{0.55}}$$

where the aspect ratio factor, F_{AR}, is a function of the blade aspect ratio:

$$F_{AR} = (C/h_B)^{0.55} \quad \text{when} \quad h_B/C \leq 0.2$$
$$F_{AR} = 1.36604 \cdot (C/h_B) \quad \text{when} \quad h_B/C > 0.2$$

3.5.2 Trailing Edge Losses

The formulation of these losses, expressed in a similar form to Ainley and Mathienson's profile losses but in terms of energy loss coefficients, is due to Kacker and Okapuu (1982):

$$\Delta E_{TE} = \Delta E_{TE}^{(\beta_{1B}=0)} + \left|\frac{\beta_{1B}}{\beta_2}\right| \cdot \left(\frac{\beta_{1B}}{\beta_2}\right) \cdot \left[\Delta E_{TE}^{(\beta_{1B}=\beta_2)} - \Delta E_{TE}^{(\beta_{1B}=0)}\right]$$

Energy losses are functions of the ratio between the trailing edge thickness and the blade pitch (t_{TE}/S):

$$\Delta E_{TE}^{(\beta_{1B}=0)} = 0.59563 \cdot \left(\frac{t_{TE}}{S} \cdot \frac{1}{\cos\beta_{2B}}\right)^2 + 0.12264 \cdot \left(\frac{t_{TE}}{S} \cdot \frac{1}{\cos\beta_{2B}}\right) - 2.2796 \times 10^{-3}$$

$$\Delta E_{TE}^{(\beta_{1B}=\beta_2)} = 0.31066 \cdot \left(\frac{t_{TE}}{S} \cdot \frac{1}{\cos\beta_{2B}}\right)^2 + 0.065617 \cdot \left(\frac{t_{TE}}{S} \cdot \frac{1}{\cos\beta_{2B}}\right) - 1.4318 \times 10^{-3}$$

In the case of perfect gases, energy losses are attributable to total pressure losses through the following expression (Fielding 2000):

$$Y_{TE} = \frac{\left[1 - \frac{k-1}{2} \cdot Ma_2^2 \cdot \left(\frac{1}{1-\Delta E_{TE}} - 1\right)\right]^{-k/(k-1)} - 1}{1 - \left(1 + \frac{k-1}{2} \cdot Ma_2^2\right)^{-k/(k-1)}}$$

In the case, however, of real fluids, the entropy increase due to these energy losses must be calculated; then, the total pressure loss coefficients can be evaluated by using this entropic increase.

With reference to Fig. 3.2, the energy losses, respectively for the stator and the rotor, are expressed by:

$$\Delta E_{TE,S} = \frac{h_{1,TE} - h_{1,is}}{\frac{1}{2} \cdot c_{1,is}^2} = \frac{h_{1,TE} - h_{1,is}}{h_{0t} - h_{1,is}}$$

$$\Delta E_{TE,R} = \frac{h_{2,TE} - h_{2',is}}{\frac{1}{2} \cdot w_{2,is}^2} = \frac{h_{2,TE} - h_{2',is}}{h_{1tr} - h_{2',is}}$$

These expressions provide the enthalpies at the stator and rotor outlet, associated with the trailing edge losses:

$$h_{1,TE} = h_{1,is} + \Delta E_{TE,S} \cdot \left(h_{0t} - h_{1,is}\right)$$

$$h_{2,TE} = h_{2',is} + \Delta E_{TE,R} \cdot \left(h_{1tr} - h_{2',is}\right)$$

and therefore the related entropies:

$$s_{1,TE} = s_{p,h}\left(p_1, h_{1,TE}\right)$$
$$s_{2,TE} = s_{p,h}\left(p_2, h_{2,TE}\right)$$

and the total and total relative pressures at the stator and rotor outlet:

$$p_{1t,TE} = p_{s,h}\left(s_{1,TE}, h_{1t}\right)$$
$$p_{2tr,TE} = p_{s,h}\left(s_{2,TE}, h_{1tr}\right)$$

By using these pressures, the pressure loss coefficients are immediately calculable in both stator and rotor.

3.5.3 Tip Clearance Losses

These losses can be calculated by using the correlations of Yaras and Sjolander (1992):

$$Y_{TC} = Y_{tip} + Y_{gap}$$

$$Y_{tip} = 1.4 \cdot K_E \cdot \sigma \cdot \frac{\tau}{h_B} \cdot \frac{\cos^2 \beta_2}{\cos^3 \beta_M} \cdot c_L^{1.5} \qquad \text{unshrouded blades}$$

$$Y_{tip} = \frac{0.37}{0.47} \cdot 1.4 \cdot K_E \cdot \sigma \cdot \frac{\tau_{eff}}{h_B} \cdot \frac{\cos^2 \beta_2}{\cos^3 \beta_M} \cdot c_L^{1.5} \qquad \text{shrouded blades}$$

$$Y_{gap} = 0.0049 \cdot K_G \cdot \sigma \cdot \frac{C}{h_B} \cdot \frac{\sqrt{c_L}}{\cos \beta_M}$$

where the theoretical blade lift coefficient, c_L, is given by:

$$c_L = \frac{2}{\sigma} \cdot \cos \beta_M \cdot (\tan \beta_1 + \tan \beta_2)$$

and β_M is that already defined in Sect. 3.5.1

For the other parameters of these expressions, we can assume:

- $K_E = 0.5$ and $K_G = 1.0$ for mid-loaded turbine blades and $K_E = 0.566$ and $K_G = 0.943$ for front-loaded or aft-loaded blades;
- tip clearances, τ and τ_{eff}, are those defined in Sect. 3.3.

3.5.4 *Stator and Stage Efficiency*

The procedure described in Sects. 3.5.1, 3.5.2 and 3.5.3 allows evaluating the pressure loss coefficients in both the stator and rotor:

$$Y_S = \frac{p_{0t} - p_{1t}}{p_{1t} - p_1}$$
$$Y_R = \frac{p_{1tr} - p_{2tr}}{p_{2tr} - p_2}$$

To assess the stator efficiency, we must first evaluate the total pressure downstream of the stator (the static pressure was calculated in the thermodynamics of the stage, Sect. 3.2):

$$p_{1t} = \frac{p_{0t} + Y_S \cdot p_1}{Y_S + 1}$$

Now, by knowing the total enthalpy downstream of the stator:

$$h_{1t} = h_{0t}$$

the thermodynamic quantities downstream of the stator (Fig. 3.3) are:

$$s_1 = s_{p,h}(p_{1t}, h_{1t})$$
$$h_1 = h_{s,p}(s_1, p_1)$$

and, therefore, the stator efficiency (Sect. 1.8.2) is:

$$\eta_S = \frac{h_{0t} - h_1}{h_{0t} - h_{1,is}}$$

Similarly, for the rotor:

$$h_{1tr} = h_1 + \frac{w_1^2}{2}$$
$$p_{1tr} = p_{s,h}(s_1, h_{1tr})$$
$$p_{2tr} = \frac{p_{1tr} + Y_R \cdot p_2}{Y_R + 1}$$
$$h_{2tr} = h_{1tr}$$
$$s_2 = s_{p,h}(p_{2tr}, h_{2tr})$$
$$h_2 = h_{s,p}(s_2, p_2)$$

Once the thermodynamics of the stage have been calculated on the basis of the stage losses, the various efficiencies can be assessed (Sect. 1.8.2):

$$\eta_{TT} = \frac{h_{0t} - h_{2t}}{h_{0t} - h_{2t,is}}$$
$$\eta_{is} = \frac{h_0 - h_2}{h_0 - h_{2,is}}$$
$$\eta_{TS} = \frac{h_{0t} - h_{2t}}{h_{0t} - h_{2,is}}$$

Remarks on input parameters

The procedure illustrated in this Sect. 3.5 does not presuppose the introduction of further input parameters, compared to those already identified in Sects. 3.1 and 3.2.

3.6 Input Parameters of the Preliminary Design Procedure

In the previous Sects. 3.1, 3.2, 3.3, 3.4 and 3.5, the procedure for the calculation of the kinematics, thermodynamics, geometry and efficiency of an axial stage was developed and the input parameters of this procedure were identified.

In this Sect. 3.6, we intend to precisely analyze these input parameters in order to suggest their numerical values (these values, however, can be reviewed in an iterative calculation, see Sect. 3.7).

First of all, let us consider the kinematics calculation procedure (Sect. 3.1).

In general, to determine the velocity triangles at the rotor inlet and outlet, we must set six independent parameters: three for the inlet triangle (station 1 of the stage) and three for the outlet triangle (station 2 of the stage).

For axial stages, however, being $u_1 = u_2 = u$ and assuming $c_{m1} = c_{m2} = c_m$, *the independent parameters are only four*. Indeed, from Sect. 3.1, we know that the absolute and relative flow angles are:

$$\alpha_1 = f(\varphi, \psi, R)$$
$$\alpha_2 = f(\varphi, \psi, R)$$
$$\beta_1 = f(\varphi, \psi, R)$$
$$\beta_2 = f(\varphi, \psi, R)$$

These angles are, therefore, functions of the following three independent parameters:

- the flow coefficient, φ
- the work coefficient, ψ
- the degree of reaction, R

As a consequence, by identifying a further independent parameter, for example, the blade speed, u, or the meridional component of the velocity, c_m, the velocity triangles are completely defined.

However, taking into account what emerged in the turbomachinery selection process (Chap. 2), it is advisable to choose the stage efficiency (isentropic or polytropic efficiency) as the additional independent parameter and not a velocity (u or c_m).

In fact, starting from the data available for the turbomachinery selection process:

- type of fluid
- fluid mass flow rate (kg/s)
- fluid inlet temperature (°C)
- fluid inlet pressure (bar)
- fluid outlet pressure (bar)

the selection process provides ranges of n ($n_{min} < n < n_{max}$) and z ($z_{min} < z < z_{max}$) compatible with the type of turbomachine chosen. Within these ranges we can choose one or more pairs of these values:

- n: rotational speed (rpm)
- z: number of stages and, consequently, the isentropic enthalpy drop in each stage, $\Delta h_{is,stage}$ (kJ/kg)

For each of these pairs, by using the stage (isoentropic or polytropic) efficiency, set as first iteration guess (Sect. 3.6.3) or calculated by the stage losses (Sect. 3.5), the adiabatic enthalpy drop of each stage can be calculated as well as the blade speed at the mean radius, after choosing the work coefficient.

Indeed, when $c_0 = c_2$ (Sect. 3.1), the blade speed is:

$$u = \sqrt{\frac{\Delta h_{t,ad,stage}}{\psi}} = \sqrt{\frac{\Delta h_{ad,stage}}{\psi}}$$

On the other hand, when $c_0 \neq c_2$ (Sect. 3.1), the work coefficient is expressed as:

$$\psi = \frac{\Delta h_{t,ad,stage}}{u^2} = \frac{\Delta h_{ad,stage}}{u^2} + \frac{c_0^2 - c_2^2}{2 \cdot u^2}$$

and, therefore, expressing the velocity c_2 as a function of the kinematic parameters, we obtain:

$$u = \sqrt{\frac{\Delta h_{ad,stage} + \frac{c_0^2}{2}}{\psi + \frac{1}{2} \cdot \left[\varphi^2 + \left(\frac{\psi}{2} + R - 1\right)^2\right]}}$$

After calculating the blade speed, the mean diameter of the rotor is:

$$D = \frac{60 \cdot u}{\pi \cdot n}$$

Alternatively, we can calculate the rotor diameter, D, from the specific diameter, D_S, and then the blade speed (Chap. 2). In this way, only three independent parameters (flow coefficient, degree of reaction and stage efficiency) must be chosen since the work coefficient is calculated by using the enthalpy drop and the blade speed. In axial stage design, however, it is preferred to assign the work coefficient, following the procedure illustrated above.

In summary, if the mean diameter and the meridional velocity are constant, we must set the following four input parameters at mean diameter in order to design preliminarily the axial stage:

- the flow coefficient, φ
- the work coefficient, ψ
- the degree of reaction, R
- the stage efficiency (isentropic or polytropic efficiency)

Taking into account what is illustrated in Sects. 3.2, 3.3, 3.4 and 3.5, these parameters, with the addition of the stator efficiency, allow to fully calculate the thermodynamics, the geometry and therefore the stage losses. Since these losses allow, in turn, to exactly evaluate the stator and stage efficiencies, the stage efficiency (in the list above) is only a starting value: the calculation will proceed iteratively for each stage until convergence for these efficiencies is achieved (Sect. 3.7).

3.6.1 Flow and Work Coefficient Selection

To choose the value of the flow and work coefficients, the Smith's diagram (Smith 1965) and its subsequent updates are widely used. This diagram (Fig. 3.6) shows the iso-efficiency lines of a stage (total-to-total efficiency) as a function of the flow and work coefficients at mean diameter, φ and ψ.

These lines were elaborated by using experimental data of 70 aeronautical axial turbines and the main assumptions are: constant meridional velocity, design conditions, degrees of reaction close to 0.5, Reynolds numbers between 10^5 and 3×10^5 and aspect ratio between 3 and 4. The Smith's diagram does not consider the effects of the Mach number variation and of the trailing edge thickness. Moreover, these efficiencies do not take into account the rotor tip clearance (RTC) losses.

Important updates of such diagram have recently been proposed in order to take into account further design parameters and the technological evolution of axial turbines (Coull and Hodson 2011 and Bertini et al. 2013), especially in the aeronautical sector.

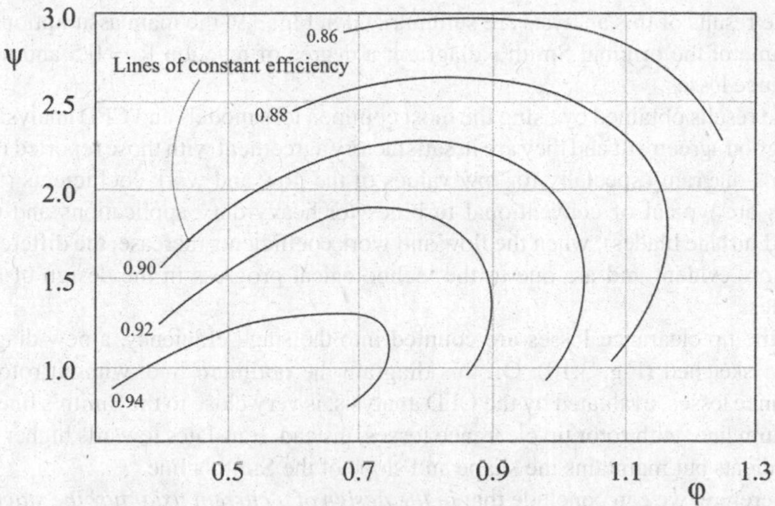

Fig. 3.6 Smith's diagram (Adapted from Korpela 2011)

Bertini et al. (2013) recognized the importance of the Smith's diagram for the preliminary design of an axial stage and visualized on this diagram some regions, for φ and ψ choice, depending on gas turbine applications (Fig. 3.7). In particular, for heavy-duty applications, *this diagram suggests a flow coefficient in the range 0.4–0.8 and a work coefficient in the range 1.0–1.4*. The diagram also highlights the variation of these coefficients with the number of stages (z) and the blade height (h_B): once the flow coefficient is chosen, when the work coefficient decreases, the efficiency increases as well as the number of stages; while, once the work coefficient is chosen, when the flow coefficient decreases, the blade height always increases but there is an optimum value of the flow coefficient. Based on these considerations, for each value of the work coefficient an optimum value of the flow coefficient can be identified (optimum line in Fig. 3.7).

Now, remembering the expressions of stator and rotor deflection angles (Sect. 3.1) as a function of φ, ψ and R, in the Smith's diagram the iso-deflection lines can be sketched; these lines are referred to different blade profiles depending on the flow and work coefficients (Fig. 3.8 is referred to $R = 0.5$ and therefore the stator and rotor deflection angles are the same, Sect. 3.1).

Bertini et al. (2013) developed an interesting comparison between the original Smith's diagram and that obtainable from up to date axial stages; the stage losses were calculated by using the currently most common loss models (already mentioned in Sect. 3.5):

- C&C model
- AMDCKO model

as well as a three-dimensional CFD (Computational Fluid Dynamics) analysis.

The results of this analysis are summarized in Fig. 3.9; the main assumptions are the same of the original Smith's diagram: a degree of reaction R = 0.5 and no tip clearance losses.

The results obtained by using the most common loss models and CFD analysis are all in good agreement and they are in satisfactory agreement with those reported in the Smith's diagram especially for low values of the flow and work coefficients (these values are typical of conventional turbines for heavy-duty applications and mid-loaded turbine blades); when the flow and work coefficients increase, the differences are more evident and are due to the technological progress in the design of axial turbines.

If the tip clearance losses are counted into the stage efficiency, a new diagram can be sketched (Fig. 3.10). On this diagram the optimum line, without rotor tip clearance losses, evaluated by the CFD analysis, is very close to the Smith's line; the optimum line, with rotor tip clearance losses, instead, translates towards higher flow coefficients but maintains the shape and slope of the Smith's line.

Therefore, we can conclude that *in the design of a current axial turbine stage the Smith's diagram can be used profitably even if slightly higher flow coefficients (an increase of about 0.1) than those suggested by Smith are recommended.*

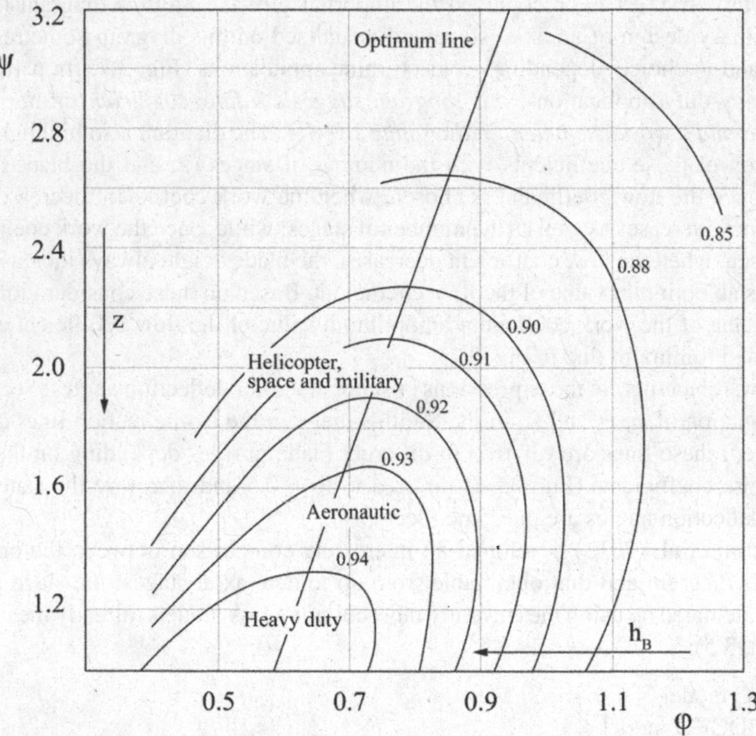

Fig. 3.7 Smith's diagram (Adapted from Bertini et al. 2013)

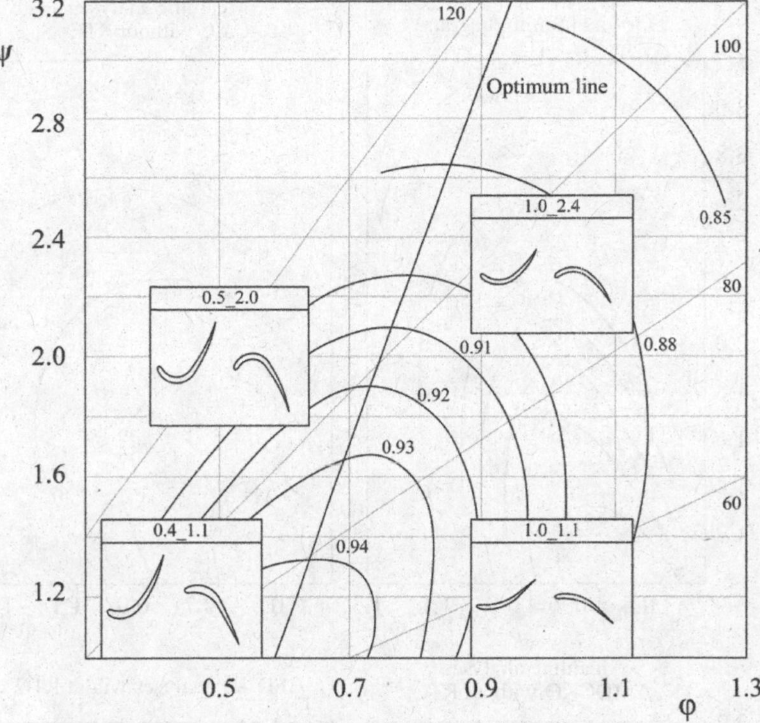

Fig. 3.8 Smith's diagram, R = 0.5 (Adapted from Bertini et al. 2013)

The previous results are referred to turbines handling conventional fluids. But also for unconventional fluids, the Smith's diagram can still be applied, if the Mach numbers remain in the typical range of conventional fluids; vice versa, the effect of the Mach number must be taken into account. For example, various authors proposed new Smith's diagrams for ORC (Organic Rankine Cycle) applications; in particular Da Lio et al. (2014) suggested adopting the work coefficient in the range 1.0–1.4 and the flow coefficient in the range 0.4-0.5 to attain the maximum efficiency.

3.6.2 Degree of Reaction Selection

With the exception of the first stages of steam turbines, where impulse stages (R = 0) are generally adopted, the typical values of the degree of reaction are in the range 0.3–0.7; optimal efficiency values are attained for degrees of reaction close to 0.5. This result is also confirmed by the aforementioned work by Bertini et al. (2013) (Fig. 3.11a, b).

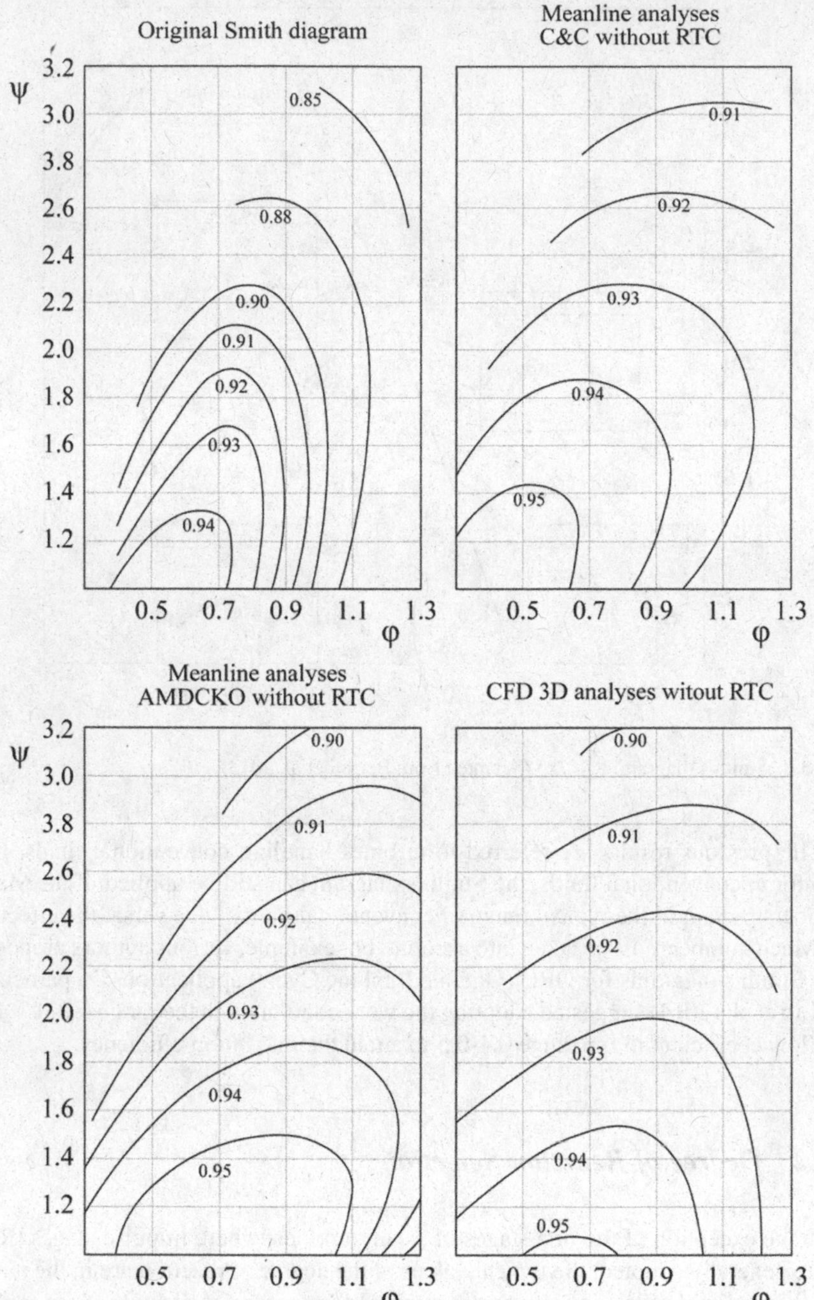

Fig. 3.9 Updated Smith's diagram (Adapted from Bertini et al. 2013)

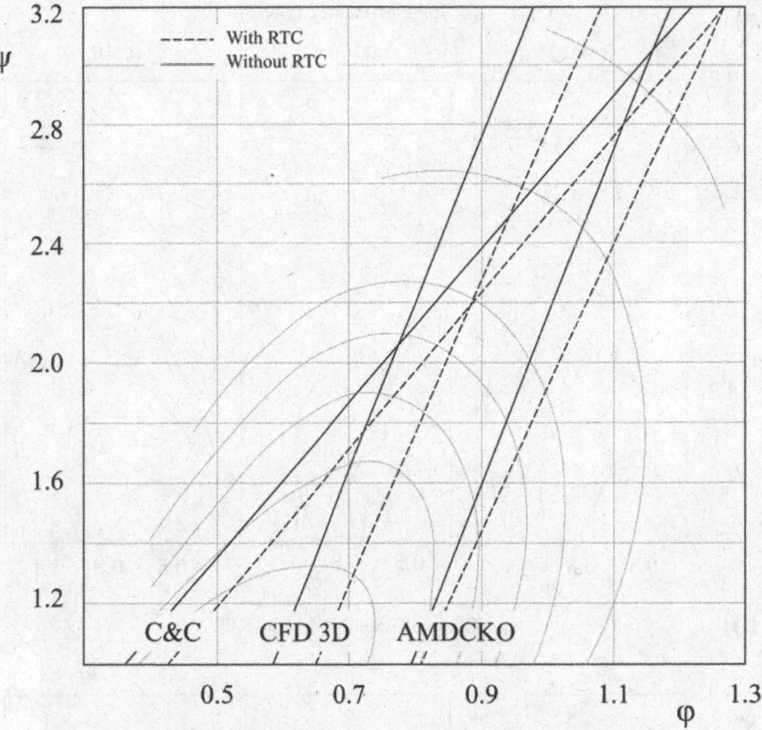

Fig. 3.10 Updated and complete Smith's diagram (Adapted from Bertini et al. 2013)

3.6.3 Starting Value of Stage Efficiency

As already illustrated, the stage efficiency is only a starting value to calculate preliminarily thermodynamics, geometry and kinematics of the stage since this parameter depends on stage losses that, in turn, depend on thermodynamics, geometry and kinematics of the stage (Sect. 3.5). The efficiency calculation proceeds iteratively for each stage until convergence is achieved (Sect. 3.7).

The starting value of the stage efficiency can be chosen on the updated Smith's diagrams, after setting the flow and work coefficients, φ and ψ.

But, in order to take into account the turbomachinery size, an effective correlation, proposed by Lozza (2016), can be used; this correlation expresses the polytropic efficiency of the stage as a function of the size parameter (SP) indicating the size of the machine:

$$\eta_p = 0.94 \cdot \left(1 - 0.02688 \cdot \log_{10}^2(SP)\right) \qquad \text{when} \quad SP < 1$$
$$\eta_p = 0.94 \qquad \text{when} \quad SP \geq 1$$

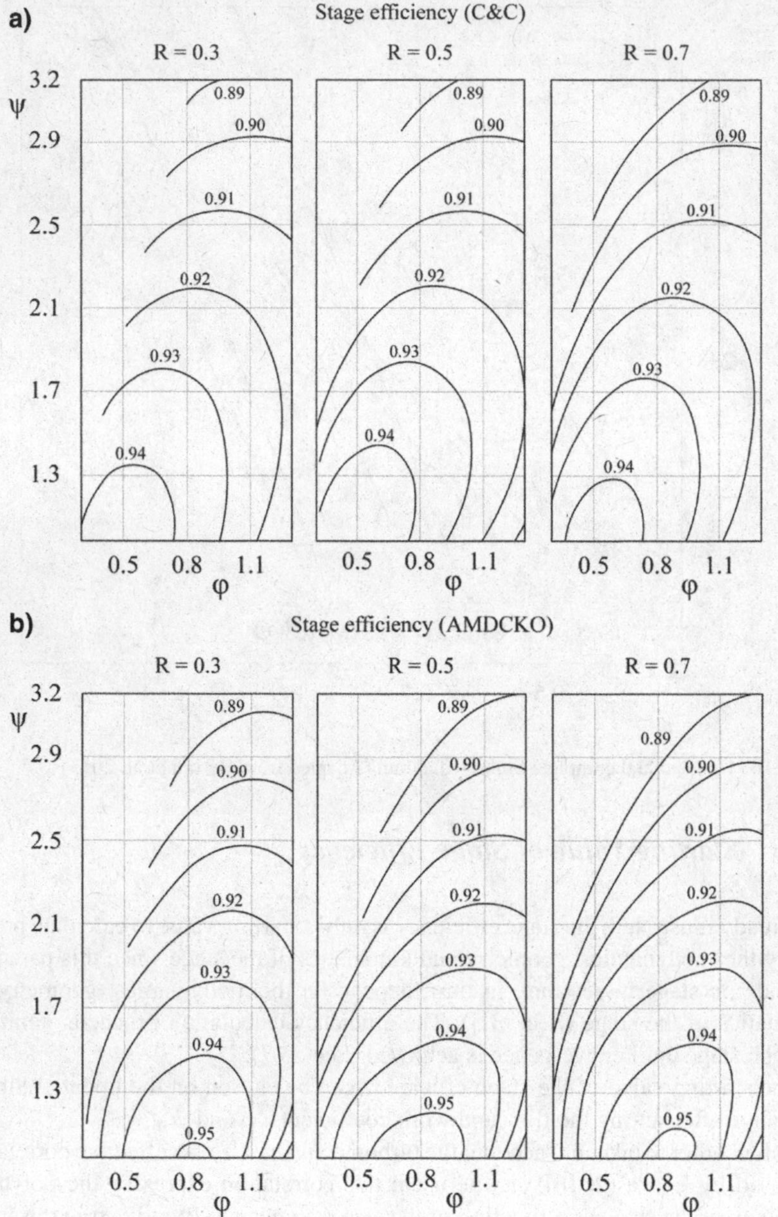

Fig. 3.11 **a** Updated Smith's diagram—C&C loss model (Adapted from Bertini et al. 2013). **b** Updated Smith's diagram—AMDCKO loss model (Adapted from Bertini et al. 2013)

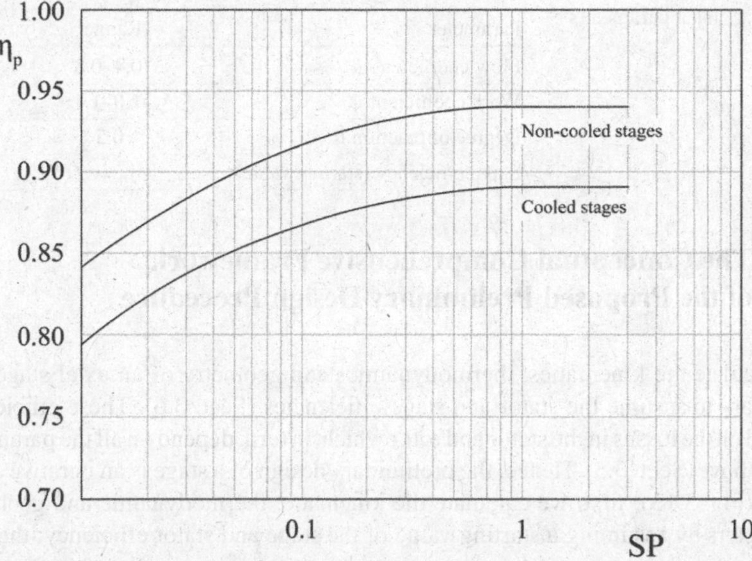

Fig. 3.12 Variation of polytropic efficiency with SP

where:

$$SP = \frac{\sqrt{V_{out}}}{\Delta h_{is,stage}^{1/4}}$$

The variation of the polytropic efficiency with SP is shown in Fig. 3.12; efficiency decreases when the absolute dimensions of the stage decrease because of a percentage increase in losses. For $SP \geq 1$, this correlation provides an efficiency value close to that of the Smith's diagrams in the recommended φ (0.4–0.8) and ψ (1.0–1.4) ranges.

By using this polytropic efficiency, the starting value of the stage isentropic efficiency can be calculated following the procedure illustrated in Sect. 1.4.2.

For the thermodynamics evaluation (Sect. 3.2), it is also necessary to choose the starting value of the stator efficiency. It can be set arbitrarily, for example, equal to the stage isentropic efficiency, since also this efficiency will be iteratively calculated by the losses in the stator (Sect. 3.6).

3.6.4 Summary of Input Parameters

All the main considerations made in the previous Sects. 3.6.1, 3.6.2, 3.6.3 and 3.6.4 can be summarized in Tab. 3.1, which reports the most suitable ranges for the input parameters of the preliminary design procedure for axial turbine stages.

Table 3.1 Input parameters

Parameter	Range
Flow coefficient φ	0.4–0.8
Work coefficient ψ	1.0–1.4
Degree of reaction R	$\approx$0.5
Stage efficiency (starting value)	$\eta_p = f(SP)$

3.7 The Conceptual Comprehensive Framework of the Proposed Preliminary Design Procedure

To calculate the kinematics, thermodynamics and geometry of an axial stage, it is necessary to assume the stator and stage efficiencies (Sect. 3.6). These efficiencies depend on the losses in the stator and rotor which, in turn, depend on all the parameters listed above (Sect. 3.5). Hence, the preliminary design of a stage is an iterative calculation (Fig. 3.13): first, we calculate the kinematic, thermodynamic and geometric parameters by assuming a starting value of the stage and stator efficiency; then, we can calculate the stator and rotor losses; at this point, the stage and stator efficiencies must be calculated by using these losses; these new efficiencies become the new input parameters of the preliminary design procedure. This iterative procedure ends when the established convergence for the stator and stage efficiencies is achieved.

The iterative procedure shown in Fig. 3.13 could be solved by using different values of the input parameters (R, φ and ψ), as well as of the geometric parameters, especially the solidity and the aspect ratio, evaluated by empirical correlations in Sect. 3.3, in order to optimize the stage performance.

For multistage turbines ($z > 1$), the same preliminary design is applied to the stages following the first one. For these stages, the only constraint is to maintain the same speed of rotation, since the input parameters R, φ and ψ could be assigned anew. If, on the other hand, an assumption is made on the characteristic diameters (constant hub or tip or mean diameter), these parameters are calculated anew according to the following procedure.

For example, if the turbomachine mean diameter is assumed constant (Fig. 3.14):

$$D_M = const$$

the blade speed is constant:

$$u_M = const$$

In addition, we can assume constant the meridional velocity:

$$c_m = const$$

while the absolute velocity angle at the stator inlet is equal to that at the upstream rotor outlet (repeating stage):

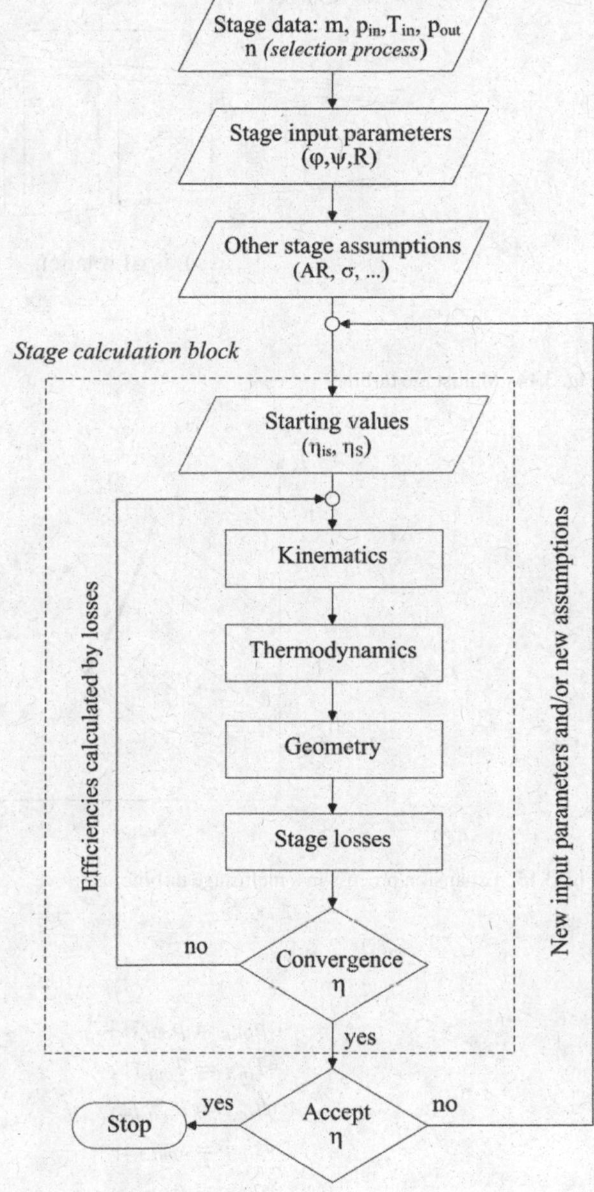

Fig. 3.13 Stage calculation block diagram

$$c_0 = c_2 \quad ; \quad \alpha_0 = \alpha_2$$

The thermodynamic parameters at the inlet and the pressure at the outlet of the stages following the first one (i = 2, z) are (Fig. 3.15):

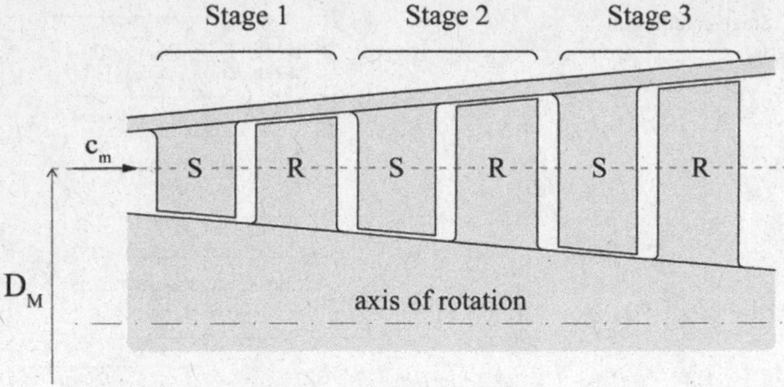

Fig. 3.14 Multistage turbine

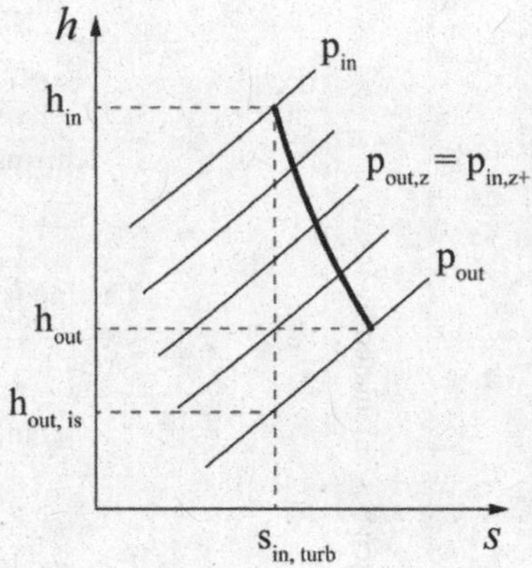

Fig. 3.15 Expansion process in a multistage turbine

$$
\begin{aligned}
p_{in,i} &= p_{out,i-1} \\
T_{in,i} &= T_{out,i-1} \\
h_{in,i} &= h_{out,i-1} \\
s_{in,i} &= s_{out,i-1} \\
\rho_{in,i} &= \rho_{out,i-1} \\
h_{out,is,i} &= h_{in} - \sum_{j=1}^{i} \Delta h_{is,j} \\
p_{out,i} &= p(h_{out,is,i}, s_{in})
\end{aligned}
$$

In the previous correlations, the isentropic enthalpy drops of each stage are evaluated according to the procedure described in Chap. 2.

The kinematics of each stage following the first one (i = 2, z) is calculated (Sect. 3.1) by considering constant the flow coefficient (according to the previous assumptions):

$$\varphi_i = \frac{c_{m,i}}{u_{M,i}} = \frac{c_m}{u_M} = const = \varphi$$

while the work coefficient must be calculated anew because now the blade speed at the mean diameter and the stage enthalpy drop are established:

$$\psi_i = \frac{\Delta h_{t,ad,i}}{u_M^2}$$

Consequently, also the degree of reaction must be recalculated:

$$R_i = 1 - \frac{\psi_i}{2} + \varphi \cdot \tan\alpha_2$$

These parameters allow the calculation of the thermodynamics (Sect. 3.2), the geometry (Sect. 3.3), the kinematics in the radial direction (Sect. 3.4) and the losses (Sect. 3.5) of each stage following the first one.

The turbine design, carried out by assigning a pair of values for the speed of rotation and the number of stages (n, z), provided by the selection process illustrated in Chap. 2, can be accomplished again (Fig. 3.16) by using a different pair of input parameters (n and z), indicated always by the turbomachinery selection process, in order to optimize the preliminary design of the turbine.

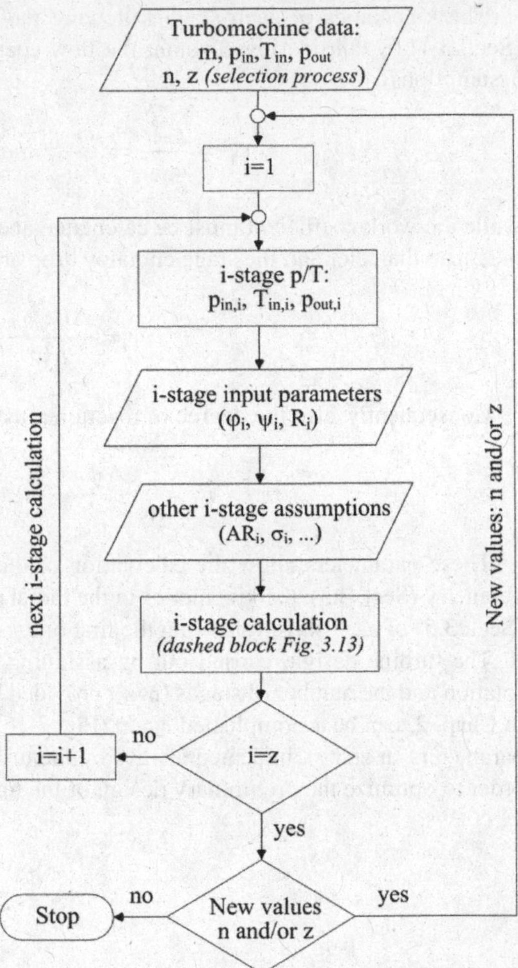

Fig. 3.16 Multistage turbine calculation block diagram

References

Ainley DG, Mathieson GCR (1951) A method of performance estimation for axial-flow turbines. British Aeronautical Research Council, Reports and Memoranda No. 2974

Aungier RH (2006) Turbine aerodynamics. Axial-flow and radial-inflow turbine design and analysis. ASME Press

Benner MW, Sjolander SA, Moustapha SH (2006a) An empirical prediction method for secondary losses in turbines—part I: a new loss breakdown scheme and penetration depth correlation. ASME, J Turbomach

Benner MW, Sjolander SA, Moustapha SH (2006b) An empirical prediction method for secondary losses in turbines—part II: a new secondary loss correlation. ASME, J Turbomach

Beschorner A, Vogeler K, Goldhahn E, Huhnd E (2013) Experimental and numerical investigations to extend the validity range of a turbine loss correlation for ultra low aspect ratios in transonic flow. In: 10th European conference on turbomachinery fluid dynamics & thermodynamics

Bertini F, Ampellio E, Marconcini M, Giovannini M (2013) A critical numerical review of loss correlation models and Smith diagram for modern low pressure turbine stages. In: Proceedings of ASME Turbo Expo

Coull JD, Hodson HP (2011) Blade loading and its application in the mean-line design of low pressure turbines. In: Proceedings of ASME Turbo Expo

Craig HRM, Cox HJA (1971) Performance estimation of axial flow turbines. In: Proceedings Institution of Mechanical Engineers

Da Lio L, Manente G, Lazzaretto A (2014) New efficiency charts for the optimum design of axial flow turbines for organic Rankine cycles. Energy

Dunham J, Came PM. (1970) Improvements to the Ainley–Mathieson method of turbine performance prediction. ASME J Eng Power

Fielding L (2000) Turbine design—the effect on axial flow turbine performance of parameter variation. ASME Press, New York

Kacker SC, Okapuu U (1982) A mean line prediction method for axial flow turbine efficiency. ASME J Eng for Power

Korpela SA (2011) Principles of turbomachinery. Wiley, New York

Krumme A (2016) Performance prediction and early design code for axial turbines and its application in research and predesign. Proceedings of ASME Turbo Expo

Lozza G (2016) Gas turbine and Combined Cycles (in Italian). Ed. Esculapio

Sammak M, Thern M, Genrup M (2013) Reduced-order through-flow design code for highly loaded, cooled axial turbines. In: Proceedings of ASME Turbo Expo

Schlichting H (1978) Boundary layer theory, 7th edn. McGraw-Hill, New York

Smith SF (1965) A simple correlation of turbine efficiency. J R Aeronaut Soc

Tournier JM, El-Genk MS (2010) Axial flow, multi stage turbine and compressor models. Energy Convers Manage

Yaras MI, Sjolander SA (1992) Prediction of tip leakage losses in axial turbines. ASME J Turbomach

Zhu J, Sjolander SA (2005) Improved profile loss and deviation correlations for axial turbine blade rows. In: Proceedings of ASME Turbo Expo

Chapter 4
Preliminary Design of Axial Flow Compressors

Abstract With reference to an axial compressor stage, a procedure for the calculation of kinematic parameters at mean diameter (Sect. 4.1), thermodynamic parameters (Sect. 4.2), geometric parameters (Sect. 4.3), parameters in the radial direction (Sect. 4.4) and stage losses (Sect. 4.5) is provided. Then, Sect. 4.6 discusses the input parameters of this procedure and suggests their numerical values to be used in the calculations. Finally, Sect. 4.7 illustrates the procedure for extending calculations to multistage compressors. The numerical application of the proposed procedure, aimed at the preliminary design of multistage axial compressors, is instead developed in Chap. 8.

4.1 Kinematic Parameters

In this section, we will deduce the kinematic correlations of an axial stage; they are correlations among kinematic parameters of a stage (degree of reaction R, absolute and relative flow angles, α and β) and dimensionless performance parameters (flow coefficient, φ, and work coefficient, ψ).

Variations of the previously introduced (Chap. 2) dimensionless factors Φ and Ψ are used in practice.

In particular, the ratio of the meridional component of the absolute velocity to the blade speed is the *flow coefficient*. Consequently, for axial stages where $u_1 = u_2$, assuming $c_m = \text{const}$, the flow coefficient is:

$$\varphi = \frac{c_m}{u}$$

Note that there is a proportionality between this flow coefficient and the flow factor, previously introduced (Chap. 2):

$$\Phi = \frac{V}{\omega \cdot D^3} \propto \frac{c_m \cdot D^2}{u \cdot D^2} = \frac{c_m}{u} = \varphi$$

M. Gambini and M. Vellini, *Turbomachinery*, Springer Tracts in Mechanical Engineering, https://doi.org/10.1007/978-3-030-51299-6_4

Analogously, the ratio of actual work exchanged between fluid flow and rotor to blade speed squared is the *work coefficient*:

$$\psi = \frac{W}{u^2} = \frac{\Delta h_{t,ad}}{u^2}$$

which is proportional to the work factor, Ψ, previously introduced (Chap. 2).

As for axial turbine stage (Fig. 3.1a), in an axial compressor stage, the fluid flow is studied (Sect. 1.7) on a cylindrical meridional section of blades developed on a plane (Fig. 4.1a). With reference to this representation, Fig. 4.1b depicts the velocity triangles of an axial compressor stage. These triangles refer to the mean line, that is, to the mean diameter of the stage (Sect. 1.7.1).

For an axial stage $u_1 = u_2 = u$ (Sect. 1.7.3). In addition, we can assume that the meridional component of the absolute velocity is constant, as well as the mean diameter:

$$c_m = const$$
$$D_M = const$$

Next, assuming the angles shown in Fig. 4.1b as positive, the energy equation applied to a compressor stage (Sect. 1.8.3) and Euler's equation (Sect. 1.6.1) provide:

$$W = \Delta h_{t,ad} = h_{3t} - h_{1t} = u \cdot \Delta c_u = u \cdot (c_{2u} - c_{1u}) = u \cdot \Delta w_u = u \cdot (w_{1u} - w_{2u})$$

The work coefficient and the degree of reaction (Sect. 1.8.3) are therefore expressed as:

$$\psi = \frac{W}{u^2} = \frac{c_{2u} - c_{1u}}{u}$$

$$R = \frac{\Delta h_R}{\Delta h_{t,ad}} = \frac{\Delta h_R}{W} = 1 - \frac{\Delta E_{c,R}}{W} = 1 - \frac{c_2^2 - c_1^2}{2 \cdot W} = 1 - \frac{c_{2u}^2 - c_{1u}^2}{2 \cdot W} = 1 - \frac{c_{2u} + c_{1u}}{2 \cdot u}$$

Now, by correlating the tangential velocity components to the angle α_1:

$$c_{1u} = c_m \cdot \tan\alpha_1 = u \cdot \varphi \cdot \tan\alpha_1$$
$$c_{2u} = u \cdot \psi + c_{1u} = u \cdot \psi + u \cdot \varphi \cdot \tan\alpha_1$$

we obtain:

$$R = 1 - \varphi \cdot \tan\alpha_1 - \frac{\psi}{2}$$

and therefore:

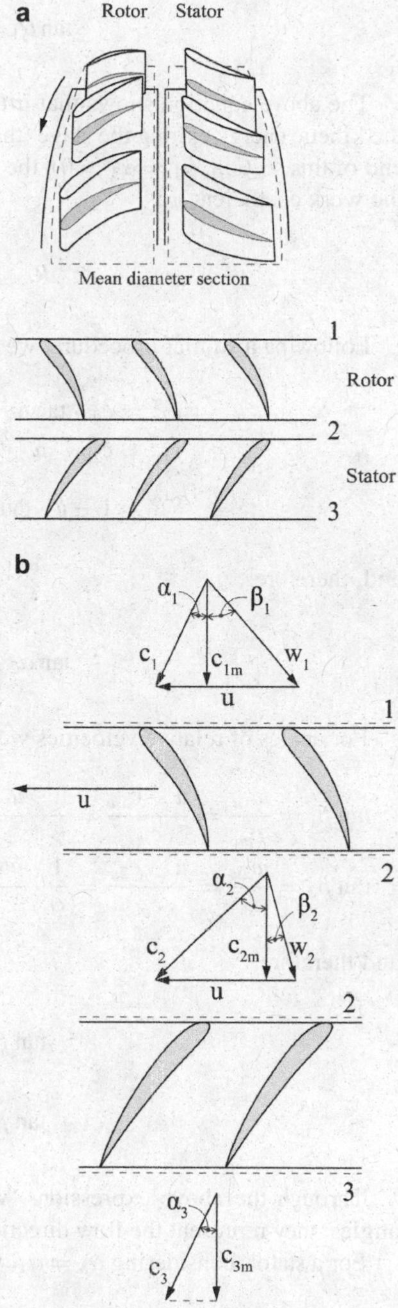

Fig. 4.1 **a** Axial compressor stage: stator and rotor representation. **b** Velocity triangles in an axial compressor stage

$$\tan\alpha_1 = \frac{1 - R - \frac{\psi}{2}}{\varphi}$$

The above equation shows that if the angle α_3 was chosen as input to minimize the kinetic energy exiting the stage (that is, $\alpha_3 = 0$ and therefore, as illustrated at the end of this section, $\alpha_1 = \alpha_3 = 0$), the degree of reaction, R, would have been set by the work coefficient ψ:

$$R_{(\alpha_1=0)} = 1 - \frac{\psi}{2}$$

Following a similar procedure, we can also obtain the angle α_2:

$$\begin{aligned} c_{2u} &= c_m \cdot \tan\alpha_2 = u \cdot \varphi \cdot \tan\alpha_2 \\ c_{1u} &= c_{2u} - u \cdot \psi = u \cdot \varphi \cdot \tan\alpha_2 - u \cdot \psi \\ R &= 1 - \varphi \cdot \tan\alpha_2 + \frac{\psi}{2} \end{aligned}$$

and, therefore:

$$\tan\alpha_2 = \frac{1 - R + \frac{\psi}{2}}{\varphi}$$

For angles of relative velocities we obtain:

$$\tan\beta_1 = \frac{w_{1u}}{c_m} = \frac{u - c_{1u}}{c_m} = \frac{1}{\varphi} \cdot \frac{u - c_{1u}}{u} = \frac{1}{\varphi} \cdot (1 - \varphi \cdot \tan\alpha_1) = \frac{1}{\varphi} - \tan\alpha_1$$
$$\tan\beta_2 = \frac{w_{2u}}{c_m} = \frac{u - c_{2u}}{c_m} = \frac{1}{\varphi} \cdot \frac{u - c_{2u}}{u} = \frac{1}{\varphi} \cdot (1 - \varphi \cdot \tan\alpha_2) = \frac{1}{\varphi} - \tan\alpha_2$$

and therefore:

$$\tan\beta_1 = \frac{R + \frac{\psi}{2}}{\varphi}$$
$$\tan\beta_2 = \frac{R - \frac{\psi}{2}}{\varphi}$$

Through the above expressions, we can obtain the stator and rotor deflection angles; they represent the flow direction changes in the stator and rotor.

For a stator, considering $\alpha_3 = \alpha_1$, we have:

$$\delta_S = \alpha_2 - \alpha_3 = \alpha_2 - \alpha_1$$

and so:

$$\tan \delta_S = \tan(\alpha_2 - \alpha_1) = \frac{\tan \alpha_2 - \tan \alpha_1}{1 + \tan \alpha_1 \cdot \tan \alpha_2}$$

$$\tan \delta_S = \frac{\frac{\psi}{\varphi}}{1 + \frac{1}{\varphi^2} \cdot \left[(1 - R)^2 - \left(\frac{\psi}{2}\right)^2\right]}$$

Similarly, the rotor deflection angle is:

$$\delta_R = \beta_1 - \beta_2$$

and so:

$$\tan \delta_R = \tan(\beta_1 - \beta_2) = \frac{\tan \beta_1 - \tan \beta_2}{1 + \tan \beta_1 \cdot \tan \beta_2}$$

$$\tan \delta_R = \frac{\frac{\psi}{\varphi}}{1 + \frac{1}{\varphi^2} \cdot \left[R^2 - \left(\frac{\psi}{2}\right)^2\right]}$$

We can observe that if R = 0.5, the stator deflection angle is equal to the rotor deflection angle.

Remarks on input parameters

From the kinematic correlations reported above, we infer that once the three independent parameters φ, ψ and R are set (Sect. 4.6), all flow angles can be calculated. Once the blade speed, u, is also established (in order to determine this speed, it is necessary to know, in addition to the work coefficient ψ, also the stage isentropic efficiency (or the polytropic efficiency, as illustrated in Sect. 4.6), the kinematics at the mean diameter of the rotor can be completely calculated:

$$c_m = u \cdot \varphi$$
$$c_{1u} = u \cdot \varphi \cdot \tan \alpha_1$$
$$c_{2u} = u \cdot \varphi \cdot \tan \alpha_2$$
$$w_{1u} = u \cdot \varphi \cdot \tan \beta_1$$
$$w_{2u} = u \cdot \varphi \cdot \tan \beta_2$$
$$c_1 = u \cdot \varphi \cdot \sqrt{1 + \tan^2 \alpha_1}$$
$$c_2 = u \cdot \varphi \cdot \sqrt{1 + \tan^2 \alpha_2}$$
$$w_1 = u \cdot \varphi \cdot \sqrt{1 + \tan^2 \beta_1}$$
$$w_2 = u \cdot \varphi \cdot \sqrt{1 + \tan^2 \beta_2}$$

In order to complete the kinematics of the stage, the velocity c_3, the absolute velocity at the stage outlet, must also be determined.

Generally (*repeating stage,* also called *normal stage*) we assume:

$$c_3 = c_1 \quad ; \quad \alpha_3 = \alpha_1$$

This assumption also allows ensuring, for multistage compressors, that the absolute velocity at the rotor inlet is equal to the absolute velocity at the upstream stage outlet.

If $c_3 \neq c_1$, the blade speed, u, can still be calculated as a function of the work coefficient, but now:

$$\Delta h_{t,ad} \neq \Delta h_{ad}$$

so we must remember that:

$$\Delta h_{t,ad} = \Delta h_{ad} + \frac{c_3^2 - c_1^2}{2}$$

4.2 Thermodynamic Parameters

The pressure p_1 and temperature T_1 at the inlet of the rotor and the pressure p_3 at the outlet of the stator are assigned based on what is illustrated in Chap. 2 (for multistage compressors, these values are calculated by the procedure described in Sect. 4.7).

Now, all the thermodynamic parameters at the rotor inlet and at the stator inlet/outlet can be calculated as follows.

If the working fluid cannot be assimilated to a perfect gas, for the following calculations, we need appropriate state functions, such as those of the NIST libraries. In the following, we will refer to a real working fluid.

Knowing p_1, T_1, we can calculate:

$$h_1 = h_{p,T}(p_1, T_1)$$
$$s_1 = s_{p,T}(p_1, T_1)$$
$$\rho_1 = \rho_{p,T}(p_1, T_1)$$

then (Sect. 1.8.1), the total quantities (Fig. 4.2):

$$h_{1t} = h_1 + \frac{c_1^2}{2}$$
$$p_{1t} = p_{h,s}(h_{1t}, s_1)$$
$$T_{1t} = T_{h,s}(h_{1t}, s_1)$$

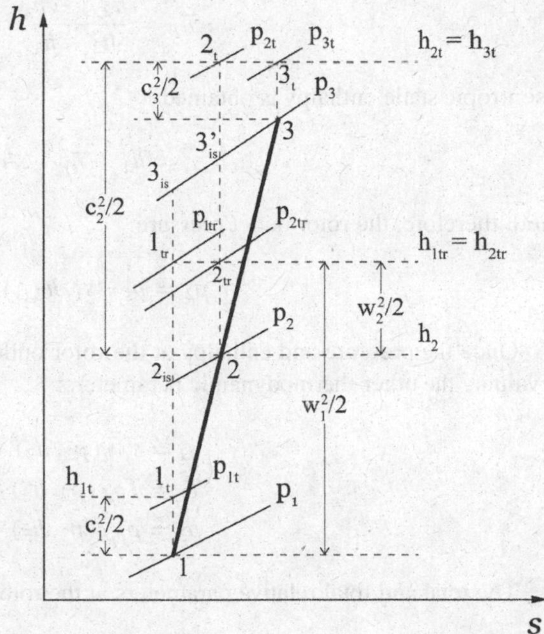

Fig. 4.2 Compression process in an axial compressor stage

$$\rho_{1t} = \rho_{h,s}(h_{1t}, s_1)$$

and the total relative quantities (Fig. 4.2):

$$h_{1tr} = h_1 + \frac{w_1^2}{2}$$

$$p_{1tr} = p_{h,s}(h_{1tr}, s_1)$$

$$T_{1tr} = T_{h,s}(h_{1tr}, s_1)$$

To calculate the thermodynamic parameters at the rotor outlet, we must know the rotor efficiency and remember the total relative enthalpy conservation in the rotor (Sect. 1.8.3):

$$h_{2tr} = h_{1tr}$$

In fact, the static enthalpy is calculated through the previous equation:

$$h_2 = h_{2tr} - \frac{w_2^2}{2}$$

while using a conventional rotor efficiency, defined by:

$$\overline{\eta}_R = \frac{h_{2,is} - h_1}{h_2 - h_1}$$

isentropic static enthalpy is obtained:

$$h_{2,is} = h_1 + \overline{\eta}_R \cdot \Delta h_R$$

and, therefore, the rotor outlet pressure:

$$p_2 = p_{s,h}(s_1, h_{2,is})$$

Once the pressure and enthalpy at the rotor outlet have been calculated, we can evaluate the other thermodynamic parameters:

$$\begin{aligned} s_2 &= s_{p,h}(p_2, h_2) \\ T_2 &= T_{p,h}(p_2, h_2) \\ \rho_2 &= \rho_{p,h}(p_2, h_2) \end{aligned}$$

The total and total relative parameters at the rotor outlet (stator inlet) are then:

$$\begin{aligned} h_{2t} &= h_2 + \frac{c_2^2}{2} \\ p_{2t} &= p_{h,s}(h_{2t}, s_2) \\ T_{2t} &= T_{h,s}(h_{2t}, s_2) \\ \rho_{2t} &= \rho_{h,s}(h_{2t}, s_2) \\ p_{2tr} &= p_{h,s}(h_{2tr}, s_2) \\ T_{2tr} &= T_{h,s}(h_{2tr}, s_2) \end{aligned}$$

Losses in the rotor are related to the entropy increase $(s_2 - s_1)$ due to irreversibilities and entail an increase in the final enthalpy $(h_2 - h_{2,is})$ and a reduction in the total relative pressure expressed by the loss coefficient (Sect. 1.9.5):

$$Y_R = \frac{p_{1tr} - p_{2tr}}{p_{1tr} - p_1}$$

At the stator outlet, the pressure is known; therefore the enthalpy, referred to the isentropic compression, can be calculated as:

$$h_{3,is} = h_{p,s}(p_3, s_1)$$

The final enthalpy in the adiabatic compression can be determined by applying the conservation of the total enthalpy in the stator (Sect. 1.8.3) and by using the velocity c_3 (Sect. 4.1):

$$h_{3t} = h_{2t}$$

$$h_3 = h_{3t} - \frac{c_3^2}{2}$$

or by applying the stage isentropic efficiency (referred to static enthalpy increases, Sect. 1.8.3):

$$\eta_{is} = \frac{h_{3,is} - h_1}{h_3 - h_1}$$

$$h_3 = h_1 + \frac{(h_{3,is} - h_1)}{\eta_{is}}$$

The calculation of the thermodynamic parameters can, therefore, be completed:

$$\begin{aligned} s_3 &= s_{p,h}(p_3, h_3) \\ T_3 &= T_{p,h}(p_3, h_3) \\ \rho_3 &= \rho_{p,h}(p_3, h_3) \\ p_{3t} &= p_{h,s}(h_{3t}, s_3) \\ T_{3t} &= T_{h,s}(h_{3t}, s_3) \\ \rho_{3t} &= \rho_{h,s}(h_{3t}, s_3) \end{aligned}$$

Losses in the stator are related to the entropy increase $(s_3 - s_2)$ due to irreversibilities and entail an increase in the final enthalpy $(h_3 - h_{3',is})$ and a reduction in the total pressure expressed by the loss coefficient (Sect. 1.9.5):

$$Y_S = \frac{p_{2t} - p_{3t}}{p_{2t} - p_2}$$

From the knowledge of the kinematics and thermodynamics of the stage, the absolute (stator inlet/outlet) and relative (rotor inlet/outlet) Mach numbers can be calculated. From the properties of the working fluid, first, the speeds of sound are calculated:

$$\begin{aligned} c_{s,1} &= c_{s,p,T}(p_1, T_1) \\ c_{s,2} &= c_{s,p,T}(p_2, T_2) \\ c_{s3} &= c_{s,p,T}(p_3, T_3) \end{aligned}$$

and therefore the Mach numbers:

$$Ma_1 = \frac{w_1}{c_{s1}}$$

$$Ma_{2,R} = \frac{w_2}{c_{s2}}$$

$$Ma_{2,s} = \frac{c_2}{c_{s2}}$$
$$Ma_3 = \frac{c_3}{c_{s3}}$$

By using the pressure and temperature conditions of the working fluid, we can calculate the fluid viscosities (necessary to assess the Reynolds numbers, Sect. 4.3):

$$\mu_1 = \mu_{p,T}(p_1, T_1)$$
$$\mu_2 = \mu_{p,T}(p_2, T_2)$$
$$\mu_3 = \mu_{p,T}(p_3, T_3)$$

Remarks on input parameters

Based on what is illustrated in this Sect. 4.2, in addition to the parameters already identified in Sect. 4.1 (φ, ψ, R e η_{is}), the conventional rotor efficiency, $\overline{\eta}_R$, must be assumed in order to calculate the thermodynamics of the stage.

4.3 Geometric Parameters

Now we will deduce the geometric correlations of an axial stage; they are correlations among geometric (blade height, tip and hub diameters, chord and blade pitch, and so on) and thermodynamic (density, volumetric flow rate) and kinematic parameters (meridional velocities, absolute velocity angles α, relative velocity angles β, stator and rotor deflection angles) of the stage.

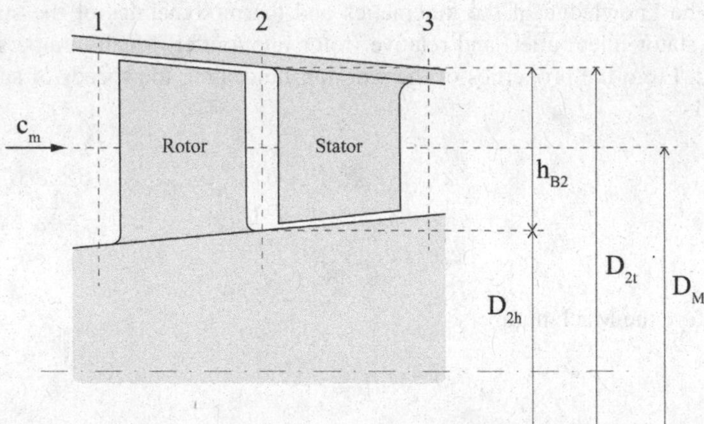

Fig. 4.3 Stage geometry (meridional section)

With reference to Fig. 4.3, it is assumed that between the stage inlet and outlet:

- the mean diameter is constant:

$$D_M = const$$

consequently, blade speed at mean diameter is constant:

$$u_M = const$$

- the meridional component of the velocity is constant:

$$c_m = const$$

The three parameters listed above depend on the kinematics of the stage (Sect. 4.1) and on the rotational speed (n, rpm) of the machine (Sect. 4.6):

$$D_M = \frac{60}{\pi} \cdot \frac{u_M}{n}$$

Volumetric flow rates in the stage (1—rotor inlet, 2—rotor outlet/stator inlet, 3—stator outlet) are a function of the assigned mass flow rate and fluid densities calculated in Sect. 4.2:

$$V_i = \frac{m}{\rho_i} \quad i = 1, 2, 3$$

By using these volumetric flow rates, we can calculate the flow areas:

$$A_i = \frac{V_i}{c_m}$$

They are correlated to the geometric parameters of a turbomachine's meridional section (tip diameter, D_t, hub diameter, D_h, and blade height, h_B) represented in Fig. 4.3.

By introducing the following geometric ratios:

$$\lambda_i = \frac{D_{h,i}}{D_{t,i}} \quad \text{and} \quad \frac{h_{B,i}}{D_M} = \frac{D_{t,i} - D_{h,i}}{2 \cdot D_M}$$

the previous geometric parameters can be expressed as a function of the mean diameter of the stage. In fact:

$$D_M = \frac{D_{t,i} + D_{h,i}}{2} = D_{t,i} \cdot \frac{1 + \lambda_i}{2} = D_{h,i} \cdot \frac{1 + \lambda_i}{2 \cdot \lambda_i}$$

and considering the expression of the flow area in the sections just upstream/downstream of the blade row (where no blade blockage occurs, Sect. 1.8), we can write:

$$A_i = \pi \cdot D_M \cdot h_{B,i} = \pi \cdot D_M \cdot \frac{D_{t,i} - D_{h,i}}{2} = \pi \cdot D_M^2 \cdot \frac{1 - \lambda_i}{1 + \lambda_i}$$

As a consequence we have:

$$h_{B,i} = \frac{A_i}{\pi \cdot D_M}$$

and:

$$\lambda_i = \frac{D_{h,i}}{D_{t,i}} = \frac{1 - \frac{h_{B,i}}{D_M}}{1 + \frac{h_{B,i}}{D_M}}$$

The *hub to tip diameter ratio*, also called *boss ratio*, λ, decreases as the blade height increases. Since, as the blade height increases, the variation of the parameters R, φ and ψ goes up between the hub and the tip of the blade, it is necessary to verify the value of λ at the hub diameter (Sect. 4.4). In fact, here, the work and flow coefficients assume the maximum values while the degree of reaction the minimum value; therefore λ must be higher than a threshold value which ensures the congruence of the parameters R, φ and ψ, chosen at the mean diameter.

After evaluating these first geometric parameters (h_B, D_h, D_t), it is necessary to calculate other geometric parameters correlated to the blade rows (Fig. 4.4), and in particular: chord (C), blade stagger angle (γ), axial chord (C_a), blade angles (α_{1B}, α_{2B}, β_{1B}, β_{2B}), blade pitch (S) and number of blades (N_B).

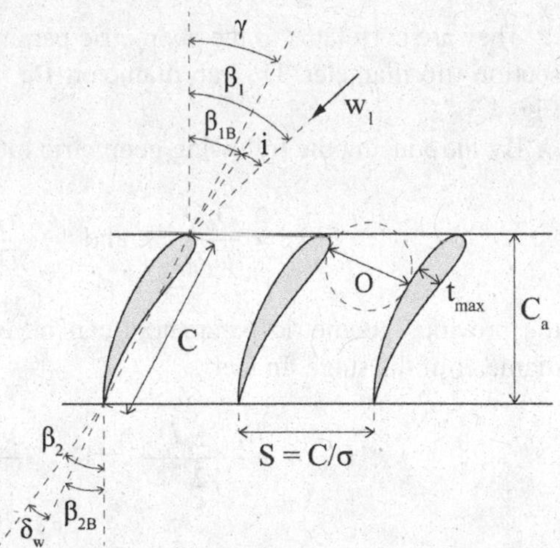

Fig. 4.4 Blade row geometry

In the following, we will refer explicitly to the rotor; the extension to the stator is obtained by replacing the angles β with the angles α and the relative velocity w with the absolute velocity c.

It is necessary to introduce two other parameters (Fig. 4.4) which will be very important in the calculation of stage losses (Sect. 4.5): the *aspect ratio*, AR, expressed as the ratio of blade height to chord, and the *blade solidity*, σ, expressed as the ratio of chord to blade pitch:

$$AR = \frac{h_B}{C} \qquad \sigma = \frac{C}{S}$$

In fact, if the blade height is established, these two parameters allow to determine the geometry of the channels and, especially through the chord, the blade surfaces related to profile losses (Sect. 1.9), and, through the blade pitch, the end wall surfaces related to end wall losses and secondary losses (Sect. 1.9).

Precisely, because of their importance in the calculation of stage losses, these two parameters can be chosen in such a way as to minimize the overall losses, even if it must be taken into account that these parameters also influence mechanical aspects and therefore they must respect structural, vibrational and acoustic constraints too. Even if their value could be determined by a loss optimization routine (Sect. 4.7), some correlations, proposed in the technical literature, may be very effective in practice. We will illustrate them hereafter.

The aspect ratio decreases with an increase of the *hub to tip diameter ratio* λ and therefore with a decrease of the blade height and volumetric flow rate in the stage.

The aspect ratio can be assigned, considering that in general it varies between 1 and 2.

Wennerstrom (1989) traced the evolution of thinking concerning appropriate aspect ratios for axial flow compressors and highlighted the increasing use of low aspect ratios (AR < 2) in design developments after 1970. To and Miller (2015) developed an in-depth study on the influence of the aspect ratio on the axial compressor performance; they determined the optimum AR values as a compromise between the profile losses and end wall losses. These authors demonstrated that the optimum aspect ratio mainly depends on the ratio between the maximum blade thickness and the blade height (t_{max}/h_B). By interpolating the results of these two authors, the following correlations are obtained:

$$AR_{opt} = 0.316 \cdot \left(\frac{t_{\max}}{h_B}\right)^{-0.416} \qquad \text{when} \qquad 0.02 \le \frac{t_{\max}}{h_B} \le 0.10$$

But another result, obtained by To and Miller, is also extremely important: by adopting the AR in the range ±20% of AR_{opt}, the efficiency variation is within 0.1%; so, an AR value can be chosen in this range without compromising stage performance but verifying always mechanical checks.

Once the AR value has been determined, the chord C can be calculated.

The solidity calculation can be carried out by following two criteria.

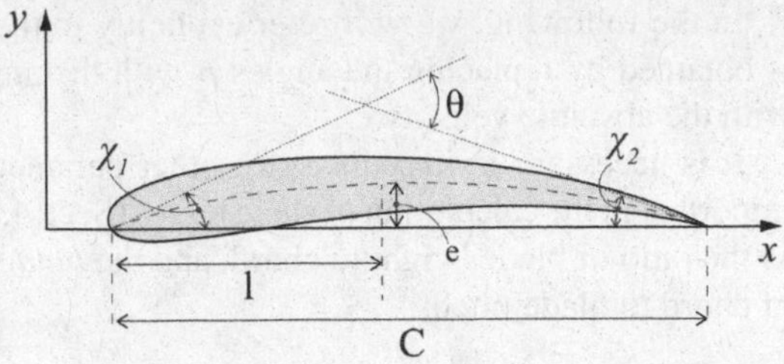

Fig. 4.5 Blade profile geometry

The first criterion is based on the diffusion coefficient DF (assumed equal to 0.45, Hall 2012):

$$\sigma = \frac{\cos\beta_1}{2\cdot\left(DF - 1 + \frac{\cos\beta_1}{\cos\beta_2}\right)}\cdot\frac{\psi}{\varphi}$$

The blade stagger angle can be approximated by:

$$\gamma = \frac{1}{2}\cdot(\beta_1 + \beta_2)$$

Alternatively, the criterion presented by Tournier and El-Genk (2010) can be used to more accurately calculate the solidity, the blade stagger angle as well as the deviation angle at the blade trailing edge.

The optimum solidity is calculated by minimizing the profile and secondary losses (Sect. 4.5):

$$(\sigma)_{opt} = A + B\cdot\beta_2^n$$

where the coefficients A, B and n are functions of the deflection angle:

$$A = -0.0197 + 0.042231\cdot(\beta_1 - \beta_2)$$
$$B = \exp\{-13.427 + (\beta_1 - \beta_2)\cdot[0.33303 - 0.002368\cdot(\beta_1 - \beta_2)]\}$$
$$n = 2.8592 - 0.04677\cdot(\beta_1 - \beta_2)$$

The equation of the optimum solidity is valid within the following limits:

$$10^\circ \le \beta_1 \le 65^\circ \qquad 0^\circ \le \beta_2 \le 55^\circ \qquad 10^\circ \le \beta_1 - \beta_2 \le 60^\circ$$

At this point we can calculate the other geometric and flow parameters (Figs. 4.4 and 4.5): blade angles, incidence angles, deviation angles and stagger angle.

By placing (Fig. 4.5):

$$\begin{aligned}\beta_{1B} &= \beta_1 - i \\ \beta_{2B} &= \beta_2 - \delta_w \\ \theta &= \beta_{1B} - \beta_{2B} = \beta_1 - \beta_2 + \delta_w - i\end{aligned}$$

the blade stagger angle can be calculated by using the following expression (Aungier 2003):

$$\gamma = \beta_1 - \left[3.6 \cdot K_t + 0.3532 \cdot \theta \cdot \left(\frac{l}{C}\right)^{0.25}\right] \cdot (\sigma)^{0.65-0.02\cdot\theta}$$

where:

$$K_t = \left(10 \cdot \frac{t_{\max}}{C}\right)^{\frac{0.28}{0.1+(t_{\max}/C)^{0.3}}}$$

In general, we can assume that:

- $\frac{l}{C} \cong 0.5$ or more generally $0.25 < \frac{l}{C} < 0.75$
- $\frac{t_{\max}}{C} \cong 0.05 \div 0.1$, and therefore the coefficient K_t takes into account values of $\frac{t_{\max}}{C}$ different from 0.1.

To calculate the deviation angle, we can apply the following expression:

$$\delta_w = \left(K_{sh} \cdot K_t' - 1\right) \cdot \delta_0^* + \frac{0.92 \cdot (l/C)^2 + 0.002 \cdot \beta_{2B}}{1 - 0.002 \cdot \theta/\sqrt{\sigma}} \cdot \frac{\theta}{\sqrt{\sigma}}$$

where K_t' is the blade thickness correction factor and it is expressed by:

$$K_t' = 6.25 \cdot \left(\frac{t_{\max}}{C}\right) + 37.5 \cdot \left(\frac{t_{\max}}{C}\right)^2$$

and δ_0^* is the base zero-camber deviation angle and it is expressed by:

$$\delta_0^* = 0.01 \cdot \sigma \cdot \beta_1 + \left(0.74 \cdot \sigma^{1.9} + 3 \cdot \sigma\right) \cdot \left(\frac{\beta_1}{90}\right)^{1.67+1.09\cdot\sigma}$$

The coefficient K_{sh} depends on the profile type (Aungier 2003) and takes values between 0.7 and 1.1. In particular: 1.0 for NACA profiles, 1.1 for C4 profiles and 0.7 for double circular arch profiles. In absence of specific information, we can assume $K_{sh} = 1$.

The calculation of γ and δ_w is therefore iterative. We set starting values of the incidence and the deviation angles (i and δ_w) and as a consequence we can calculate the first value of the blade camber angle, θ:

$$\theta = \beta_1 - \beta_2 + \delta_w - i$$

and also the blade stagger angle (once the value l/C is set).

Then, once the value l/C has been set, the ratio e/C is calculated (Fig. 4.5) by using (Aungier 2003):

$$\frac{e}{C} = \frac{\sqrt{1 + (4 \cdot \tan\theta)^2 \cdot \left[l/C - (l/C)^2 - 3/16\right]} - 1}{4 \cdot \tan\theta}$$

and therefore the blade angles measured from the chord line, χ_1 and χ_2 (Fig. 4.5):

$$\tan\chi_1 = \frac{e/C}{l/C - 1/4}$$
$$\tan\chi_2 = \frac{e/C}{3/4 - l/C}$$

By using these angles, the blade angles can be calculated by:

$$\beta_{1B} = \gamma + \chi_1$$
$$\beta_{2B} = \gamma - \chi_2$$

and therefore the new incidence angle:

$$i = \beta_1 - \beta_{1B}$$

and also the deviation angle through the expression previously proposed. These calculations are repeated until convergence for the incidence and deviation angles is achieved.

At the end of this iterative procedure, the axial chord (C_a), the blade pitch (S) and the number of blades (N_B) can be calculated:

$$C_a = C \cdot \cos\gamma$$
$$S = \frac{C}{\sigma}$$
$$N_B = \text{int}\left(\frac{\pi \cdot D_M}{S}\right)$$

The *blade distance* at the leading edge, O (Fig. 4.4), can be assessed by:

$$O \cong S \cdot \cos\beta_{1B}$$

by assuming as a right triangle the triangle having a hypotenuse equal to S, a base equal to O and an angle between them equal to β_{1B}. This parameter O also represents the flow area width perpendicular to the fluid velocity, assuming the flow angle very close to the blade angle ($\beta_1 \cong \beta_{1B}$).

The ratio between the blade trailing edge thickness and the blade pitch can be assumed as:

$$\frac{t_{TE}}{S} \cong 0.0005$$

Regarding the blade tip clearance (necessary for calculating the tip clearance losses, Sect. 4.5), we can assume that:

$$\frac{\tau}{h_B} = 0.6\% \div 1.5\%$$

but the minimum values must be not lower than $\tau = 0.3\ \text{mm} \div 0.5\ \text{mm}$.

From the knowledge of the kinematics, thermodynamics and geometry of the stage, the absolute (stator inlet/outlet) and relative (rotor inlet/outlet) Reynolds numbers can be calculated:

$$\text{Re}_1 = \frac{\rho_1 \cdot w_1 \cdot C_R}{\mu_1}$$

$$\text{Re}_{2,R} = \frac{\rho_2 \cdot w_2 \cdot C_R}{\mu_2}$$

$$\text{Re}_{2,S} = \frac{\rho_2 \cdot c_2 \cdot C_S}{\mu_2}$$

$$\text{Re}_3 = \frac{\rho_3 \cdot c_3 \cdot C_S}{\mu_3}$$

Once the Reynolds numbers have been calculated, considerations on surface roughness can be made according to Sect. 1.9.1.

In particular, the *admissible surface roughness*, $k_{s,adm}$, can be assessed; it is the limit roughness below which there are no effects on losses allowing the walls to be considered smooth. This condition occurs when the surface roughness is contained within the laminar sub-layer of the boundary layer.

This roughness (Sect. 1.9.1) can be evaluated, for the stator and the rotor, by:

$$\frac{k_{s,adm,R}}{C_R} = \frac{100}{\text{Re}_1}$$

$$\frac{k_{s,adm,S}}{C_S} = \frac{100}{\text{Re}_{2,S}}$$

The calculation of $k_{s,adm}$ assumes considerable importance as it allows to establish the finishing level of the channel surfaces in order to contain dissipations.

Taking into account that turbomachinery surface roughness generally falls within the range:

$$5\ \mu\text{m} < k_s < 50\ \mu\text{m}$$

the knowledge of $k_{s,adm}$ allows to evaluate a possible, and economically convenient, surface treatment aimed at respecting this admissible roughness value as well as not resorting to very high finish levels where not required.

In other words, if the admissible surface roughness, $k_{s,adm}$, is lower than the minimum roughness that is technologically and economically achievable, $k_{s,min}$, we assume:

$$k_s = k_{s,\min}$$

otherwise:

$$k_{s,min} \le k_s \le k_{s,adm}$$

In particular, in the latter case, a roughness lower than the admissible one can be assumed in order to take into account that the roughness tends to increase with the turbomachine operation.

Remarks on input parameters

The procedure illustrated in this Sect. 4.3 does not presuppose any further input parameters, with respect to those already identified in Sects. 4.1 and 4.2, taking into account the assumptions on the aspect ratio, AR, the solidity, σ, the ratios t_{TE}/S, $t_{\max}/C$, τ/h_B, k_s/C, and that the mean diameter, D_M, depends on the blade speed, u_M, and on the speed of rotation, n (Sect. 4.6).

4.4 Kinematic Parameters in the Radial Direction

After evaluating the blade heights at the rotor and stator inlet/outlet, all the kinematic parameters previously calculated at the mean diameter can be assessed along these blade heights, and in particular at the hub and tip diameters.

In this regard, the free-vortex flow is adopted (Sect. 1.7.4); it ensures that work transfer is equal at all radii:

$$\frac{\partial h_t}{\partial r} = 0 \quad ; \quad \frac{\partial c_m}{\partial r} = 0$$

from these equations we obtain:

$$c_{u(r)} \cdot r = const$$
$$c_{m(r)} = const$$

The velocity triangles along the blade height, and consequently the Mach and Reynolds numbers, can be deduced by calculating the flow and work coefficients and the degrees of reaction along the blade height. With reference to the blade hub and tip, we have:

$$\varphi_t = \varphi_M \cdot \frac{D_M}{D_t} \quad ; \quad \varphi_h = \varphi_M \cdot \frac{D_M}{D_h}$$

$$\psi_t = \psi_M \cdot \left(\frac{D_M}{D_t}\right)^2 \quad ; \quad \psi_h = \psi_M \cdot \left(\frac{D_M}{D_h}\right)^2$$

$$R_t = 1 - (1 - R_M) \cdot \frac{D_M^2}{D_t^2} \quad ; \quad R_h = 1 - (1 - R_M) \cdot \frac{D_M^2}{D_h^2}$$

By using these parameters and the correlations deduced in Sect. 4.1, the absolute and relative flow angles along the blade height can be calculated, and in particular at the blade hub and tip:

$$\tan \alpha_{1,t} = \frac{1 - R_t - \frac{\psi_t}{2}}{\varphi_t} \qquad \tan \alpha_{1,h} = \frac{1 - R_h - \frac{\psi_h}{2}}{\varphi_h}$$

$$\tan \alpha_{2,t} = \frac{1 - R_t + \frac{\psi_t}{2}}{\varphi_t} \qquad \tan \alpha_{2,h} = \frac{1 - R_h + \frac{\psi_h}{2}}{\varphi_h}$$

$$\tan \beta_{1,t} = \frac{R_t + \frac{\psi_t}{2}}{\varphi_t} \qquad \tan \beta_{1,h} = \frac{R_h + \frac{\psi_h}{2}}{\varphi_h}$$

$$\tan \beta_{2,t} = \frac{R_t - \frac{\psi_t}{2}}{\varphi_t} \qquad \tan \beta_{2,h} = \frac{R_h - \frac{\psi_h}{2}}{\varphi_h}$$

Consequently, the kinematics and geometry of the blade profiles, previously calculated at the mean diameter (Sects. 4.1 and 4.3), can also be completely evaluated in the radial direction.

Since the maximum values of the flow and work coefficients and the minimum degree of reaction occur at the blade hub, a positive degree of reaction at the hub imposes a minimum value on the hub to tip ratio (Tournier and El-Genk 2010).

Taking up the expression of the degree of reaction at the hub and imposing:

$$R_{h,\min} = 1 - (1 - R_M) \cdot \frac{D_M^2}{D_{h,\min}^2} = 0$$

introducing:

$$\lambda_{\min} = \frac{D_{h,\min}}{D_t}$$

and therefore:

$$D_M = \frac{D_t}{2} \cdot (1 + \lambda_{\min})$$

we obtain:

$$(1 - R_M) \cdot \left[\frac{1}{2} \cdot \left(\frac{1 + \lambda_{\min}}{\lambda_{\min}}\right)\right]^2 = 1$$

and as a consequence:

$$\lambda = \frac{D_h}{D_t} \geq \lambda_{\min} = \left(\frac{2}{\sqrt{1 - R_M}} - 1\right)^{-1}$$

Then, by imposing that the flow and work coefficients at the hub are lower than one, to ensure surge stability, the most general expression is obtained:

$$\lambda = \frac{D_h}{D_t} \geq \lambda_{\min} = \left(\frac{2}{MAX\left(\sqrt{1 - R_M}, \sqrt{\psi_M}, \varphi_M\right)} - 1\right)^{-1}$$

If this condition is not verified, the hub to tip diameter ratio, λ, must be increased by choosing a different mean-diameter degree of reaction and/or a different flow coefficient in order to reduce the blade height.

Remarks on input parameters

This procedure does not presuppose the introduction of further input parameters, compared to those already identified in the previous sections.

4.5 Stage Losses and Efficiency

The loss sources and the loss models in the different types of turbomachinery are extensively described in Sect. 1.9. In this section, we will provide specific correlations for loss models able to quantify the losses introduced in Sect. 1.9.

Losses in the stator and rotor of an axial stage (Sect. 1.9) depend on the kinematic (Sects. 4.1 and 4.4), thermodynamic (Sect. 4.2) and geometric (Sect. 4.3) parameters of the stage.

Here we will refer to the Lieblein model (1959), with the modifications and additions provided by Koch and Smith (1976) and Aungier (2003); in particular, reference will be made to the complete and in-depth analysis carried out by Tournier and El-Genk (2010).

The pressure loss coefficients are respectively for the rotor and the stator (Sect. 1.9.5):

$$Y_R = \frac{p_{1tr} - p_{2tr}}{p_{1tr} - p_1}$$
$$Y_S = \frac{p_{2t} - p_{3t}}{p_{2t} - p_2}$$

In the following, we will refer explicitly to the rotor; the extension to the stator is obtained by replacing the relative angles, β, with the absolute angles, α, the relative velocities, w, with the absolute velocities, c, and taking into account what will be expressly illustrated for the stator.

The pressure loss coefficient is composed of (Sect. 1.9.4):

$$Y = Y_p + Y_s + Y_{EW} + Y_{shock} + Y_{TC}$$

where:

- Y_p represents the profile losses;
- Y_s represents the secondary losses;
- Y_{EW} represents the end wall losses;
- Y_{shock} represents the shock losses
- Y_{TC} represents the tip clearance losses.

4.5.1 Profile Losses

Lieblein (1959) proposed the following expression to calculate the blade-profile pressure loss coefficient:

$$Y_p = 2 \cdot \left(\frac{\theta_2}{C}\right) \cdot \frac{\sigma}{\cos\beta_2} \cdot \left(\frac{\cos\beta_1}{\cos\beta_2}\right)^2 \cdot \left(\frac{2 \cdot H_{TE}}{3 \cdot H_{TE} - 1}\right) \cdot \left[1 - \left(\frac{\theta_2}{C}\right) \cdot \frac{\sigma \cdot H_{TE}}{\cos\beta_2}\right]^{-3}$$

The profile losses depend on the boundary-layer momentum thickness at the blade outlet, θ_2, and on the boundary layer trailing-edge shape factor, H_{TE} (that is, the ratio of the boundary layer displacement thickness to the momentum thickness, Sect 1.9.1), in addition to the flow angles, the chord and the solidity (Sects. 4.1 and 4.3).

These parameters were provided by Koch and Smith (1976):

$$\frac{\theta_2}{C} = \left(\frac{\theta_2^0}{C}\right) \cdot \zeta_M \cdot \zeta_H \cdot \zeta_{Re}$$
$$H_{TE} = H_{TE}^0 \cdot \xi_M \cdot \xi_H \cdot \xi_{Re}$$

The values of θ_2^0 / C and H_{TE}^0 are for inlet Mach numbers lower than 0.05, no contraction in the height of the flow annulus, an inlet Reynolds number equal to 10^6 and hydraulically smooth blades. By using experimental data obtained under these conditions, Koch and Smith provided the following correlations as a function of the equivalent diffusion ratio $\mathrm{DF_{eq}}$ (see below for its expression):

$$\frac{\theta_2^0}{C} = 2.644 \cdot 10^{-3} \cdot DF_{eq} - 1.519 \cdot 10^{-4} + \frac{6.713 \cdot 10^{-3}}{2.60 - DF_{eq}}$$

$$H_{TE}^0 = \frac{\delta_{TE}^*}{\theta_2^0} = \left(0.91 + 0.35 \cdot DF_{eq}\right) \cdot \left[1 + 0.48 \cdot \left(DF_{eq} - 1\right)^4 + 0.21 \cdot \left(DF_{eq} - 1\right)^6\right] \quad \text{when} \quad DF_{eq} \leq 2$$

$$H_{TE}^0 = 2.7209 \quad \text{when} \quad DF_{eq} > 2$$

The correction factor of θ_2^0 / C for inlet Mach number is ζ_M and it can be expressed as:

$$\zeta_M = 1.0 + \left(0.11757 - 0.16983 \cdot DF_{eq}\right) \cdot Ma_1^n$$
$$n = 2.853 + DF_{eq} \cdot \left(-0.97747 + 0.19477 \cdot DF_{eq}\right)$$

The correction factor of θ_2^0 / C for the flow area contraction is ζ_H and it is given by:

$$\zeta_H = 0.53 \cdot \frac{h_{B,1}}{h_{B,2}} + 0.47$$

The correction factor of θ_2^0 / C for the Reynolds number and the roughness surface is ζ_{Re}; for this correction factor Koch and Smith provided a diagram that can be described analytically through the procedure illustrated in Sect. 1.9.1. A critical Reynolds number, above which the effects of roughness become significant (Sect. 1.9.1), is introduced:

$$\mathrm{Re}_{cr} = 100 \cdot \frac{C}{k_s}$$

where k_s is the roughness surface, evaluated in Sect. 4.3.

If $\mathrm{Re}_1 \leq \mathrm{Re}_{cr}$ we can assume:

$$\zeta_{\mathrm{Re}} = \left(\frac{10^6}{\mathrm{Re}_1}\right)^{0.166} \quad \text{when} \quad \mathrm{Re}_1 \geq 2 \cdot 10^5$$

$$\zeta_{\mathrm{Re}} = 1.30626 \cdot \left(\frac{2 \cdot 10^5}{\mathrm{Re}_1}\right)^{0.5} \quad \text{when} \quad \mathrm{Re}_1 < 2 \cdot 10^5$$

If $\mathrm{Re}_1 > \mathrm{Re}_{cr}$ we can assume:

$$\zeta_{\text{Re}} = \left(\frac{10^6}{\text{Re}_{cr}}\right)^{0.166} \qquad \text{when} \quad \text{Re}_{cr} \geq 2 \cdot 10^5$$

$$\zeta_{\text{Re}} = 1.30626 \cdot \left(\frac{2 \cdot 10^5}{\text{Re}_{cr}}\right)^{0.5} \qquad \text{when} \quad \text{Re}_{cr} < 2 \cdot 10^5$$

The correction factors of H^0_{TE} can be assessed by the following expressions:

$$\xi_M = 1.0 + \left[1.0725 + DF_{eq} \cdot (-0.8671 + 0.18043 \cdot DF_{eq})\right] \cdot Ma_1^{1.8}$$

$$\xi_H = 1.0 + \left(\frac{h_{B,1}}{h_{B,2}} - 1\right) \cdot (0.0026 \cdot DF^8_{eq} - 0.024)$$

$$\xi_{\text{Re}} = \left(\frac{10^6}{\text{Re}_1}\right) \qquad \text{when} \quad \text{Re}_1 < \text{Re}_{cr}$$

$$\xi_{\text{Re}} = \left(\frac{10^6}{\text{Re}_{cr}}\right)^{0.06} \qquad \text{when} \quad \text{Re}_1 \geq \text{Re}_{cr}$$

The equivalent diffusion ratio, DF_{eq}, is expressed by (Koch e Smith 1976):

$$DF_{eq} = \frac{w_1}{w_2} \cdot \left(1 + K_3 \cdot \left(\frac{t_{\max}}{C}\right) + K_4 \cdot \Gamma\right) \cdot \sqrt{(sen\beta_1 - K_1 \cdot \sigma \cdot \Gamma)^2 + \left(\frac{\cos\beta_1}{A^*_{throat} \cdot \rho_{throat}/\rho_1}\right)^2}$$

The contraction ratio is given by:

$$A^*_{throat} = \left[1.0 - K_2 \cdot \sigma \cdot \left(\frac{t_{\max}}{C}\right) \Big/ \cos\big((\beta_1 + \beta_2)/2\big)\right] \cdot \frac{A_{throat}}{A_1}$$

By assuming that the cascade throat area occurs at one-third of the axial chord, we can write:

$$A_{throat} = A_1 - \frac{1}{3} \cdot (A_1 - A_2)$$

The fluid density at the throat is calculated as:

$$\frac{\rho_{throat}}{\rho_1} = 1 - \frac{Ma^2_{x1}}{1 - Ma^2_{x1}} \cdot \left(1 - A^*_{throat} - K_1 \cdot \sigma \cdot \Gamma \cdot \frac{\tan\beta_1}{\cos\beta_1}\right)$$

where Ma_{x1} is the Mach number calculated by using the meridional velocity.

The constants that appear in the above expressions were obtained experimentally by Koch and Smith and they are:

$$K_1 = 0.2445$$
$$K_2 = 0.4458$$
$$K_3 = 0.7688$$
$$K_4 = 0.6024$$

Finally, the dimensionless blade circulation parameter, Γ, is given by:

$$\Gamma = \frac{D_{2M} \cdot c_{2u} - D_{1M} \cdot c_{1u}}{\sigma \cdot w_1 \cdot (D_{1M} + D_{2M})/2}$$

With the hypothesis of constant mean diameter and therefore constant blade speed at the mean diameter, the above parameter becomes:

$$\Gamma = (\tan \beta_1 - \tan \beta_2) \cdot \frac{\cos \beta_1}{\sigma}$$

Summarising, in order to calculate the profile losses, we must first assess the circulation parameter, Γ; then, by using the definition of A_{throat}, we can calculate A^*_{throat} and ρ_{throat}. All these quantities allow to evaluate the equivalent diffusion ratio, DF_{eq}, and as a consequence all the other parameters appearing in the expression of the profile losses.

4.5.2 Secondary Losses

Secondary losses are calculated through the correlation proposed by Howell (1947):

$$Y_s = 0.018 \cdot \sigma \cdot \frac{\cos^2 \beta_1}{\cos^3 \beta_M} \cdot c_L^2$$

where the blade lift coefficient, c_L, is expressed as:

$$c_L = \frac{2}{\sigma} \cdot \cos \beta_M \cdot (\tan \beta_1 - \tan \beta_2)$$

and the mean flow angle as:

$$\tan \beta_M = \frac{(\tan \beta_1 + \tan \beta_2)}{2}$$

4.5.3 End Wall Losses

End wall losses are calculated by using the correlation proposed by Aungier (2003):

$$Y_{EW} = 0.0146 \cdot \left(\frac{C}{h_B}\right) \cdot \left(\frac{\cos\beta_1}{\cos\beta_2}\right)^2$$

4.5.4 Shock Wave Losses

Shock wave losses occur in transonic and supersonic compressors (Ma > 1). Koch and Smith (1976) sketched on a graph the shock losses of a series of compressors as a function of the relative Mach number (at the rotor inlet). We interpolated these data and as a consequence we elaborated the following expressions:

$$Y_{shock} = 0.32 \cdot Ma_1^2 - 0.62 \cdot Ma_1 + 0.30 \qquad \text{when} \quad \mathrm{Ma}_1 \geq 1$$
$$Y_{shock} = 0 \qquad \text{when} \quad \mathrm{Ma}_1 < 1$$

4.5.5 Tip Clearance Losses

These losses can be calculated by using the correlations of Yaras and Sjolander (1992):

$$Y_{TC} = Y_{tip} + Y_{gap}$$
$$Y_{tip} = 1.4 \cdot K_E \cdot \sigma \cdot \left(\frac{\tau}{h_B}\right) \cdot \frac{\cos^2\beta_1}{\cos^3\beta_M} \cdot c_L^{1.5}$$
$$Y_{gap} = 0.0049 \cdot K_G \cdot \sigma \cdot \frac{C}{h_B} \cdot \frac{\sqrt{c_L}}{\cos\beta_M}$$

where the theoretical blade lift coefficient, c_L, and the mean flow angle, β_M,are already defined (Sect. 4.5.2) as well as the tip clearance, τ (Sect. 4.3).

For the other parameters of these expressions, we can assume (mid-loaded compressor blades l/C = 0.5):

- $K_E = 0.5$
- $K_G = 1.0$

4.5.6 Rotor and Stage Efficiency

The procedure described in Sects. 4.5.1, 4.5.2, 4.5.3, 4.5.4 and 4.5.5 allows evaluating the pressure loss coefficients in both the stator and rotor:

$$Y_R = \frac{p_{1tr} - p_{2tr}}{p_{1tr} - p_1}$$
$$Y_S = \frac{p_{2t} - p_{3t}}{p_{2t} - p_2}$$

To assess the rotor efficiency, we must first evaluate the total relative pressure downstream of the rotor (the total relative pressure upstream of the rotor was calculated in the thermodynamics of the stage, Sect. 4.2):

$$p_{2tr} = p_{1tr} - Y_R \cdot (p_{1tr} - p_1)$$

Now, by knowing the total relative enthalpy downstream of the rotor:

$$h_{2tr} = h_{1tr}$$

the thermodynamic quantities downstream of the rotor (Fig. 4.3) are:

$$s_2 = s_{p,h}(p_{2tr}, h_{2tr})$$
$$h_2 = h_{p,s}(p_2, s_2)$$

and, therefore, the rotor efficiency is:

$$\overline{\eta}_R = \frac{h_{2,is} - h_1}{h_2 - h_1}$$

Similarly, for the stator we have:

$$h_{2t} = h_2 + \frac{c_2^2}{2}$$
$$p_{2t} = p_{s,h}(s_2, h_{2t})$$
$$p_{3t} = p_{2t} - Y_S \cdot (p_{2t} - p_2)$$
$$h_{3t} = h_{2t}$$
$$s_3 = s_{p,h}(p_{3t}, h_{3t})$$
$$h_3 = h_{s,p}(s_3, p_3)$$

Once the thermodynamics of the stage have been calculated on the basis of the stage losses, the various efficiencies can be assessed (Sect. 1.8.3):

$$\eta_{TT} = \frac{h_{3t,is} - h_{1t}}{h_{3t} - h_{1t}}$$
$$\eta_{is} = \frac{h_{3,is} - h_1}{h_3 - h_1}$$
$$\eta_{TS} = \frac{h_{3,is} - h_{1t}}{h_{3t} - h_{1t}}$$

Remarks on input parameters

The procedure illustrated in this Sect. 4.5 does not presuppose the introduction of further input parameters, compared to those already identified in Sects. 4.1 and 4.2.

4.6 Input Parameters of the Preliminary Design Procedure

In the previous Sects. 4.1, 4.2, 4.3, 4.4 and 4.5, the procedure for the calculation of the kinematics, thermodynamics, geometry and efficiency of an axial stage was developed and the input parameters of this procedure were identified.

In this Sect. 4.6 we intend to precisely analyze these input parameters in order to suggest their numerical values (these values, however, can be reviewed in an iterative calculation, see Sect. 4.7).

First of all, let us consider the kinematics calculation procedure (Sect. 4.1).

In general, to determine the velocity triangles at the rotor inlet and outlet, we must set six independent parameters: three for the inlet triangle (station 1 of the stage) and three for the outlet triangle (station 2 of the stage).

For axial stages, however, being $u_1 = u_2 = u$ and assuming $c_{m1} = c_{m2} = c_m$, *the independent parameters are only four*. Indeed, from Sect. 4.1, we know that the absolute and relative flow angles are:

$$\alpha_1 = f(\varphi, \psi, R)$$
$$\alpha_2 = f(\varphi, \psi, R)$$
$$\beta_1 = f(\varphi, \psi, R)$$
$$\beta_2 = f(\varphi, \psi, R)$$

These angles are, therefore, functions of the following three independent parameters:

- the flow coefficient, φ
- the work coefficient, ψ
- the degree of reaction, R

As a consequence, by identifying a further independent parameter, for example, the blade speed, u, or the meridional component of the velocity, c_m, the velocity triangles are completely defined.

However, taking into account what emerged in the turbomachinery selection process (Chap. 2), it is advisable to choose the stage efficiency (isentropic or polytropic efficiency) as the additional independent parameter and not a velocity (u or c_m).

In fact, starting from the data available for the turbomachinery selection process:

- type of fluid
- fluid mass flow rate (kg/s)
- fluid inlet temperature (°C)
- fluid inlet pressure (bar)
- fluid outlet pressure (bar)

the selection process provides ranges of n ($n_{min} < n < n_{max}$) and z ($z_{min} < z < z_{max}$) compatible with the type of turbomachine chosen. Within these ranges we can choose one or more pairs of these values:

- n: rotational speed (rpm)
- z: number of stages and, consequently, the isentropic enthalpy increase in each stage, $\Delta h_{is,stage}$ (kJ/kg)

For each of these pairs, by using the stage (isoentropic or polytropic) efficiency, set as first iteration guess (Sect. 4.6.3) or calculated by the stage losses (Sect. 4.5), the adiabatic enthalpy increase of each stage can be calculated as well as the blade speed at the mean radius, after choosing the work coefficient.

Indeed, when $c_3 = c_1$ (Sect. 4.1), the blade speed is:

$$u = \sqrt{\frac{\Delta h_{t,ad,stage}}{\psi}} = \sqrt{\frac{\Delta h_{ad,stage}}{\psi}}$$

On the other hand, when $c_3 \neq c_1$ (Sect. 4.1), the work coefficient is expressed as:

$$\psi = \frac{\Delta h_{t,ad,stage}}{u^2} = \frac{\Delta h_{ad,stage}}{u^2} + \frac{c_3^2 - c_1^2}{2 \cdot u^2}$$

and, therefore, expressing the velocity c_1 as a function of the kinematic parameters, we obtain:

$$u = \sqrt{\frac{\Delta h_{ad,stage} + \frac{c_3^2}{2}}{\psi + \frac{1}{2} \cdot \left[\varphi^2 + \left(1 - R - \frac{\psi}{2}\right)^2\right]}}$$

After calculating the blade speed, the mean diameter of the rotor is:

$$D = \frac{60 \cdot u}{\pi \cdot n}$$

Alternatively, we can calculate the rotor diameter, D, from the specific diameter, D_S, and then the blade speed (Chap. 2). In this way, only three independent parameters (flow coefficient, degree of reaction and stage efficiency) must be chosen since the work coefficient is calculated by using the enthalpy increase and the blade speed. In axial stage design, however, it is preferred to assign the work coefficient, following the procedure illustrated above.

In summary, if the mean diameter and the meridional velocity are constant, we must set the following four input parameters at mean diameter in order to design preliminarily the axial stage:

- the flow coefficient, φ
- the work coefficient, ψ
- the degree of reaction, R
- the stage efficiency (isentropic or polytropic efficiency)

Taking into account what is illustrated in Sects. 4.2, 4.3, 4.4 and 4.5, these parameters, with the addition of the rotor efficiency, allow to fully calculate the thermodynamics, the geometry and therefore the stage losses. Since these losses allow, in turn, to exactly evaluate the rotor and stage efficiencies, the stage efficiency (in the list above) is only a starting value: the calculation will proceed iteratively for each stage until convergence for these efficiencies is achieved (Sect. 4.7).

4.6.1 Flow and Work Coefficient Selection

In analogy to the turbine stages, Smith-type diagrams have also been proposed for the axial compressor stages over the years.

In Fig. 4.6 the diagram, proposed by Lewis (1996) and based on the assessments by Casey (1987), is sketched for three degrees of reaction (0.5, 0.7 and 0.9). These diagrams, compared with Smith's diagrams for the turbines (Sect. 3.6.1), highlight that the work coefficient of an axial compression stage is much lower than that of a turbine stage, while the flow coefficient range is nearly the same (0.4–0.8). To attain high efficiencies (not less than 90%), in fact, the work coefficient must be in the range 0.2–0.45 (that is, 4–5 times lower than that of a turbine stage). This is due to the limitations to the flow deflection and the diffusion limits in typical compressor cascades.

Higher values of the work coefficient could be advantageous to reduce the number of stages (and therefore the compressor weight and cost), but these values necessarily entail an efficiency decay. Concerning that point, Dickens and Day (2009) studied the design of highly loaded compressors (high work coefficient). Figure 4.7, proposed by Dickens and Day (2009) and based on the 1D correlations of Wright and Miller (1991), shows the conventional design range for the flow coefficient (0.4–0.6) and work coefficient (0.3–0.45) as well as the design ranges for highly loaded compressors: in this last case, the work coefficient can vary between 0.6–0.75, while the flow coefficient range is the same. The study by Dickens and Day showed that if the

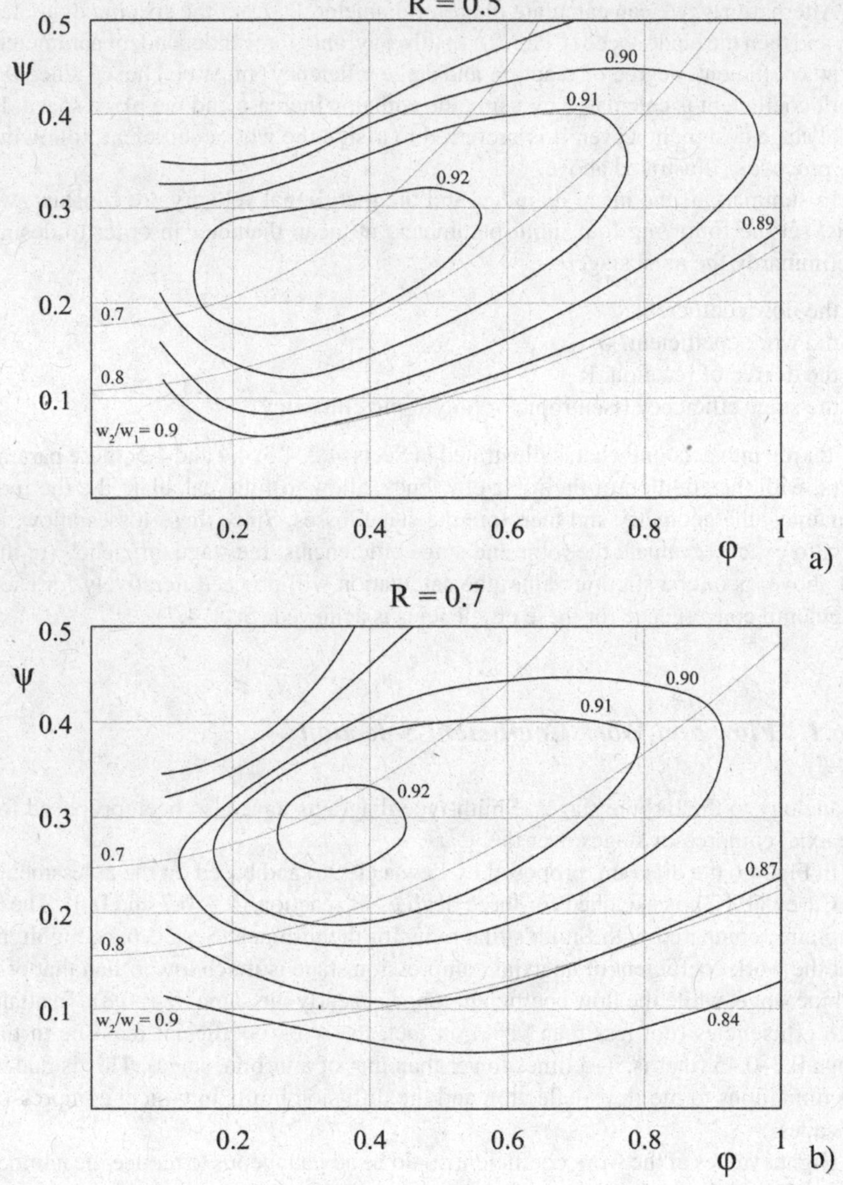

Fig. 4.6 Smith-type diagram for an axial compressor stage (**a** R = 0.5, **b** R = 0.7, **c** R = 0.9). Adapted from Lewis (1996)

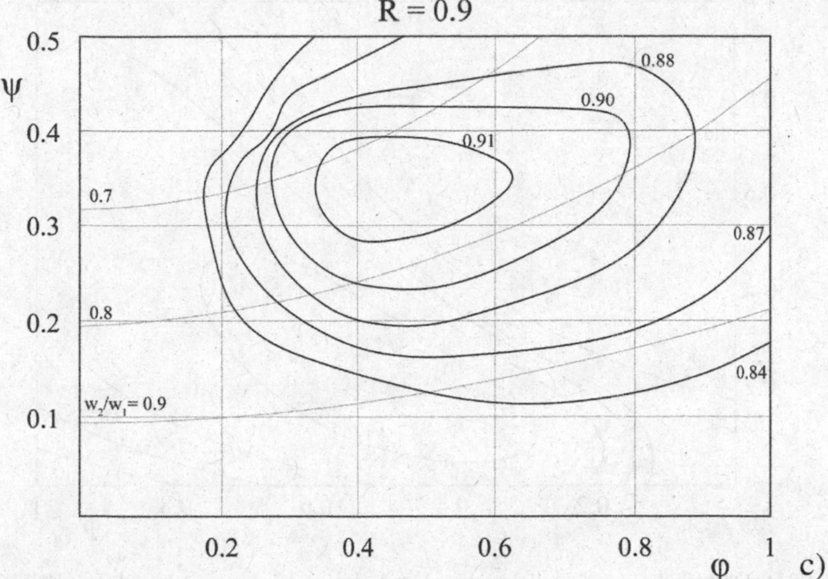

Fig. 4.6 (continued)

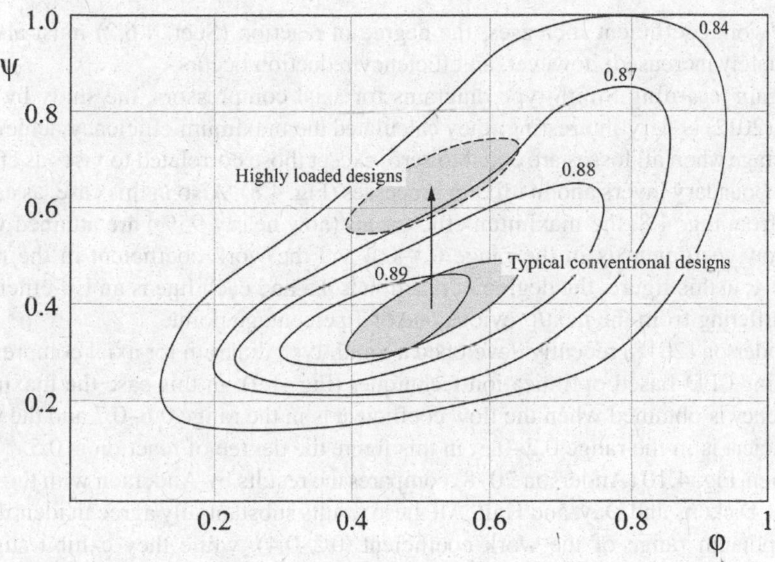

Fig. 4.7 Smith-type diagram for axial stages of conventional and highly loaded compressors. Adapted from Dickens and Day (2009)

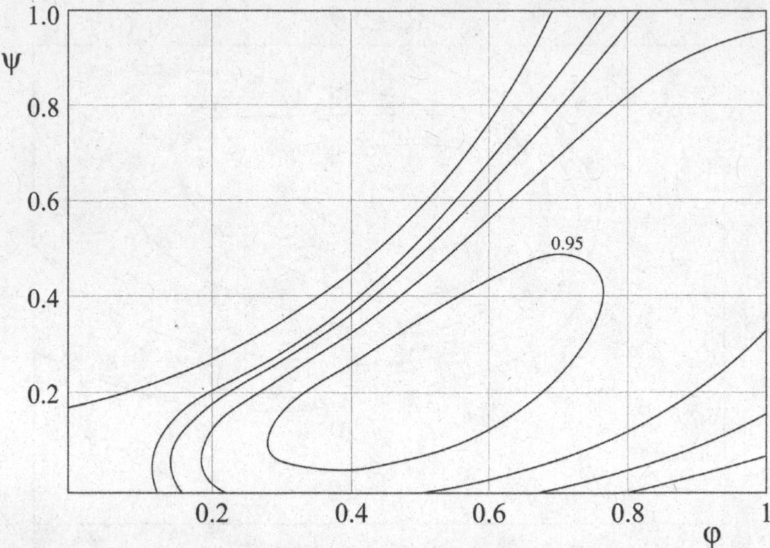

Fig. 4.8 Smith-type diagram for axial stages where losses are minimum. Adapted from Hall et al. (2012)

stage work coefficient increases, the degree of reaction (Sect. 4.6.2) must also be adequately increased; however, an efficiency reduction occurs.

Again regarding Smith-type diagrams for axial compressors, the study by Hall et al. (2012) is very interesting: they calculated the maximum efficiency achievable by a stage when all losses are equal to zero, except those correlated to viscous effects in the boundary layers and in mixing processes (Fig. 4.8). Also in this case, as can be seen from Fig. 4.8, the maximum efficiencies (now nearly 95%) are attained when the flow coefficient is in the range 0.4–0.6 and the work coefficient in the range 0.2–0.4; in this figure, the degree of reaction is 0.5 and each line is an iso-efficiency line differing from the next/previous one of 1 percentage point.

Anderson (2018) recently developed a Smith-type diagram for axial compressors by using CFD-based optimization techniques (Fig. 4.9): in this case the maximum efficiency is obtained when the flow coefficient is in the range 0.6–0.7 and the work coefficient is in the range 0.2–0.3; in this figure the degree of reaction is 0.5.

Then, Fig. 4.10 (Anderson 2018) compares the results by Anderson with those by Casey, Dickens and Day, and Hall. All these results substantially agree in identifying the optimum range of the work coefficient (0.2–0.4), while they exhibit slightly different values of the flow coefficient: the values proposed by Casey and Dickens are lower than those by Anderson, while those proposed by Hall are similar. This difference is certainly due to the application of different loss models, considering that Anderson's analysis used the most advanced CFD-based optimization techniques.

Based on the above observations, we can conclude that *in the preliminary design of axial stages of conventional compressors, Smith-type diagrams can be effectively*

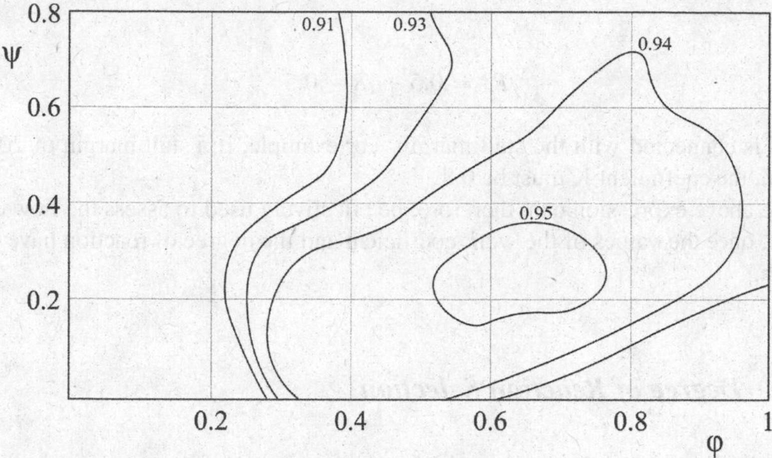

Fig. 4.9 Smith-type diagram for axial stages elaborated by using CFD-based optimization techniques. Adapted from Anderson (2018)

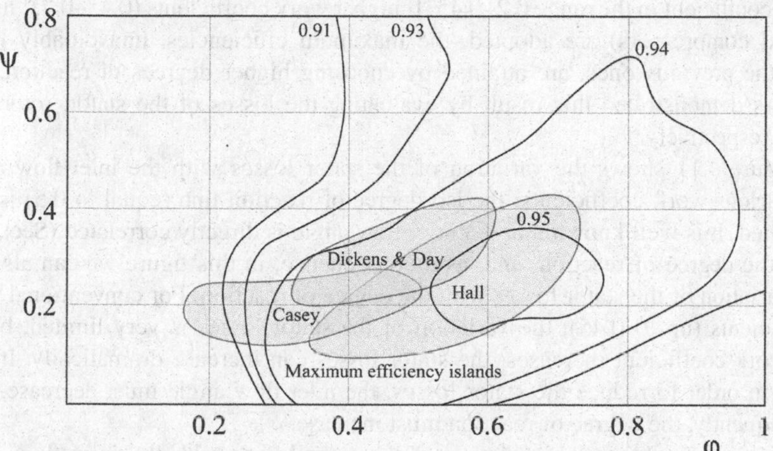

Fig. 4.10 Smith-type diagram comparison. Adapted from Anderson (2018)

applied, and, therefore, a work coefficient in the range 0.2–0.45 and a flow coefficient in the range 0.4–0.8 can be adopted.

However, a further clarification, regarding the flow coefficient, is needed. To ensure a margin to stall conditions, Aungier (2003) developed the following empirical expression, which correlates the flow and work coefficients and the degree of reaction:

$$\varphi = \frac{\pi}{4} \cdot \left[\frac{\psi / K - R^* \cdot 6/17}{0.86} \cdot \left(\frac{R^*}{0.5} \right)^{1.18} \right]^{\frac{1}{2.0+1/R^*}}$$

where

$$R^* = 0.5 + |R - 0.5|$$

and K is connected with the stall margin. For example, if a stall margin of 20% is chosen, the coefficient K must be 0.8.

The above expression can, therefore, be effectively used to assess the flow coefficient, once the values of the work coefficient and the degree of reaction have been chosen.

4.6.2 Degree of Reaction Selection

Generally, the degree of reaction, which ensure optimum efficiency, is very close to 0.5.

As previously mentioned, Dickens and Day (2009) stated that this degree of reaction guarantees maximum efficiency for a conventional compressor having the work coefficient in the range 0.2–0.45. If higher work coefficients (0.45–0.75, highly loaded compressors) are adopted, the maximum efficiencies, unavoidably lower than the previous ones, are attained by choosing higher degrees of reaction. The authors demonstrated this result by evaluating the losses of the stator, rotor, and stage respectively.

Figure 4.11 shows the variation of the stator losses with the inlet flow angle for various work coefficients; the iso-degree of reaction line (equal to 0.5) is also sketched. It is well known that the inlet flow angle is directly correlated (Sect. 4.1) with the degree of reaction, and, as a consequence, in this figure we can also see the variation of the stator losses with the degree of reaction. For conventional work coefficients (up to 0.45), the variation of the stator losses is very limited, but, if the work coefficient increases, the stator losses can increase dramatically. In this case, in order to reduce the stator losses, the inlet flow angle must decrease, and, consequently, the degree of reaction must increase.

Figure 4.12 shows the variation of the rotor losses with the inlet flow angle for various work coefficients; the iso-degree of reaction line (equal to 0.5) is also sketched. In this case, the variation of the rotor losses with the degree of reaction is much less accentuated than that of the stator losses.

Dickens and Day, then, calculated the stage loss variation with the inlet flow angle; they highlighted that, for each work coefficient, there is an optimum degree of reaction, R, which maximizes efficiency (locus of peak efficiency in Fig. 4.13): the optimum R increases as the work coefficient rises.

The study by Dickens and Day, therefore, can be effectively used to evaluate the optimum degree of reaction, according to the chosen work coefficient.

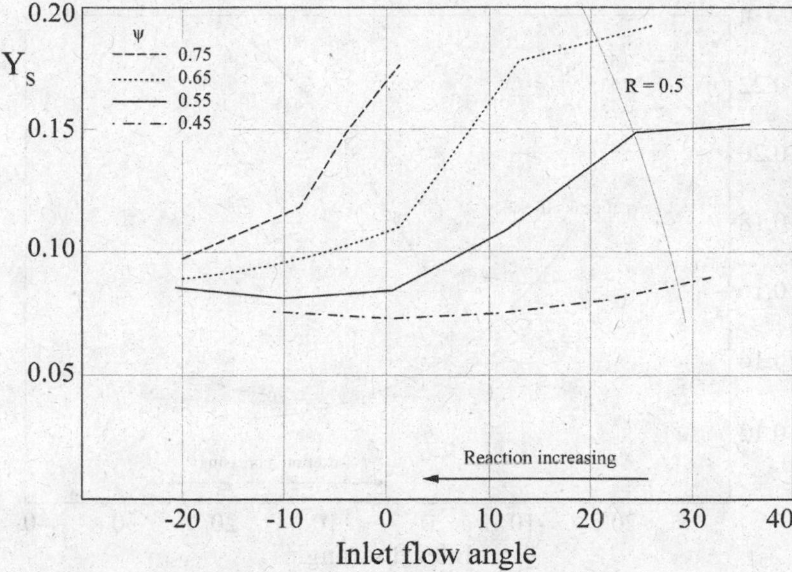

Fig. 4.11 Stator losses. Adapted from Dickens and Day (2009)

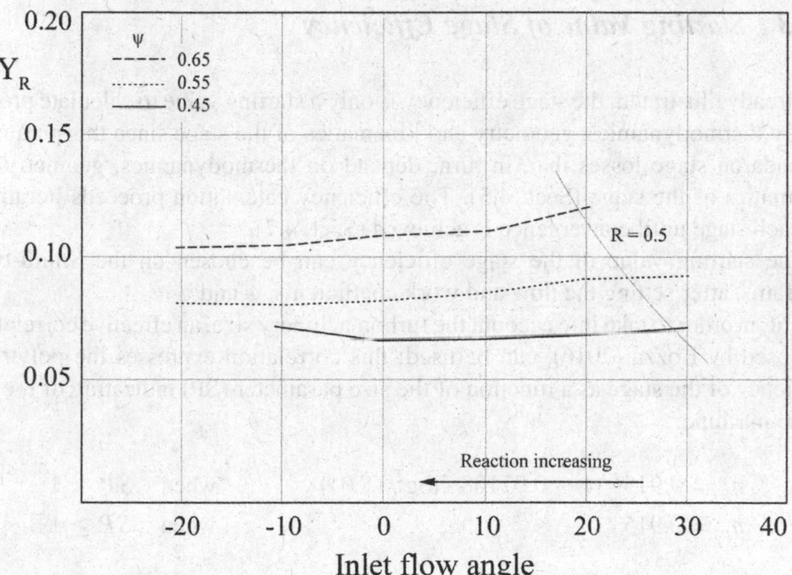

Fig. 4.12 Rotor losses Adapted from Dickens and Day (2009)

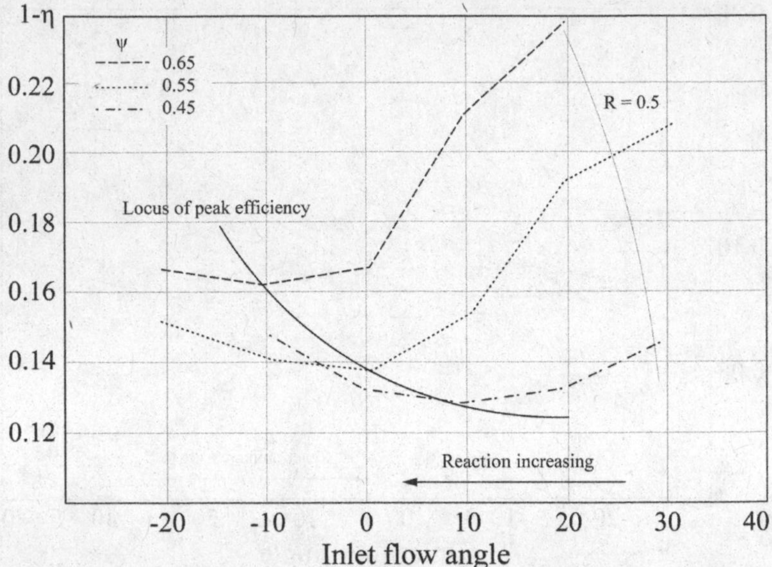

Fig. 4.13 Stage losses Adapted from Dickens and Day (2009)

4.6.3 Starting Value of Stage Efficiency

As already illustrated, the stage efficiency is only a starting value to calculate preliminarily thermodynamics, geometry and kinematics of the stage since this parameter depends on stage losses that, in turn, depend on thermodynamics, geometry and kinematics of the stage (Sect. 4.5). The efficiency calculation proceeds iteratively for each stage until convergence is achieved (Sect. 4.7).

The starting value of the stage efficiency can be chosen on the Smith-types diagrams, after setting the flow and work coefficients, φ and ψ.

But, in order to take into account the turbomachinery size, an effective correlation, proposed by Lozza (2016), can be used; this correlation expresses the polytropic efficiency of the stage as a function of the size parameter (SP) indicating of the size of the machine:

$$\eta_p = 0.915 \cdot \left(1 - 0.07108 \cdot \log_{10}^2(SP)\right) \qquad \text{when} \quad SP < 1$$

$$\eta_p = 0.915 \qquad \text{when} \quad SP \geq 1$$

where:

$$SP = \frac{\sqrt{V_{in}}}{\Delta h_{is,stage}^{1/4}}$$

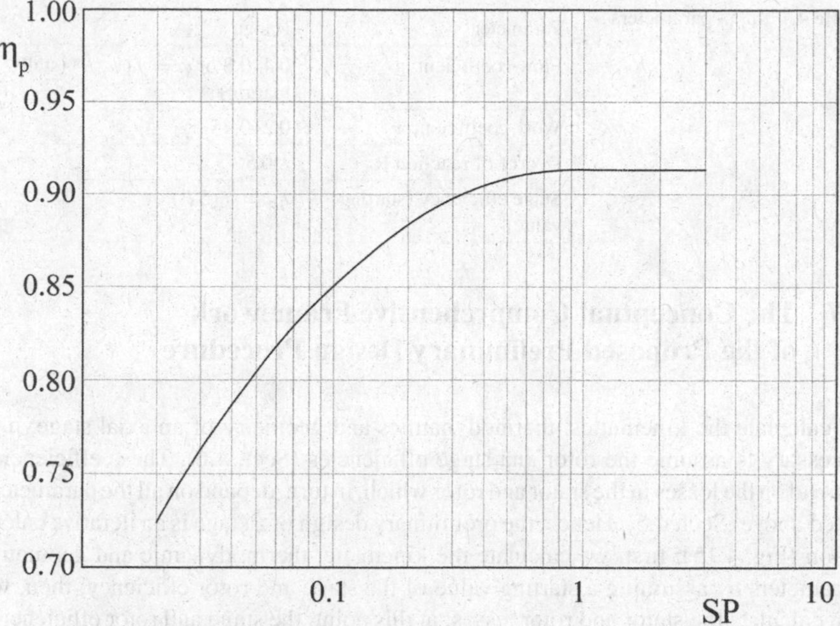

Fig. 4.14 Variation of polytropic efficiency with SP

The variation of the polytropic efficiency with SP is shown in Fig. 4.14; efficiency decreases when the absolute dimensions of the stage decrease because of a percentage increase in losses. For $SP \geq 1$, this correlation provides an efficiency value close to that of the Smith-type diagrams in the recommended φ (0.4–0.8) and ψ (0.2–0.45) ranges.

By using this polytropic efficiency, the starting value of the stage isentropic efficiency can be calculated following the procedure illustrated in Sect. 1.4.2.

For thermodynamics evaluation (Sect. 4.2), it is also necessary to choose the starting value of the rotor efficiency. It can be set arbitrarily, for example equal to the stage isentropic efficiency, since also this efficiency will be iteratively calculated by the losses in the rotor (Sect. 4.6).

4.6.4 Summary of Input Parameters

All the main considerations made in the previous Sects. 4.6.1, 4.6.2 and 4.6.3 can be summarized in Table 4.1, which reports the most suitable ranges for the input parameters of the preliminary design procedure for axial compressor stages.

Table 4.1 Input parameters

Parameter	Range
Flow coefficient φ	0.4–0.8 or $\varphi = f(\psi, R)$ (stall margin)
Work coefficient ψ	0.2–0.45
Degree of reaction R	$\approx$0.5
Stage efficiency (starting value)	$\eta_p = f(SP)$

4.7 The Conceptual Comprehensive Framework of the Proposed Preliminary Design Procedure

To calculate the kinematics, thermodynamics and geometry of an axial stage, it is necessary to assume the rotor and stage efficiencies (Sect. 4.6). These efficiencies depend on the losses in the stator and rotor which, in turn, depend on all the parameters listed above (Sect. 4.5). Hence, the preliminary design of a stage is an iterative calculation (Fig. 4.15): first, we calculate the kinematic, thermodynamic and geometric parameters by assuming a starting value of the stage and rotor efficiency; then, we can calculate the stator and rotor losses; at this point, the stage and rotor efficiencies must be calculated by using these losses; these new efficiencies become the new input parameters of the preliminary design procedure. This iterative procedure ends when the established convergence for the rotor and stage efficiencies is achieved.

The iterative procedure shown in Fig. 4.15 could be solved by using different values of the input parameters (R, φ and ψ), as well as of the geometric parameters, especially the solidity and the aspect ratio, evaluated by empirical correlations in Sect. 4.3, in order to optimize the stage performance.

For multistage compressors (z > 1), the same preliminary design is applied to the stages following the first one. For these stages, the only constraint is to maintain the same speed of rotation, since the input parameters R, φ and ψ could be assigned anew. If, on the other hand, an assumption is made on the characteristic diameters (constant hub or tip or mean diameter), these parameters are calculated anew according to the following procedure.

For example, if the turbomachine mean diameter is assumed constant (Fig. 4.16):

$$D_M = const$$

the blade speed is constant:

$$u_M = const$$

In addition, we can assume constant the meridional velocity:

$$c_m = const$$

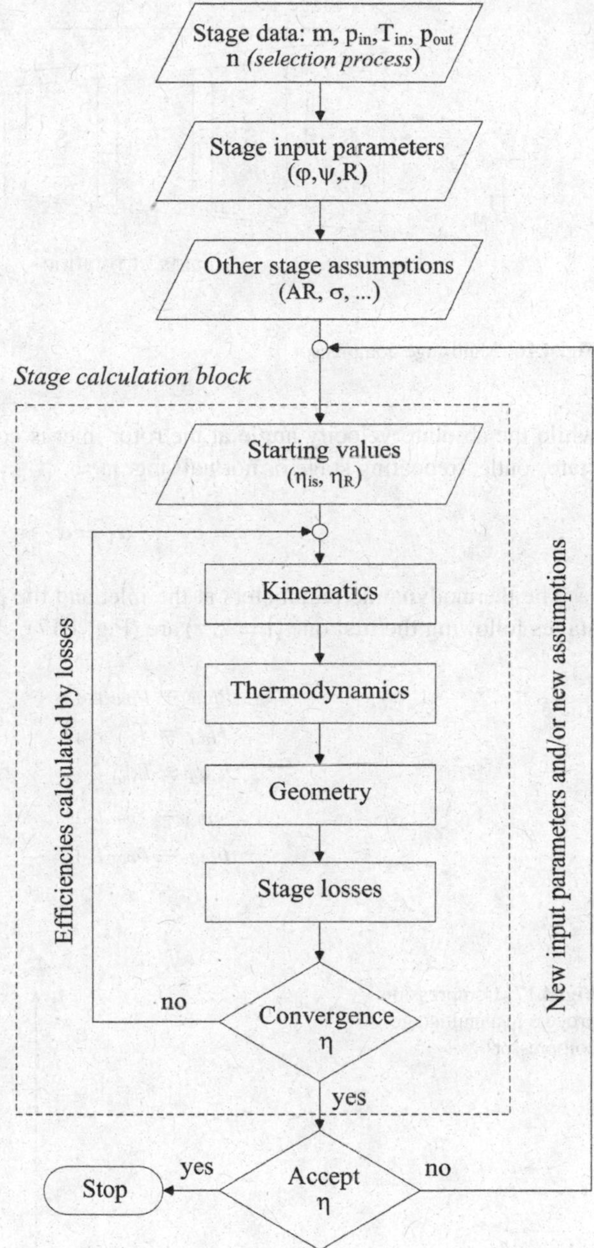

Fig. 4.15 Stage calculation block diagram

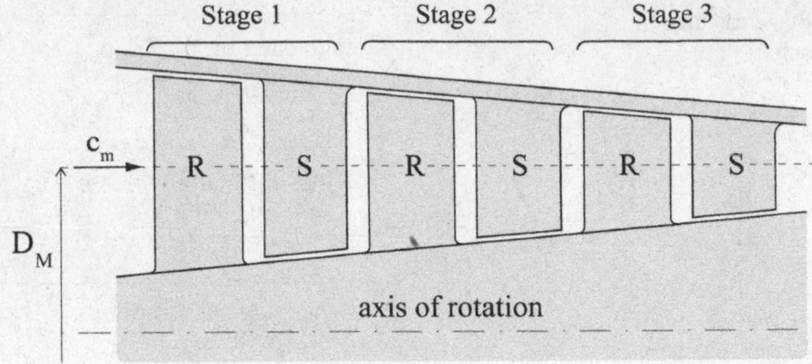

Fig. 4.16 Multistage compressor

while the absolute velocity angle at the rotor inlet is equal to that at the upstream stator outlet (repeating stage or normal stage):

$$c_1 = c_3 \quad ; \quad \alpha_1 = \alpha_3$$

The thermodynamic parameters at the inlet and the pressure at the outlet of the stages following the first one (i = 2, z) are (Fig. 4.17):

$$\begin{aligned} p_{in,i} &= p_{out,i-1} \\ T_{in,i} &= T_{out,i-1} \\ h_{in,i} &= h_{out,i-1} \\ s_{in,i} &= s_{out,i-1} \\ \rho_{in,i} &= \rho_{out,i-1} \end{aligned}$$

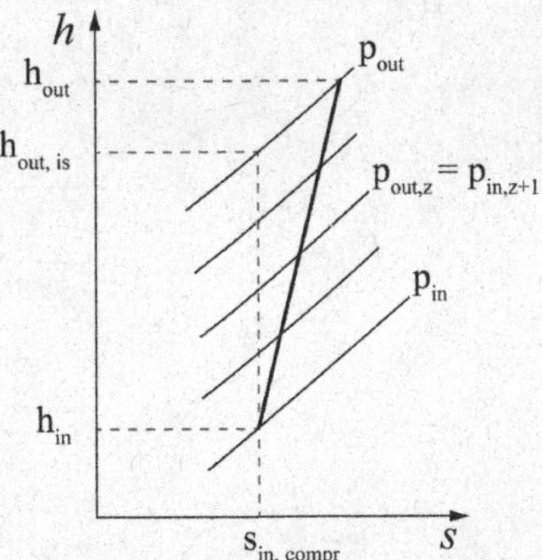

Fig. 4.17 Compression process in a multistage compressor

$$h_{out,is,i} = h_{in} + \sum_{j=1}^{i} \Delta h_{is,j}$$

$$p_{out,i} = p(h_{out,is,i}, s_{in})$$

In the previous correlations, the isentropic enthalpy increases of each stage are evaluated according to the procedure described in Chap. 2.

The kinematics of each stage following the first one (i = 2, z) is calculated (Sect. 4.1) by considering constant the flow coefficient (according to the previous assumptions):

$$\varphi_i = \frac{c_{m,i}}{u_{M,i}} = \frac{c_m}{u_M} = const = \varphi$$

while the work coefficient must be calculated anew because now the blade speed at the mean diameter and the stage enthalpy increase are established:

$$\psi_i = \frac{\Delta h_{t,ad,i}}{u_M^2}$$

Consequently, also the degree of reaction must be recalculated:

$$R_i = 1 - \frac{\psi_i}{2} - \varphi \cdot \tan \alpha_1$$

These parameters allow the calculation of the thermodynamics (Sect. 4.2), the geometry (Sect. 4.3), the kinematics in the radial direction (Sect. 4.4) and the losses (Sect. 4.5) of each stage following the first one.

The compressor design, carried out by assigning a pair of values for the speed of rotation and the number of stages (n, z), provided by the selection process illustrated in Chap. 2, can be accomplished again (Fig. 4.18) by using a different pair of input parameters (n and z), indicated always by the turbomachinery selection process, in order to optimize the preliminary design of the compressor.

Fig. 4.18 Multistage compressor calculation block diagram

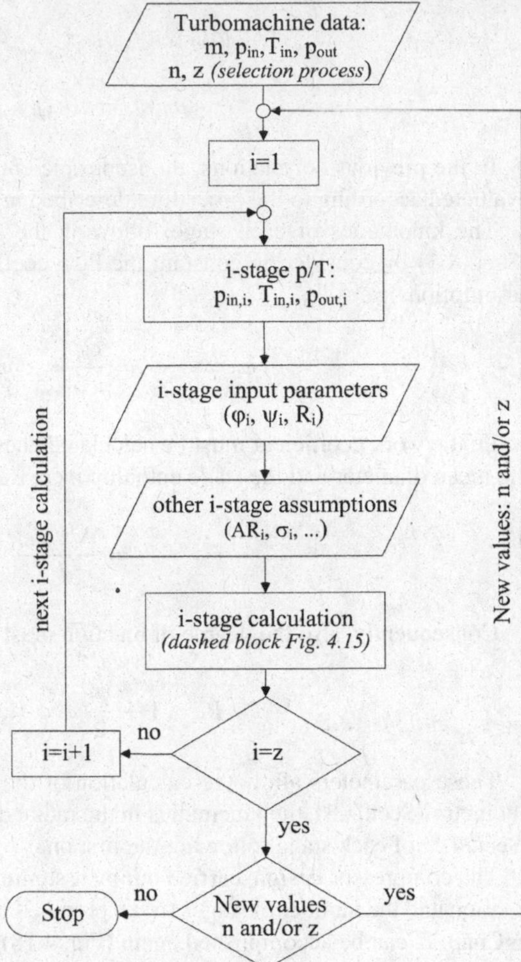

References

Anderson MR (2018) Improved smith chart for axial compressor design. In: Proceedings of ASME Turbo Expo GT2018

Aungier RH (2003) Axial-flow compressors. A strategy for aerodynamic design and analysis. ASME Press

Casey M (1987) A mean-line prediction method for estimating the performance characteristic of an axial compressor stage. In: International conference on turbomachinery: efficiency prediction and improvement. Inst. Mech. Eng., Cambridge

Dickens T, Day I (2009) The design of highly loaded axial compressors. In: Proceedings of ASME Turbo Expo GT2009

Hall DK, Greitzer EM, Tan CS (2012) Performance limits of axial compressor stages. In: Proceedings of ASME Turbo Expo GT2012

Howell AR (1947) Fluid dynamics of axial compressors. Lectures on the Development of the British Gas Turbine Jet Unit. Institution of Mechanical Engineers

Koch CC, Smith LH (1976) Loss Sources and magnitudes in axial-flow compressors. J. Eng. Power

Lewis RI (1996) Turbomachinery performance analysis. Elsevier

Lieblein S (1959) Loss and stall analysis in compressor cascades. J Basic Eng

Lozza G (2016) Gas turbine and combined cycles (in Italian). Ed. Esculapio

To HO, Miller RJ (2015) The effect of aspect ratio on compressor performance. In: Proceedings of ASME Turbo Expo GT2015

Tournier JM, El-Genk MS (2010) Axial flow, multi stage turbine and compressor models. Energy Convers Manage

Wennerstrom AJ (1989) Low aspect ratio axial flow compressors: why and what it means. J Turbomach

Wright PI, Miller DC (1991) An Improved compressor performance prediction method. In: Proceedings of the European conference on turbomachinery (J Turbomach)

Yaras MI, Sjolander SA (1992) Prediction of tip-leakage losses in axial turbines. J Turbomach

Chapter 5
Preliminary Design of Radial Inflow Turbines

Abstract With reference to a radial inflow turbine stage, a procedure for the calculation of kinematic parameters (Sect. 5.1), thermodynamic parameters (Sect. 5.2), geometric parameters (Sect. 5.3) and stage losses (Sect. 5.4) is provided. Then, Sect. 5.5 discusses the input parameters of this procedure and suggests their numerical values to be used in the calculations. Finally, Sect. 5.6 illustrates the procedure for extending calculations to multistage turbines. The numerical application of the proposed procedure, aimed at the preliminary design of radial turbines, is instead developed in Chap. 8.

5.1 Kinematic Parameters

In this section, we will deduce the kinematic correlations of a radial inflow stage; they are correlations among kinematic parameters of a stage (degree of reaction R, absolute and relative flow angles, α and β) and dimensionless performance parameters (flow coefficient, φ, and work coefficient, ψ).

Variations of the previously introduced (Chap. 2) dimensionless factors Φ and Ψ are used in practice.

In particular, the ratio of the meridional velocity at the rotor exit, c_{2m}, to the blade speed at the rotor inlet, u_1, is the *flow coefficient*:

$$\varphi = \frac{c_{2m}}{u_1}$$

Note that there is a proportionality between this flow coefficient and the flow factor, previously introduced (Chap. 2):

$$\Phi = \frac{V}{\omega \cdot D^3} \propto \frac{c_{2m} \cdot D_1^2}{u_1 \cdot D_1^2} = \frac{c_{2m}}{u_1} = \varphi$$

Analogously, the ratio of actual work exchanged between fluid flow and rotor to blade speed (at rotor inlet) squared is the *work coefficient*:

M. Gambini and M. Vellini, *Turbomachinery*, Springer Tracts in Mechanical Engineering, https://doi.org/10.1007/978-3-030-51299-6_5

$$\psi = \frac{W}{u_1^2} = \frac{\Delta h_{t,ad}}{u_1^2}$$

which is proportional to the work factor, Ψ, previously introduced (Chap. 2).

In a radial stage (Fig. 5.1a), the fluid flow is studied (Sect. 1.7) on two stage sections: one perpendicular to the axis of rotation (Fig. 5.1b, left) and the other one containing the axis of rotation (meridional section, Fig. 5.1b right). With reference to these two sections, Fig. 5.1b depicts the velocity triangles of a radial turbine stage.

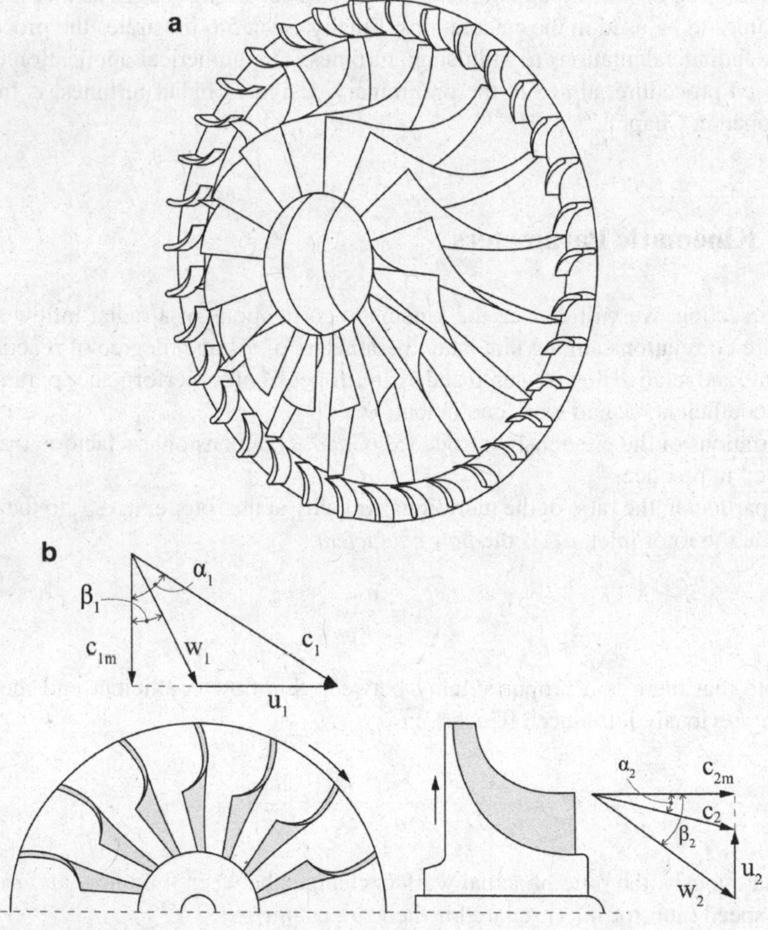

Fig. 5.1 **a** A radial turbine stage. **b** Velocity triangles in a radial turbine stage

In the following, the kinematic correlations will be expressed by using $\alpha_2 \neq 0$, even if, in a radial stage, the outlet absolute velocity is generally axial ($\alpha_2 = 0$).

Two parameters are then introduced: they are the *rotor meridional velocity ratio*, ξ, and the *rotor tip diameter ratio*, δ_t:

$$\xi = \frac{c_{1m}}{c_{2m}}$$

$$\delta_t = \frac{D_{2t}}{D_1}$$

At the stage exit, in the radial direction, the meridional absolute velocity can be considered constant and the free-vortex flow can be assumed (Sect. 1.7.4):

$$c_{2m(r)} = const$$

$$c_{2u(r)} \cdot r = const$$

and as a consequence:

$$u_{2(r)} \cdot c_{2u(r)} = const$$

Therefore it is possible to express the work transfer by using indifferently the velocity triangles at the tip or the hub of the rotor outlet.

Referring to the tip, the rotor tip diameter ratio, δ_t, can be expressed through the blade speed ratio:

$$\delta_t = \frac{D_{2t}}{D_1} = \frac{u_{2t}}{u_1} = \frac{u_2}{u_1}$$

Next, assuming the angles shown in Fig. 5.1b as positive (i.e. c_{1u} and c_{2u} are in opposite directions), the energy equation applied to a turbine stage (Sect. 1.8.2) and Euler's equation (Sect. 1.6.1) provide:

$$W = \Delta h_{t,ad} = h_{0t} - h_{2t} = u_1 \cdot c_{1u} + u_2 \cdot c_{2u}$$

The work coefficient can be therefore expressed as:

$$\psi = \frac{W}{u_1^2} = \frac{c_{1u}}{u_1} + \delta_t \cdot \frac{c_{2u}}{u_1}$$

The angle α_2 is:

$$\tan \alpha_2 = \frac{c_{2u}}{c_{2m}}$$

Since this angle can be assumed as an input parameter (generally, as already mentioned, we can assume $\alpha_2 = 0$), it is convenient to express all the other flow angles as a function of α_2 and of the kinematic parameters introduced above.

Following this approach, the rotor inlet absolute flow angle is expressed as:

$$\tan\alpha_1 = \frac{c_{1u}}{c_{1m}} = \frac{c_{1u}}{u_1}\cdot\frac{u_1}{c_{2m}}\cdot\frac{c_{2m}}{c_{1m}} = \frac{c_{1u}}{u_1}\cdot\frac{1}{\varphi\cdot\xi} = \frac{1}{\varphi\cdot\xi}\cdot(\psi - \varphi\cdot\delta_t\cdot\tan\alpha_2)$$

and thus we obtain:

$$\tan\alpha_1 = \frac{\psi}{\xi\cdot\varphi} - \frac{\delta_t}{\xi}\cdot\tan\alpha_2$$

The rotor relative flow angles are expressed as:

$$\tan\beta_1 = \frac{w_{1u}}{c_{1m}} = \frac{c_{1u} - u_1}{c_{1m}} = \tan\alpha_1 - \frac{1}{\varphi\cdot\xi}$$

$$\tan\beta_2 = \frac{w_{2u}}{c_{2m}} = \frac{u_2 + c_{2u}}{c_{2m}} = \frac{\delta_t}{\varphi} + \tan\alpha_2$$

and thus we obtain:

$$\tan\beta_1 = \frac{1}{\varphi\cdot\xi}\cdot(\psi - 1) - \frac{\delta_t}{\xi}\cdot\tan\alpha_2$$

$$\tan\beta_2 = \frac{\delta_t}{\varphi} + \tan\alpha_2$$

The degree of reaction is given by:

$$R = \frac{\Delta h_R}{\Delta h_{t,ad}} = \frac{\Delta h_R}{\psi\cdot u_1^2} = 1 - \frac{\Delta E_{c,R}}{\psi\cdot u_1^2} = 1 - \frac{c_1^2 - c_2^2}{2\cdot\psi\cdot u_1^2} = 1 - \frac{c_{1u}^2 + c_{1m}^2 - c_{2u}^2 - c_{2m}^2}{2\cdot\psi\cdot u_1^2}$$
$$= 1 - \frac{c_{1m}^2\cdot(1 + \tan^2\alpha_1) - c_{2m}^2\cdot(1 + \tan^2\alpha_2)}{2\cdot\psi\cdot u_1^2}$$

and thus it is:

$$R = 1 - \frac{\psi}{2} + \frac{\varphi^2}{2\cdot\psi}\cdot\left[(1 - \xi^2) + \tan^2\alpha_2\cdot(1 - \delta_t^2)\right] + \varphi\cdot\delta_t\cdot\tan\alpha_2$$

From this equation we infer that, unlike what happens in axial stages, in radial stages, the angle α_2 can be assumed as input parameter (for example, $\alpha_2 = 0$ to minimize the kinetic energy exiting the stage), keeping R, φ and ψ (or ξ, φ and ψ) as independent parameters.

The previous equation also allows to express ξ as a function of R:

$$\xi = \sqrt{1 - \frac{2 \cdot \psi}{\varphi^2} \cdot \left(R + \frac{\psi}{2} - 1\right) + \tan^2 \alpha_2 \cdot (1 - \delta_t^2) + \frac{2 \cdot \psi \cdot \delta_t}{\varphi} \cdot \tan \alpha_2}$$

Consequently there is a constraint on the maximum value of the degree of reaction. For example, if $\alpha_2 = 0$, it must be:

$$R < 1 + \frac{\varphi^2}{2 \cdot \psi} - \frac{\psi}{2}$$

Remarks on input parameters

From the kinematic correlations reported above, we infer that once the *five independent parameters* φ, ψ, R (or ξ), δ_t and α_2 are set (Sect. 5.5), all flow angles, α_1, β_1, β_2, can be calculated. Once the blade speed, u_1, is also established (in order to determine this speed, it is necessary to know, in addition to the work coefficient ψ, also the stage isentropic efficiency (or the polytropic efficiency, as illustrated in Sect. 5.5)), the kinematics of the rotor can be completely calculated:

$$\begin{aligned}
c_{1m} &= u_1 \cdot \xi \cdot \varphi \\
c_{1u} &= u_1 \cdot \xi \cdot \varphi \cdot \tan \alpha_1 \\
c_{2m} &= u_1 \cdot \varphi \\
c_{2u} &= u_1 \cdot \varphi \cdot \tan \alpha_2 \\
w_{1u} &= u_1 \cdot \xi \cdot \varphi \cdot \tan \beta_1 \\
w_{2u} &= u_1 \cdot \varphi \cdot \tan \beta_2 \\
c_1 &= u_1 \cdot \xi \cdot \varphi \cdot \sqrt{1 + \tan^2 \alpha_1} \\
c_2 &= u_1 \cdot \varphi \cdot \sqrt{1 + \tan^2 \alpha_2} \\
w_1 &= u_1 \cdot \xi \cdot \varphi \cdot \sqrt{1 + \tan^2 \beta_1} \\
w_2 &= u_1 \cdot \varphi \cdot \sqrt{1 + \tan^2 \beta_2} \\
u_2 &= u_1 \cdot \delta_t
\end{aligned}$$

In order to complete the kinematics of the stage, the velocity c_0, the absolute velocity at the stator inlet, must also be determined.

Generally we assume:

$$c_0 = c_2$$

and radial inlet:

$$\alpha_0 = 0$$

If (Sect. 5.3) this assumption results in a blade height at the stator inlet (b_0) lower than a preset minimum value, or if b_0 is assigned as an input value, c_0 must be recalculated. In this case, the blade speed, u_1, can still be calculated as a function of the work coefficient, but now:

$$\Delta h_{t,ad} \neq \Delta h_{ad}$$

so we must remember that:

$$\Delta h_{t,ad} = \Delta h_{ad} + \frac{c_0^2 - c_2^2}{2}$$

5.2 Thermodynamic Parameters

The pressure p_0 and temperature T_0 at the stator inlet and the pressure p_2 at the rotor outlet are assigned based on what is illustrated in Chap. 2 (for multistage turbines, these values are calculated by the procedure described in Sect. 5.6).

Now, all the thermodynamic parameters at the stator inlet and at the rotor inlet/outlet can be calculated as follows.

If the working fluid cannot be assimilated to a perfect gas, for the following calculations we need appropriate state functions, such as those of the NIST libraries. In the following, we will refer to a real working fluid.

Knowing p_0, T_0, we can calculate:

$$h_0 = h_{p,T}(p_0, T_0)$$
$$s_0 = s_{p,T}(p_0, T_0)$$
$$\rho_0 = \rho_{p,T}(p_0, T_0)$$

Once c_0 is established (Sect. 5.1) or calculated (Sect. 5.3), the following total quantities (Sect. 1.8.1) can be evaluated (Fig. 5.2):

$$h_{0t} = h_0 + \frac{c_0^2}{2}$$
$$p_{0t} = p_{h,s}(h_{0t}, s_0)$$
$$T_{0t} = T_{h,s}(h_{0t}, s_0)$$
$$\rho_{0t} = \rho_{h,s}(h_{0t}, s_0)$$

To calculate the thermodynamic parameters at the stator outlet, we must know the stator efficiency and remember the total enthalpy conservation in the nozzle (Sect. 1.8.2):

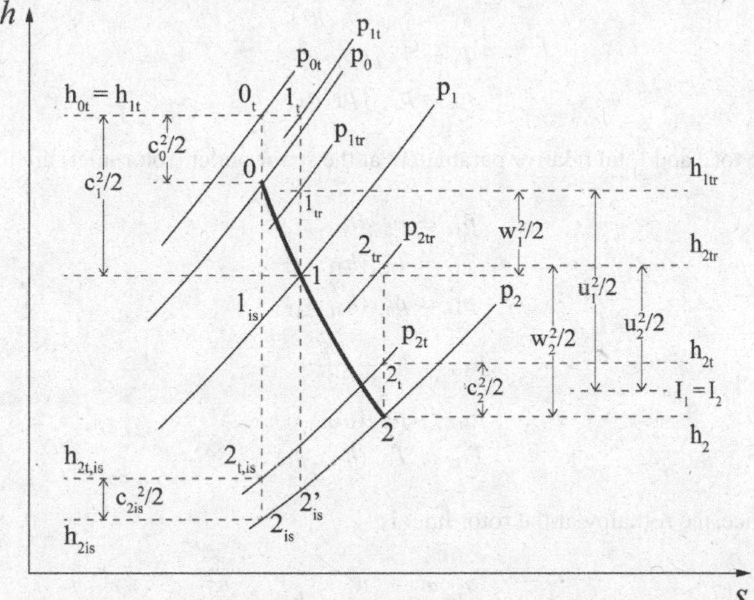

Fig. 5.2 Expansion process in a radial turbine stage

$$h_{1t} = h_{0t}$$

In fact, the static enthalpy is calculated through the previous equation:

$$h_1 = h_{1t} - \frac{c_1^2}{2}$$

while using the expression of the stator efficiency (Sect. 1.8.2):

$$\eta_S = \frac{h_{0t} - h_1}{h_{0t} - h_{1,is}}$$

isentropic static enthalpy is obtained:

$$h_{1,is} = h_{0t} - \frac{h_{0t} - h_1}{\eta_S}$$

and, therefore, the stator outlet pressure:

$$p_1 = p_{s,h}(s_0, h_{1,is})$$

Once the pressure and enthalpy at the stator outlet have been calculated, we can evaluate the other thermodynamic parameters:

$$s_1 = s_{p,h}(p_1, h_1)$$
$$T_1 = T_{p,h}(p_1, h_1)$$
$$\rho_1 = \rho_{p,h}(p_1, h_1)$$

The total and total relative parameters at the stator outlet (rotor inlet) are then:

$$p_{1t} = p_{h,s}(h_{1t}, s_1)$$
$$T_{1t} = T_{h,s}(h_{1t}, s_1)$$
$$\rho_{1t} = \rho_{h,s}(h_{1t}, s_1)$$
$$h_{1tr} = h_1 + \frac{w_1^2}{2}$$
$$p_{1tr} = p_{h,s}(h_{1tr}, s_1)$$
$$T_{1tr} = T_{h,s}(h_{1tr}, s_1)$$

Hence, the rothalpy at the rotor inlet is

$$I_1 = h_1 + \frac{w_1^2}{2} - \frac{u_1^2}{2} = h_{1tr} - \frac{u_1^2}{2}$$

Losses in the stator are related to the entropy increase $(s_1 - s_0)$ due to irreversibilities and entail an increase in the final enthalpy $(h_1 - h_{1,is})$ and a reduction in the total pressure expressed by the loss coefficient (Sect. 1.9.5):

$$Y_S = \frac{p_{0t} - p_{1t}}{p_{1t} - p_1}$$

At the rotor outlet, the pressure is known; therefore the enthalpy, referred to the isentropic expansion, can be calculated as:

$$h_{2,is} = h_{p,s}(p_2, s_0)$$

The final enthalpy in the adiabatic expansion can be determined by applying the conservation of the rothalpy in the rotor (Sect. 1.8.2):

$$I_2 = I_1$$

$$h_2 = I_2 + \frac{u_2^2}{2} - \frac{w_2^2}{2}$$

or by applying the stage isentropic efficiency (referred to static enthalpy drops, Sect. 1.8.2):

$$\eta_{is}' = \frac{h_0 - h_2}{h_0 - h_{2,is}}$$

$$h_2 = h_0 - \eta_{is} \cdot (h_0 - h_{2,is})$$

The calculation of the thermodynamic parameters can, therefore, be completed:

$$s_2 = s_{p,h}(p_2, h_2)$$
$$T_2 = T_{p,h}(p_2, h_2)$$
$$\rho_2 = \rho_{p,h}(p_2, h_2)$$
$$h_{2t} = h_2 + \frac{c_2^2}{2}$$
$$p_{2t} = p_{h,s}(h_{2t}, s_2)$$
$$T_{2t} = T_{h,s}(h_{2t}, s_2)$$
$$\rho_{2t} = \rho_{h,s}(h_{2t}, s_2)$$
$$h_{2tr} = h_2 + \frac{w_2^2}{2}$$
$$p_{2tr} = p_{h,s}(h_{2tr}, s_2)$$
$$T_{2tr} = T_{h,s}(h_{2tr}, s_2)$$

Losses in the rotor are related to the entropy increase $(s_2 - s_1)$ due to irreversibilities and entail an increase in the final enthalpy $(h_2 - h_{2',is})$ and a reduction in the total relative pressure expressed by the loss coefficient (Aungier 2006):

$$Y_R = \frac{p_{2tr,is} - p_{2tr}}{p_{2tr} - p_2}$$

The rothalpy conservation in the rotor allows to calculate the total relative isentropic pressure, necessary to calculate the pressure losses in the stage (Sect. 5.4), according to the previous expression Y_R.

By applying the rothalpy conservation along an ideal (isentropic) and a real (adiabatic) process, we have:

$$dh + w\,dw = u\,du$$

$$\left(h_{2',is} + \frac{w_{2',is}^2}{2}\right) - \left(h_1 + \frac{w_1^2}{2}\right) = \frac{u_2^2 - u_1^2}{2}$$

$$\left(h_2 + \frac{w_2^2}{2}\right) - \left(h_1 + \frac{w_1^2}{2}\right) = \frac{u_2^2 - u_1^2}{2}$$

and as a consequence:

$$h_{2'tr,is} = h_{2tr}$$

Now it is possible to calculate the total relative isentropic pressure and temperature as:

$$p_{2tr,is} = p_{h,s}\left(h_{2'tr,is}, s_1\right)$$
$$T_{2tr,is} = T_{h,s}\left(h_{2'tr,is}, s_1\right)$$

From the knowledge of the kinematics and thermodynamics of the stage, the absolute (stator inlet/outlet) and relative (rotor inlet/outlet) Mach numbers can be calculated. From the properties of the working fluid, first, the speeds of sound are calculated:

$$c_{s,0} = c_{s,p,T}(p_0, T_0)$$
$$c_{s,1} = c_{s,p,T}(p_1, T_1)$$
$$c_{s2} = c_{s,p,T}(p_2, T_2)$$

and therefore the Mach numbers:

$$Ma_0 = \frac{c_0}{c_{s0}}$$
$$Ma_{1,S} = \frac{c_1}{c_{s1}}$$
$$Ma_{1,R} = \frac{w_1}{c_{s1}}$$
$$Ma_2 = \frac{w_2}{c_{s2}}$$

By using the pressure and temperature conditions of the working fluid, we can calculate the fluid viscosities (necessary to assess the Reynolds numbers, Sect. 5.3):

$$\mu_0 = \mu_{p,T}(p_0, T_0)$$
$$\mu_1 = \mu_{p,T}(p_1, T_1)$$
$$\mu_2 = \mu_{p,T}(p_2, T_2)$$

Remarks on input parameters

Based on what is illustrated in this Sect. 5.2, in addition to the parameters already identified in Sect. 5.1 (φ, ψ, R, δ_t, α_2 and η_{is}), the stator efficiency must be assumed in order to calculate the thermodynamics of the stage.

5.3 Geometric Parameters

Now we will deduce the geometric correlations of a radial stage; they are correlations among geometric (rotor and stator diameters, blade height, chord and blade pitch, etc.), thermodynamic (density, volumetric flow rate) and kinematic parameters (meridional velocities, absolute velocity angles α, relative velocity angles β) of the stage.

The rotor inlet diameter (D_1) is determined by the kinematics of the stage (Sect. 5.1), and in particular by the knowledge of the blade speed at the rotor inlet (u_1) and the speed of rotation (n, rpm) (Sect. 5.5):

$$D_1 = \frac{60}{\pi} \cdot \frac{u_1}{n}$$

Before evaluating the rotor and stator geometry, we must highlight the differences between the geometric parameters in the radial and axial stages. In particular, in the axial stages, after determining the blade height through the equation of continuity, the geometry of the rotor and stator channels (Chaps. 3 and 4) was calculated by using two parameters: the aspect ratio and the solidity. In the radial stages, on the other hand, it is not simple the definition of a rotor chord; moreover, the blade heights and the rotor diameters change, sometimes even a lot, between the rotor inlet and outlet. For these reasons, in order to evaluate the stage losses, an equivalent rotor channel, having a circular section of diameter D_{hyd} (hydraulic diameter) and length L_{hyd} (hydraulic length), is used (Sect. 5.4).

The hydraulic diameter is defined as:

$$D_{hyd} = 4 \cdot \frac{Cross\,sectional\,area}{Wetted Perimeter}$$

To define the hydraulic length, first, we must introduce the channel length on the meridional plane, that is, the plane containing the axis of rotation:

$$L_m = Meridional\,length$$

The hydraulic length is correlated with the meridional length through the angle β_M (for the rotor or α_M for the stator) representing the mean inclination of the rotor/stator channel on a plane perpendicular to the axis of rotation; these mean angles are similar to the stagger angle in the axial stages:

$$L_{hyd} = \frac{L_m}{\cos \beta_M}$$

For the stator, even if a chord can be defined as for the axial stages, for homogeneity of the discussion with the rotor, we define the same parameters: the hydraulic diameter and hydraulic length. For the stator, we will also use the solidity, σ, since the hydraulic length substantially corresponds to the chord, but we must specify the station where the blade pitch is calculated, as it is variable between the stator inlet and outlet.

Now, we can illustrate the geometry evaluation of the rotor and stator for a radial turbine stage.

Rotor geometry

The volumetric flow rates in the rotor (1 - rotor inlet, 2- rotor outlet) are functions of the assigned mass flow rate and fluid densities calculated in Sect. 5.2:

$$V_i = \frac{m}{\rho_i} \qquad i = 1, 2$$

By using these volumetric flow rates, knowing the meridional velocities c_{1m} and c_{2m}, we can calculate the flow areas:

$$A_i = \frac{V_i}{c_{im}}$$

and, therefore, considering a section just upstream of the rotor inlet (where no blade blockage occurs, Sect. 1.8), the blade height at the rotor inlet (Fig. 5.3):

$$b_1 = \frac{A_1}{\pi \cdot D_1}$$

The flow area at the rotor outlet can be correlated to the rotor diameter ratios; in fact we have (Fig. 5.4):

$$A_2 = \frac{\pi}{4} \cdot \left(D_{2t}^2 - D_{2h}^2\right) = \frac{\pi}{4} \cdot D_1^2 \cdot \left(\delta_t^2 - \delta_h^2\right)$$

and, as a consequence, we can obtain the rotor tip diameter ratio:

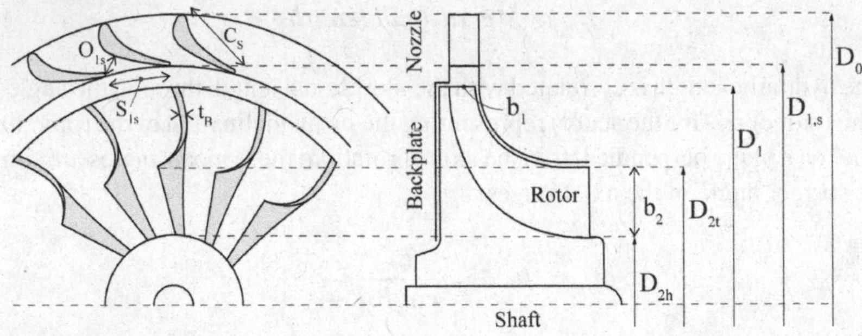

Fig. 5.3 Stage geometry

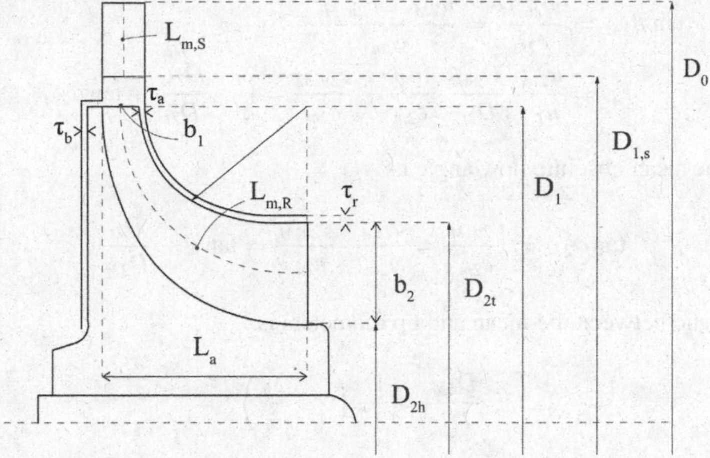

Fig. 5.4 Stage geometry in the meridional section

$$\delta_t = \sqrt{\delta_h^2 + \frac{4 \cdot A_2}{\pi \cdot D_1^2}}$$

since the rotor hub diameter ratio is defined as:

$$\delta_h = \frac{D_{2h}}{D_1}$$

After choosing δ_h, for example Aungier (2006) suggested:

$$\delta_h = 0.185$$

or δ_t (Sect. 5.5), the rotor outlet geometry is determined:

$$D_{2h} = \delta_h \cdot D_1$$

$$D_{2t} = \delta_t \cdot D_1$$

$$b_2 = \frac{D_{2t} - D_{2h}}{2} = \frac{D_1}{2} \cdot (\delta_t - \delta_h)$$

By using these parameters, we can calculate the outlet mean flow angle, that is, the flow angle at the outlet mean diameter:

$$D_{2M} = \frac{D_{2h} + D_{2t}}{2}$$

In fact, remembering that the kinematics at the rotor outlet (Sect. 5.1) is evaluated at the tip diameter, the mean relative flow angle is:

$$\tan \beta_{2M} = \frac{w_{2u,M}}{c_{2m}} = \frac{u_{2M} + c_{2u,M}}{c_{2m}}$$
$$= \frac{u_{2t}}{u_1} \cdot \frac{D_{2M}}{D_{2t}} \cdot \frac{u_1}{c_{2m}} + \frac{c_{2u,M}}{c_{2m}} = \frac{\delta_t}{\varphi} \cdot \frac{D_{2M}}{D_{2t}} + \tan \alpha_{2M}$$

where, the mean absolute flow angle is:

$$\tan \alpha_{2M} = \frac{c_{2u,M}}{c_{2m}} = \frac{c_{2u,t}}{c_{2m}} \cdot \frac{c_{2u,M}}{c_{2u,t}} = \tan \alpha_2 \cdot \frac{D_{2t}}{D_{2M}}$$

and the ratio between the mean and tip diameters is:

$$\frac{D_{2M}}{D_{2t}} = \frac{1}{2} \cdot \left(1 + \frac{\delta_h}{\delta_t}\right)$$

As a consequence, we can write:

$$\tan \beta_{2M} = \frac{1}{2 \cdot \varphi} \cdot (\delta_t + \delta_h) + \frac{2 \cdot \delta_t \cdot \tan \alpha_2}{\delta_t + \delta_h}$$

Therefore, the mean relative velocity is:

$$w_{2M} = \frac{c_{2m}}{\cos \beta_{2M}}$$

In order to calculate the incidence losses (Sect. 5.4), it is important to determine the optimum incidence. In this regard, Dixon and Hall (2014) resumed the study by Whitfield (1990), and introduced the following *incidence factor*:

$$IF = \frac{c_{1u,opt}}{u_1}$$

using the correlation proposed by Stanitz (1952):

$$IF = 1 - \frac{0.63 \cdot \pi}{N_{B,R}} = 1 - \frac{1.98}{N_{B,R}}$$

Through this equation, we can calculate the optimum incidence angle:

$$\tan \beta_{1,opt} = \frac{w_{1u,opt}}{c_{1m}} = \frac{c_{1u,opt} - u_1}{c_{1m}} = \frac{u_1}{c_{1m}} \cdot (IF - 1) = -\frac{u_1}{c_{1m}} \cdot \frac{1.98}{N_{B,R}}$$
$$\tan \beta_{1,opt} = -\frac{1}{\xi \cdot \varphi} \cdot \frac{1.98}{N_{B,R}}$$

Spence and Artt (1998) developed an experimental validation of this optimum incidence angle.

To calculate the number of blades, we can use the equation proposed by Glassman (1976), and adopted by Ventura et al. (2012):

$$N_{B,R} = \text{int}\left(\frac{\pi \cdot (110 - \alpha_1) \cdot \tan \alpha_1}{30}\right)$$

Regarding the blade thickness, we can assume:

$$\frac{t_B}{D_1} \cong 0.02 \qquad \textit{at the rotor inlet}$$
$$\frac{t_B}{D_1} \cong 0.01 \qquad \textit{at the rotor outlet}$$

The ratio between the blade thickness at the trailing edge and the outlet blade height can be assumed as:

$$\frac{t_{TE}}{b_2} \cong 0.04$$

This parameter will be necessary for calculating the trailing edge losses (Sect. 5.4).

Regarding clearances (necessary for calculating the clearance losses, Sect. 5.5), we can assume that:

$$\frac{\tau}{b_1} = 2.0\text{–}5.0\%$$

but the minimum values must be not lower than $\tau = 0.3–0.5$ mm.

In absence of specific data, we can choose:

$$\frac{\tau_a}{b_1} = \frac{\tau_r}{b_1} = \frac{\tau_b}{b_1} = 0.05$$

where (Fig. 5.4):

- τ_a is the axial clearance (blade at rotor inlet)
- τ_r is the radial clearance (blade at rotor outlet)
- τ_b is the backface clearance (on the back of the rotor disc)

The blade pitch at the rotor inlet is:

$$S_{1,R} = \frac{\pi \cdot D_1}{N_{B,R}}$$

and at the rotor outlet (considering the mean diameter) is:

$$S_{2,R} = \frac{\pi \cdot D_{2M}}{N_{B,R}} = \frac{\pi \cdot (D_{2t} + D_{2h})}{2 \cdot N_{B,R}}$$

Now, the rotor geometry must be completed calculating the hydraulic diameter, D_{hyd}, and the hydraulic length, L_{hyd}.

As previously mentioned, the hydraulic diameter is defined as:

$$D_{hyd} = 4 \cdot \frac{Cross\ sectional\ area}{Wetted\ Perimeter}$$

To evaluate the wetted perimeter, and the cross sectional area delimited by this perimeter, it is necessary to calculate the blade distance at the leading edge, O_1, and at the trailing edge, O_2, following what was done for the axial stages.

By assuming as right triangles the triangles having a hypotenuse equal to S, a base equal to O and an angle between them equal to β, we obtain:

$$O_1 = S_{1,R} \cdot \cos \beta_1$$

$$O_2 = S_{2,R} \cdot \cos \beta_{2M}$$

The cross sectional areas and the wetted perimeters are therefore:

$$A_{w1} = O_1 \cdot b_1$$

$$P_{w1} = 2 \cdot (O_1 + b_1)$$

$$A_{w2} = O_2 \cdot b_2$$

$$P_{w2} = 2 \cdot (O_2 + b_2)$$

and the hydraulic diameters at the rotor inlet and outlet become:

$$D_{hyd,1} = 2 \cdot \frac{O_1 \cdot b_1}{O_1 + b_1}$$

$$D_{hyd,2} = 2 \cdot \frac{O_2 \cdot b_2}{O_2 + b_2}$$

The rotor hydraulic diameter is the mean hydraulic diameter between the rotor inlet and outlet:

$$D_{hyd,R} = \frac{D_{hyd,1} + D_{hyd,2}}{2} = \frac{O_1 \cdot b_1}{O_1 + b_1} + \frac{O_2 \cdot b_2}{O_2 + b_2}$$

To calculate the hydraulic length L_{hyd}, as previously illustrated, first, we must assess the channel length, L_m, on the meridional plane. Following the assumption by Glassman (1976), this length is equivalent to a quarter of an ellipse having the following semi-axes (Fig. 5.4):

$$a = L_a - \frac{b_1}{2}$$

$$b = \frac{D_1 - D_{2M}}{2} = \frac{D_1 - D_{2t} + b_2}{2}$$

and, therefore, we obtain:

$$L_{m,R} \cong \frac{\pi}{2} \cdot \sqrt{\frac{a^2 + b^2}{2}} = \frac{\pi}{4} \cdot \sqrt{\frac{(2 \cdot L_a - b_1)^2 + (D_1 - D_{2t} + b_2)^2}{2}}$$

Alternatively, similarly to the centrifugal compressors (Chap. 6), the meridional length can be considered equivalent to a quarter circle having an equivalent radius equal to:

$$r_{eq} = \frac{a + b}{2} = \frac{L_a - \frac{b_1}{2} + \frac{D_1 - D_{2t} + b_2}{2}}{2}$$

and, as a consequence, we obtain:

$$L_{m,R} \cong \frac{\pi}{2} \cdot r_{eq} = \frac{\pi}{8} \cdot (2 \cdot L_a - b_1 + D_1 - D_{2t} + b_2)$$

In the previous equations, L_a is the rotor axial length. This parameter, influencing the meridional length, assumes considerable importance in the loss calculations (Sect. 5.4). According to Glassman (1976), on the meridional plane, the shroud can be considered circular, so we obtain (neglecting the clearances):

$$L_a = \frac{D_1 - D_{2t}}{2} + b_1$$

Alternatively, the rotor axial length can be calculated by using the following correlation proposed by Aungier (2006):

$$L_a = 1.5 \cdot b_2$$

After assessing the meridional length, the hydraulic length is:

$$L_{hyd,R} = \frac{L_{m,R}}{\cos \beta_M} = \frac{L_{m,R}}{\cos\left(\frac{\beta_{2M} + \beta_1}{2}\right)}$$

Stator geometry

Baines (2003) and Simpson et al. (2008) suggested adopting a distance between the stator and rotor equal to (Fig. 5.4):

$$\Delta D_{S,R} = D_{1,S} - D_1 = 4 \cdot b_1 \cdot \cos \alpha_1$$

Glassman (1976) proposed the following equation for calculating the stator inlet diameter; this equation was also adopted by Ventura et al. (2012):

$$D_0 = 1.25 \cdot D_1 \cdot \left(1 + 4 \cdot \frac{b_1}{D_1} \cdot \cos\alpha_1\right)$$

The volumetric flow rate entering the stator is a function of the assigned mass flow rate and fluid density calculated in Sect. 5.2:

$$V_0 = \frac{m}{\rho_0}$$

By using this volumetric flow rate and assuming a purely meridional inlet velocity equal to c_2 ($c_0 = c_2$), we can calculate the inlet flow area:

$$A_0 = \frac{V_0}{c_0}$$

and, as a consequence, the blade height:

$$b_0 = \frac{A_0}{\pi \cdot D_0}$$

If b_0 is lower than a preset minimum value, or if b_0 is assigned (for example, $b_0 = b_{1,s} = b_1$), we must recalculate c_0:

$$c_0 = c_{0m} = \frac{V_0}{\pi \cdot D_0 \cdot b_0}$$

and, consequently, the stage thermodynamics (Sect. 5.2).

Since the stator channel develops in a completely radial direction, the meridional length on the meridional plane is:

$$L_{m,S} = \frac{D_0 - D_{1,S}}{2} = 0.125 \cdot D_1 + 0.5 \cdot b_1 \cdot \cos\alpha_1$$

Assuming a radial inlet in the stator, the hydraulic length (which, in this case, corresponds to the chord) is:

$$L_{hyd,S} = C_S = \frac{L_{m,S}}{\cos\alpha_M} \cong \frac{L_{m,S}}{\cos\left(\frac{\alpha_1}{2}\right)}$$

The number of blades is calculated by using the optimum solidity suggested by Simpson et al. (2008):

$$\sigma_S = \frac{C_S}{S_{1,S}} = 1.25$$

The solidity and the chord are correlated with the blade pitch at the stator outlet:

$$S_{1,S} = \frac{C_S}{\sigma_S}$$

and, therefore, the number of blades is:

$$N_{B,S} = \text{int}\left(\frac{\pi \cdot D_{1,S}}{S_{1,S}}\right)$$

Then, the blade pitch at the stator inlet is:

$$S_{0,S} = \frac{\pi \cdot D_0}{N_{B,S}}$$

To evaluate the hydraulic diameter, it is necessary to calculate the blade distance at the leading edge, O_0, and at the trailing edge, O_{1S}. By assuming as right triangles the triangles having a hypotenuse equal to S, a base equal to O and an angle between them equal to α, we obtain:

$$O_0 = S_{0,S} \cdot \cos\alpha_0$$

$$O_{1S} = S_{1,S} \cdot \cos\alpha_1$$

Therefore, the cross sectional areas and the wetted perimeters are:

$$A_{w0} = O_0 \cdot b_0$$

$$P_{w0} = 2 \cdot (O_0 + b_0)$$

$$A_{w1S} = O_{1S} \cdot b_1$$

$$P_{w1S} = 2 \cdot (O_{1S} + b_1)$$

and the hydraulic diameters at the stator inlet and outlet become:

$$D_{hyd,0} = 2 \cdot \frac{O_0 \cdot b_0}{O_0 + b_0}$$

$$D_{hyd,1S} = 2 \cdot \frac{O_{1S} \cdot b_1}{O_{1S} + b_1}$$

The stator hydraulic diameter is the mean hydraulic diameter between the stator inlet and outlet:

$$D_{hyd,S} = \frac{D_{hyd,0} + D_{hyd,1S}}{2} = \frac{O_0 \cdot b_0}{O_0 + b_0} + \frac{O_{1S} \cdot b_1}{O_{1S} + b_1}$$

From the knowledge of the kinematics, thermodynamics and geometry of the stage, the absolute (stator inlet/outlet) and relative (rotor inlet/outlet) Reynolds numbers can be calculated:

$$\mathrm{Re}_0 = \frac{\rho_0 \cdot c_0 \cdot D_{hyd,S}}{\mu_0}$$

$$\mathrm{Re}_{1,S} = \frac{\rho_1 \cdot c_1 \cdot D_{hyd,S}}{\mu_1}$$

$$\mathrm{Re}_{1,R} = \frac{\rho_1 \cdot w_1 \cdot D_{hyd,R}}{\mu_1}$$

$$\mathrm{Re}_2 = \frac{\rho_2 \cdot w_2 \cdot D_{hyd,R}}{\mu_2}$$

In these equations, the characteristic length is the hydraulic diameter; if the Reynolds numbers must be calculated by using another characteristic dimension, these new equations will be explicitly expressed.

Once the Reynolds numbers have been calculated, considerations on surface roughness can be made according to Sect. 1.9.1.

In particular, the *admissible surface roughness*, $k_{s,adm}$, can be assessed; it is the limit roughness below which there are no effects on losses allowing the walls to be considered smooth. This condition occurs when the surface roughness is contained within the laminar sub-layer of the boundary layer.

This roughness (Sect. 1.9.1) can be evaluated, for the stator and the rotor, by:

$$\frac{k_{s,adm,S}}{D_{hyd,S}} = \frac{100}{\mathrm{Re}_{1,S}}$$

$$\frac{k_{s,adm,R}}{D_{hyd,R}} = \frac{100}{\mathrm{Re}_2}$$

The calculation of $k_{s,adm}$ assumes considerable importance as it allows to establish the finishing level of the channel surfaces in order to contain dissipations.

Taking into account that turbomachinery surface roughness generally falls within the range:

$$5\ \mu\mathrm{m} < k_s < 50\ \mu\mathrm{m}$$

the knowledge of $k_{s,adm}$ allows to evaluate possible, and economically convenient, surface treatment aimed at respecting this admissible roughness value as well as not resorting to very high finish levels where not required.

In other words, if the admissible surface roughness, $k_{s,adm}$, is lower than the minimum roughness that is technologically and economically achievable, $k_{s,min}$, we assume:

$$k_s = k_{s,\mathrm{min}}$$

otherwise:

$$k_{s,min} \leq k_s \leq k_{s,adm}$$

In particular, in the latter case, a roughness lower than the admissible one can be assumed in order to take into account that the roughness tends to increase with the turbomachine operation.

Remarks on input parameters

The procedure illustrated in this Sect. 5.3 does not presuppose any further input parameters, with respect to those already identified in Sects. 5.1 and 5.2, taking into account the assumptions, defining the stator and rotor geometry, adopted in this section.

5.4 Stage Losses and Efficiency

The loss sources and the loss models in the different types of turbomachinery are extensively described in Sect. 1.9. In this section we will provide specific correlations for loss models able to quantify the losses introduced in Sect. 1.9.

Losses occurring in the stator and rotor of a radial stage depend on the kinematic (Sect. 5.1), thermodynamic (Sect. 5.2) and geometric (Sect. 5.3) parameters of the stage.

In the literature there are some loss models based on pressure loss coefficients and other loss models based on energy losses, expressed in terms of enthalpy increase compared to isentropic expansion.

Both loss models will be illustrated below.

In particular, regarding the models based on energy losses, widely proposed in the most recent literature (Suhrmann et al. 2010; Ghosh et al. 2011; Ventura et al. 2012; Rahbar et al. 2015; Jung and Krumdieck 2016; Qi et al. 2017; Cho et al. 2018; Gao and Petrie-Repar 2018; Meroni et al. 2018; Persky and Sauret 2019; Alshammari et al. 2020), for each type of loss we will report the most frequently used correlations in the literature (Sect. 5.4.1.2); then, after provided the loss sets proposed by some authors, we will propose (Sect. 5.4.1.3) the set considered by us the optimum one.

5.4.1 Loss Models Based on Energy Losses

5.4.1.1 Stator Losses

Several authors (Ventura et al. 2012; Jung and Krumdieck 2016; Gao and Petrie-Repar 2018) considered the stator losses negligible compared to the rotor ones. Other authors (Rodgers 1987, Ghosh et al. 2011, Deng et al. 2018, Meroni et al. 2018) instead proposed the following correlation of stator passage loss:

$$\Delta h_S = \frac{c_1^2}{2} \cdot \frac{0.05}{\mathrm{Re}_1^{0.2}} \cdot \left(\frac{3 \cdot \tan \alpha_1}{(S_{1,S}/C_S)} + \frac{S_{1,S}}{b_1} \cdot \cos \alpha_1 \right)$$

where the Reynolds number is calculated as:

$$\mathrm{Re}_1 = \frac{\rho_1 \cdot c_1 \cdot b_1}{\mu_1}$$

and all the other parameters have already been defined in the previous sections.

5.4.1.2 Rotor Losses

The losses in the rotor (Sect. 1.9) are essentially divided into six types of losses:

$$\Delta h_R = \Delta h_{incidence} + \Delta h_{passage} + \Delta h_{TE} + \Delta h_{CL} + \Delta h_{exit} + \Delta h_{windage}$$

where:

$\Delta h_{incidence}$ represents the incidence losses
$\Delta h_{passage}$ represents the passage losses
Δh_{TE} represents the trailing edge losses
Δh_{CL} represents the clearance losses
Δh_{exit} represents the exit energy losses
$\Delta h_{windage}$ represents the windage losses
Δh_R represents the sum of internal and external losses (for example windage losses):

$$\Delta h_R = \Delta h_{R,\mathrm{int}} + \Delta h_{R,ext}$$

and therefore it is necessary to take into account what is illustrated in Sect. 1.9.6 regarding the calculation of the isentropic efficiency (Sect. 5.4.3).

Incidence losses

Futral and Wasserbauer (1965) proposed to calculate the incidence losses as the kinetic energy associated with the tangential relative velocity at the rotor inlet:

$$\Delta h_{incidence} = \frac{w_1^2}{2} \cdot sen^2 \beta_1$$

While, according to the formulation by Wasserbauer and Glassman (1975), these losses are correlated with the difference between the inlet relative flow angle and the optimum one:

$$\Delta h_{incidence} = \frac{w_1^2}{2} \cdot sen^2 (\beta_1 - \beta_{1,opt})$$

where $\beta_{1,opt}$ is the optimum incidence flow angle, already obtained in Sect. 5.3.

Passage losses

Passage losses include profile losses and secondary losses (Sect. 1.9).

Again according to the formulation by Wasserbauer and Glassman (1975), these losses can be expressed through:

$$\Delta h_{passage} = \frac{1}{2} \cdot K \cdot \left[w_1^2 \cdot \cos^2 (\beta_1 - \beta_{1,opt}) + w_{2M}^2 \right]$$

where

- K is a constant coefficient that can be assumed equal to 0.3;
- w_{2M} is the relative velocity at the rotor outlet, evaluated at the mean diameter (Sect. 5.3)

Several authors (Suhrmann et al. 2010; Ventura et al. 2012; Alshammari et al. 2020) recommend instead a formulation similar to that proposed for passage losses in centrifugal compressors (Musgrave 1980); this equation is a function of the friction coefficient and the characteristic geometric parameters of the rotor (the hydraulic length L_{hyd}, and the hydraulic diameter D_{hyd}, already defined in Sect. 5.3):

$$\Delta h_{passage} = 2 \cdot f_t \cdot \frac{L_{hyd,R}}{D_{hyd,R}} \cdot \bar{w}^2$$

where:

- the mean velocity $\bar{w}$ is an average between inlet and outlet (at the mean diameter) relative velocity:

$$\bar{w} = \frac{w_1 + w_{2M}}{2}$$

- the skin friction coefficient, f_t, is:

$$f_t = f_C \cdot \left[\text{Re} \cdot \left(\frac{D_1}{2 \cdot r_C} \right)^2 \right]^{0.05}$$

In this equation we have:

$$\text{Re} = \frac{\rho_1 \cdot u_1 \cdot (D_1/2)}{\mu_1}$$

$$r_C = \frac{D_1 - D_{2t}}{2}$$

$$f_C = f \cdot \left(1 + 0.075 \cdot \text{Re}^{0.25} \cdot \sqrt{\frac{D_{hyd,R}}{2 \cdot r_C}} \right)$$

and the coefficient f depends on the flow regime. If the flow is laminar (Re < 2,000), it is:

$$f = f_l = \frac{16}{\text{Re}_2}$$

while if Re > 4,000 it is:

$$f = f_{tr}$$

$$\frac{1}{\sqrt{f_{tr}}} = -4 \cdot \log_{10} \left[\frac{(k_s / D_{hyd,R})}{3.7} + \frac{1.256}{\text{Re}_2 \cdot \sqrt{f_{tr}}} \right]$$

where k_s is the superficial roughness defined in Sect. 5.3.

If 2000 < Re < 4000, the boundary layer is in transition from laminar to turbulent flow; the skin friction coefficient in this transition zone is well approximated by a weighted average of the laminar and turbulent values:

$$f = f_{l,(\text{Re}=2000)} - \left(f_{l,(\text{Re}=2000)} - f_{tr,(\text{Re}=4000)} \right) \cdot \left(\frac{\text{Re}_2}{2000} - 1 \right)$$

Baines (2003) proposed the following formulation of the passage losses:

$$\Delta h_{passage} = K_p \cdot \left\{ \left(\frac{L_{hyd,R}}{D_{hyd,R}} \right) + 0.68 \cdot \left[1 - \left(\frac{D_{2M}}{D_1} \right)^2 \right] \cdot \frac{\cos \beta_{2M}}{(b_2 / L_{hyd,R})} \right\} \cdot \frac{w_1^2 + w_{2M}^2}{2}$$

This equation depends on the characteristic geometric parameters of the rotor (the hydraulic length, L_{hyd}, and the hydraulic diameter, D_{hyd}, already defined in Sect. 5.3), while K_p is a coefficient equal to:

$$K_p = 0.11 \qquad \text{when} \qquad \frac{D_1 - D_{2t}}{2 \cdot b_2} \geq 0.2$$

$$K_p = 0.22 \qquad \text{when} \qquad \frac{D_1 - D_{2t}}{2 \cdot b_2} < 0.2$$

Ventura et al. (2012), in accordance with Suhrmann et al. (2010), proposed to calculate the sum of the passage and secondary losses through a combination of the Musgrave (1980) passage losses and the secondary losses as follows:

$$\Delta h_{passage} + \Delta h_s = \frac{1}{2} \cdot \left(2 \cdot f_t \cdot \frac{L_{hyd,R}}{D_{hyd,R}} \cdot \bar{w}^2 + \frac{2 \cdot c_1^2 \cdot D_1}{N_{B,R} \cdot r_C} \right)$$

and all the parameters appearing in the previous equation have already been defined. Suhrmann et al. (2010), however, concluded that in this formulation the secondary losses are overestimated. As a consequence, Alshammari et al. (2020) proposed a different combination of Musgrave (1980) passage losses and secondary losses; this new equation significantly attenuates the secondary losses:

$$\Delta h_{passage} + \Delta h_s = f_t \cdot \frac{L_{hyd,R}}{D_{hyd,R}} \cdot \bar{w}^2 + \frac{1}{4} \cdot \frac{c_1^2 \cdot D_1}{N_{B,R} \cdot r_C}$$

Trailing edge losses

Meroni et al. (2018) adopted the Glassman's correlation (1995):

$$\Delta h_{TE} = \frac{1}{2} \cdot w_{2M}^2 \cdot \left(\frac{N_{B,R} \cdot t_{TE}}{\pi \cdot D_{2M} \cdot \cos \beta_{2M}} \right)^2 \cdot \left(1 + \frac{k_2 - 1}{2} \cdot Ma_2^2 \right)^{\frac{k_2}{1-k_2}}$$

where

- k_2 is the ratio between the specific heats, c_p and c_v, at the rotor outlet;
- Ma_2 is the relative Mach number at the rotor outlet;
- t_{TE} is the blade thickness at the trailing edge defined in Sect. 5.3.

Qi et al. (2017) proposed a simplification of the aforementioned correlation:

$$\Delta h_{TE} = \frac{1}{2} \cdot w_{2M}^2 \cdot \left(\frac{N_{B,R} \cdot t_{TE}}{\pi \cdot D_{2M} \cdot \cos \beta_2} \right)^2$$

Clearance losses

According to the loss model proposed by Baines (2003), these losses are:

$$\Delta h_{CL} = \frac{u_1^3 \cdot N_{B,R}}{8 \cdot \pi} \cdot \left(K_a \cdot \tau_a \cdot C_a + K_r \cdot \tau_r \cdot C_r + K_{a,r} \cdot \sqrt{\tau_a \cdot C_a \cdot \tau_r \cdot C_r} \right)$$

where

- the coefficients K assume the following values:

$$K_a = 0.4;\;\; K_r = 0.75;\;\; K_{a,r} = -0.3$$

- the parameters τ represent the axial and radial clearances already defined in Sect. 5.3
- the parameters C are:

$$C_a = \frac{1 - \frac{D_{2t}}{D_1}}{c_{1m} \cdot b_1} = \frac{1 - \delta_t}{c_{1m} \cdot b_1}$$

$$C_r = 2 \cdot \delta_t \cdot \frac{L_a - b_1}{c_{2m} \cdot D_{2M} \cdot b_2}$$

where L_a represents the axial length of the rotor already defined in Sect. 5.3.

Rodgers (1987) proposed a simplified correlation:

$$\Delta h_{CL} = 0.4 \cdot \frac{\tau_r}{b_1} \cdot c_{1u}^2$$

Qi et al. (2017) and Persky and Sauret (2019) then proposed the following correlations derived from the interpolations of experimental data provided by Futral and Holeski (1969):

$$\Delta h_{CL} = \Delta h_{t,ad} \cdot \left(-0.09678 \cdot A - 1.69997 \cdot R + 0.096844 \cdot R^2 - 0.03379 \cdot A \cdot R \right) \cdot \frac{1}{100}$$

$$\Delta h_{CL} = \Delta h_{t,ad} \cdot \left(1.355 - 0.08371 \cdot A - 1.772 \cdot R - 0.2285 \cdot A \cdot R + 0.9725 \cdot R^2 \right) \cdot \frac{1}{100}$$

where the coefficients A and R:

$$A = 100 \cdot \frac{\tau_a}{b_1};\;\; \mathrm{R} = 100 \cdot \frac{\tau_r}{b_1}$$

For the radial and axial clearances can be assumed the numerical values previously reported.

Exit energy losses

These losses appear in single-stage turbines or in the last stage of multistage turbines. They are equal to the kinetic energy (calculated at the mean diameter) at the rotor outlet:

$$\Delta h_{exit} = \frac{c_{2M}^2}{2}$$

For turbines equipped with a final diffuser, these losses are expressed as (Erbas et al. 2013):

$$\Delta h_{exit} = (1 - C_D) \cdot \frac{c_{2M}^2}{2}$$

where C_D is the diffuser efficiency, assessable by:

$$C_D = -3.25 \cdot 10^{-4} \cdot |\alpha_{2M}|^2 + 6.25 \cdot 10^{-3} \cdot |\alpha_{2M}| + 0.57$$

Windage losses

They represent losses due to disk friction; Baines (2003), starting from the correlation proposed by Daily and Nece (1960), suggested the following equation:

$$\Delta h_{windage} = K_f \cdot \frac{\bar{\rho}}{2} \cdot \frac{u_1^3 \cdot D_1^2}{8 \cdot m}$$

where

- m is the mass flow rate handled by the rotor;
- $\bar{\rho}$ is a mean average density defined as:

$$\bar{\rho} = \frac{\rho_1 + \rho_2}{2}$$

- the coefficient K_f can be expressed as:

$$K_f = \frac{3.7 \cdot (2 \cdot \tau_b / D_1)^{0.1}}{\mathrm{Re}^{0.5}} \qquad \text{when} \qquad \mathrm{Re} < 10^5$$

$$K_f = \frac{0.102 \cdot (2 \cdot \tau_b / D_1)^{0.2}}{\mathrm{Re}^{0.2}} \qquad \text{when} \qquad \mathrm{Re} > 10^5$$

and

$$\mathrm{Re} = \frac{\rho_1 \cdot u_1 \cdot (D_1/2)}{\mu_1}$$
$$\tau_b = \tau_a.$$

5.4.1.3 Rotor Loss Model Proposals

As illustrated at the beginning of Sect. 5.4, in the literature there are several loss models for the rotor; the various correlations proposed for each type of loss were reported in previous sections.

Only the most recent loss models are listed below:

Meroni et al. (2018):

$$\Delta h_{incidence} = \frac{w_1^2}{2} \cdot sen^2(\beta_1 - \beta_{1,opt})$$

$$\Delta h_{passage} = K_p \cdot \left\{ \left(\frac{L_{hyd,R}}{D_{hyd,R}} \right) + 0.68 \cdot \left[1 - \left(\frac{D_{2M}}{D_1} \right)^2 \right] \cdot \frac{\cos \beta_{2M}}{(b_2/L_{hyd,R})} \right\} \cdot \frac{w_1^2 + w_{2M}^2}{2}$$

$$\Delta h_{TE} = \frac{1}{2} \cdot w_{2M}^2 \cdot \left(\frac{N_{B,R} \cdot t_{TE}}{\pi \cdot D_{2M} \cdot \cos \beta_{2M}} \right)^2 \cdot \left(1 + \frac{k_2 - 1}{2} \cdot Ma_2^2 \right)^{\frac{k_2}{1-k_2}}$$

$$\Delta h_{CL} = \frac{u_1^3 \cdot N_{B,R}}{8 \cdot \pi} \cdot \left(K_a \cdot \tau_a \cdot C_a + K_r \cdot \tau_r \cdot C_r + K_{a,r} \cdot \sqrt{\tau_a \cdot C_a \cdot \tau_r \cdot C_r} \right)$$

$$\Delta h_{windage} = K_f \cdot \frac{\bar{\rho}}{2} \cdot \frac{u_1^3 \cdot D_1^2}{8 \cdot m}$$

In this loss model, exit energy loss is not assessed but authors calculated the rotor post-expansion losses, applying the correlation proposed by Aungier (2006).

Persky and Sauret (2019):

$$\Delta h_{incidence} = \frac{w_1^2}{2} \cdot sen^2(\beta_1 - \beta_{1,opt})$$

$$\Delta h_{passage} = \frac{1}{2} \cdot K \cdot \left[w_1^2 \cdot \cos^2(\beta_1 - \beta_{1,opt}) + w_{2M}^2 \right]$$

$$\Delta h_{TE} = \frac{1}{2} \cdot w_{2M}^2 \cdot \left(\frac{N_{B,R} \cdot t_{TE}}{\pi \cdot D_{2M} \cdot \cos \beta_{2M}} \right)^2$$

$$\Delta h_{CL} = \Delta h_{t,ad} \cdot \left(1.355 - 0.08371 \cdot A - 1.772 \cdot R - 0.2285 \cdot A \cdot R + 0.9725 \cdot R^2 \right) \cdot \frac{1}{100}$$

$$\Delta h_{exit} = (1 - C_D) \cdot \frac{c_{2M}^2}{2}$$

$$\Delta h_{windage} = K_f \cdot \frac{\bar{\rho}}{2} \cdot \frac{u_1^3 \cdot D_1^2}{8 \cdot m}$$

Alshammari et al. (2020):

$$\Delta h_{incidence} = \frac{w_1^2}{2} \cdot sen^2(\beta_1 - \beta_{1,opt})$$

$$\Delta h_{passage} + \Delta h_s = f_t \cdot \frac{L_{hyd,R}}{D_{hyd,R}} \cdot \bar{w}^2 + \frac{1}{4} \cdot \frac{c_1^2 \cdot D_1}{N_{B,R} \cdot r_C}$$

$$\Delta h_{CL} = \frac{u_1^3 \cdot N_{B,R}}{8 \cdot \pi} \cdot \left(K_a \cdot \tau_a \cdot C_a + K_r \cdot \tau_r \cdot C_r + K_{a,r} \cdot \sqrt{\tau_a \cdot C_a \cdot \tau_r \cdot C_r} \right)$$

$$\Delta h_{exit} = \frac{c_{2M}^2}{2}$$

$$\Delta h_{windage} = K_f \cdot \frac{\bar{\rho}}{2} \cdot \frac{u_1^3 \cdot D_1^2}{8 \cdot m}$$

We propose the following loss model:

$$\Delta h_{incidence} = \frac{w_1^2}{2} \cdot sen^2(\beta_1 - \beta_{1,opt})$$

$$\Delta h_{passage} = 2 \cdot f_t \cdot \frac{L_{hyd,R}}{D_{hyd,R}} \cdot \bar{w}^2$$

$$\Delta h_{TE} = \frac{1}{2} \cdot w_{2M}^2 \cdot \left(\frac{N_{B,R} \cdot t_{TE}}{\pi \cdot D_{2M} \cdot \cos \beta_{2M}} \right)^2$$

$$\Delta h_{CL} = \frac{u_1^3 \cdot N_{B,R}}{8 \cdot \pi} \cdot \left(K_a \cdot \tau_a \cdot C_a + K_r \cdot \tau_r \cdot C_r + K_{a,r} \cdot \sqrt{\tau_a \cdot C_a \cdot \tau_r \cdot C_r} \right)$$

$$\Delta h_{exit} = (1 - C_D) \cdot \frac{c_{2M}^2}{2}$$

$$\Delta h_{windage} = K_f \cdot \frac{\bar{\rho}}{2} \cdot \frac{u_1^3 \cdot D_1^2}{8 \cdot m}$$

This loss model is very similar to those proposed by the authors previously cited but for the passage losses we believe that the best correlation is that proposed by Musgrave (1980) because it takes directly into account the Reynolds number, the type of the working fluid and the surface roughness. This approach is fundamental especially for the analysis of turbomachines handling unconventional fluids (Chap. 8).

5.4.2 Loss Models Based on Pressure Loss Coefficients

The pressure loss model was proposed by Aungier (2006).

The main characteristic of this model is the passage loss calculation in the stator and rotor through the evaluation of the boundary layers on the blade surfaces and on the end wall contours.

The inlet, midpassage and exit stations of the stator/rotor channel are indicated with the subscripts in, mid and out respectively; Aungier proposed the following expressions of the boundary layer momentum thickness (Sect. 1.9) at the component discharge:

$$\theta = f_{sf} \cdot \bar{\rho} \cdot \left[\left(\frac{v_{in}}{v_{out}} \right)^5 + 2 \cdot \left(\frac{v_{mid}}{v_{out}} \right)^5 + 1 \right] \cdot \frac{L_{hyd}}{8 \cdot \rho_{out}}$$

where

- v is the fluid flow velocity relative to wall, that is the absolute velocity, c, in the stator and the relative velocity, w, in the rotor. In particular, v_{in} and v_{out} are the absolute/relative velocity of the fluid flow entering and exiting the channel while v_{mid} is the velocity at the midpassage. To calculate this last velocity we must distinguish the end wall boundary layer and the blade surface boundary layer; in the first case v_{mid} is the mean velocity in the channel while in the second case this mean velocity must be corrected for blade loading effects (this specialization for the stator and for the rotor will be described below);
- $\bar{\rho}$ is the mean density defined as:

$$\bar{\rho} = \frac{\rho_{in} + 2 \cdot \rho_{mid} + \rho_{out}}{4}$$

- L_{hyd} is the hydraulic length of the channel;
- f_{sf} is the skin friction coefficient. For this coefficient, Aungier proposed a formulation completely analogous to that used for radial compressors (Chap. 6), which, for convenience, is illustrated below.

  Let us define the Reynolds number as:

$$\mathrm{Re}_D = \frac{\rho_{out} \cdot v_{out} \cdot b_{out}}{\mu_{out}}$$

If the flow is laminar ($\mathrm{Re_D} < 2000$), the skin friction coefficient is:

$$f_{sf} = f_l = \frac{16}{\mathrm{Re}_D}$$

while if $\mathrm{Re_D} > 4000$, the skin friction coefficient is a weighted average between the skin friction coefficient for smooth walls (f_{ts}) and for fully rough walls (f_{tr}):

$$\frac{1}{\sqrt{f_{ts}}} = -4 \cdot \log_{10}\left(\frac{1.255}{\mathrm{Re}_D \cdot \sqrt{f_{ts}}} \right)$$

$$\frac{1}{\sqrt{f_{tr}}} = -4 \cdot \log_{10}\left(\frac{1}{3.71} \cdot \frac{k_s}{D_{hyd,R}}\right)$$

where k_s is the superficial roughness defined in Sect. 5.3

To calculate this weighted average, the following Reynolds number is introduced:

$$\text{Re}_e = (\text{Re}_D - 2000) \cdot \frac{k_s}{D_{hyd,R}}$$

and by using this parameter we can calculate the skin friction coefficient when $\text{Re}_\text{D} > 4{,}000$. In fact we can apply:

$$f_{sf} = f_t = f_{ts} \qquad \text{when} \quad \text{Re}_e < 60$$

$$f_{sf} = f_t = f_{ts} + (f_{tr} - f_{ts}) \cdot \left(1 - \frac{60}{\text{Re}_e}\right) \qquad \text{when} \quad \text{Re}_e \geq 60$$

If $2000 < \text{Re}_\text{D} < 4000$, the boundary layer is in transition from laminar to turbulent flow; the skin friction coefficient in this transition zone is well approximated by a weighted average of the laminar and turbulent values:

$$f = f_{l,(\text{Re}_D=2000)} - \left(f_{l,(\text{Re}_D=2000)} - f_{t,(\text{Re}_D=4000)}\right) \cdot \left(\frac{\text{Re}_D}{2000} - 1\right)$$

Subsequently, the displacement thickness of the boundary layer is evaluated (Sect. 1.9):

$$\delta^* = \theta \cdot H$$

where H is the boundary layer shape factor and it can be assumed equal to 1.2857.

These parameters (θ and δ^*), which represent the reduction of momentum and mass because of the boundary layer, are then normalized through the width, b, of the channel:

$$\Theta = \sum \frac{\theta}{b}$$

$$\Delta = \sum \frac{\delta^*}{b}$$

The aforementioned summations are the addition of all the boundary layers in the channel: blade pressure and blade suction sides and end walls. Considering the boundary layers on the two end walls (blade hub and tip) equal and instead distinguishing the boundary layers on the blade pressure and suction sides, the above summations can be expressed as:

$$\Theta = 1 - \left[1 - 2 \cdot \left(\frac{\theta_{EW}}{b_{EW}}\right)\right] \cdot \left[1 - \left(\frac{\theta_{B,SS}}{b_{B,SS}} + \frac{\theta_{B,PS}}{b_{B,PS}}\right)\right]$$

$$\Delta = 1 - \left[1 - 2 \cdot \left(\frac{\delta^*_{EW}}{b_{EW}}\right)\right] \cdot \left[1 - \left(\frac{\delta^*_{B,SS}}{b_{B,SS}} + \frac{\delta^*_{B,PS}}{b_{B,PS}}\right)\right]$$

where the subscript EW is referred to the end wall parameters and the subscripts B,SS and B,PS to the parameters at the suction side and pressure side of the blades.

In the following these correlations will be further specialized for stator and rotor channels.

By using these normalized parameters, the passage losses are expressed as:

$$Y_p = \frac{2 \cdot \Theta + \Delta^2}{(1 - \Delta)^2}$$

5.4.2.1 Stator Losses

The total pressure loss coefficient in the stator is expressed as (Sect. 5.2):

$$Y_S = \frac{p_{0t} - p_{1t}}{p_{1t} - p_1}$$

Stator losses include passage losses and incidence losses. In design conditions the latter ones can be neglected, so:

$$Y_S = Y_p + Y_{inc} \cong Y_p = \frac{2 \cdot \Theta + \Delta^2}{(1 - \Delta)^2}$$

To calculate the passage losses, the correlations provided above are specialized as:

$$\theta_{EW} = f_{sf,EW} \cdot \bar{\rho}_S \cdot \left[\left(\frac{c_0}{c_1}\right)^5 + 2 \cdot \left(\frac{\bar{c}_{EW}}{c_1}\right)^5 + 1\right] \cdot \frac{L_{hyd,S}}{8 \cdot \rho_1}$$

$$\theta_{B,SS} = f_{sf,B} \cdot \bar{\rho}_S \cdot \left[\left(\frac{c_0}{c_1}\right)^5 + 2 \cdot \left(\frac{\bar{c}_{SS}}{c_1}\right)^5 + 1\right] \cdot \frac{L_{hyd,S}}{8 \cdot \rho_1}$$

$$\theta_{B,PS} = f_{sf,B} \cdot \bar{\rho}_S \cdot \left[\left(\frac{c_0}{c_1}\right)^5 + 2 \cdot \left(\frac{\bar{c}_{PS}}{c_1}\right)^5 + 1\right] \cdot \frac{L_{hyd,S}}{8 \cdot \rho_1}$$

where, in addition to the parameters already defined in the previous sections, the velocities at the midpassage appear. In order to take into account the blade loading effects, Aungier suggested to evaluate the velocity difference between blade suction and pressure sides by means of:

$$\Delta c = \frac{2 \cdot \pi \cdot (D_1 \cdot c_{1u} - D_0 \cdot c_{0u})}{N_{B,S} \cdot L_{hyd,S}}$$

Following this suggestion, the velocities at the midpassage become:

$$\bar{c}_{EW} = \frac{c_1 + c_0}{2}$$
$$\bar{c}_{B,SS} = \bar{c}_{EW} + \frac{\Delta c}{2}$$
$$\bar{c}_{B;PS} = \bar{c}_{EW} - \frac{\Delta c}{2}$$

The passage widths are:

$$b_{B,SS} = b_{B,PS} = O_{1S}$$

$$b_{EW} = b_1$$

Finally, as regards the skin friction coefficients on the end walls ($f_{sf,EW}$) and blade surfaces ($f_{sf,B}$), the procedure previously described is applied by using the previous passage widths in the Reynolds numbers.

5.4.2.2 Rotor Losses

The relative total pressure loss coefficient in the rotor is expressed as (Sect. 5.2):

$$Y_R = \frac{p_{2tr,is} - p_{2tr}}{p_{2tr} - p_2}$$

The losses in the rotor can be divided into:

$$Y_R = Y_{inc} + Y_p + Y_{BL} + Y_{HS} + Y_{CL}$$

where

- Y_{inc} represents the incidence losses;
- Y_p represents the passage losses;
- Y_{BL} represents the blade loading losses;
- Y_{HS} represents the hub to shroud losses;
- Y_{CL} represents the blade clearance losses.

Incidence losses

They are given by:

$$Y_{inc} = sen^2(\alpha_1 - \alpha_1^*) \cdot \frac{p_{1tr} - p_1}{p_{2tr} - p_2}$$

Aungier suggested to calculate the optimum flow angle α_1^* at the rotor inlet by using a procedure very similar to that used for centrifugal compressors, based on the *slip factor*. He introduced the slip factor as:

$$SF = \frac{c_{1u}^*}{c_{1u}}$$

By using this slip factor, α_1^* may be expressed as:

$$\tan \alpha_1^* = \frac{c_{1u}^*}{c_{1m}} = \frac{SF \cdot c_{1u}}{c_{1m}} = SF \cdot \tan \alpha_1$$

In order to calculate the slip factor, we can use the correlation proposed by Wiesner (1967):

$$SF = 1 - \frac{\sqrt{\cos \beta_1}}{N_{B,R}^{0.7}}$$

with the correction suggested by Aungier (2006) when the rotor mean diameter ratio, defined as:

$$\delta_M = \frac{D_{2M}}{D_1} = \frac{\delta_t + \delta_h}{2}$$

exceeds the following limit value:

$$\delta_{M,\lim} = \frac{SF - SF^*}{1 - SF^*}$$

where:

$$SF^* = \sin[19° + 0.2 \cdot (90° - \beta_1)]$$

Hence, when $\delta_M > \delta_{M,\lim}$, Aungier (2006) suggested the following expression of the slip factor, correcting the previous correlation proposed by Wiesner:

$$SF_{cor} = SF \cdot \left[1 - \left(\frac{\delta_M - \delta_{M,\lim}}{1 - \delta_{M,\lim}}\right)^{\sqrt{(90-\beta_1)/10}}\right]$$

Passage losses

They are evaluated by using a procedure completely analogous to that followed for the stator, that is:

$$Y_p = \frac{2 \cdot \Theta + \Delta^2}{(1 - \Delta)^2}$$

To calculate these losses, the correlations provided above are specialized as:

$$\theta_{EW} = f_{sf,EW} \cdot \bar{\rho}_R \cdot \left[\left(\frac{w_1}{w_2} \right)^5 + 2 \cdot \left(\frac{\bar{w}_{EW}}{w_2} \right)^5 + 1 \right] \cdot \frac{L_{hyd,R}}{8 \cdot \rho_2}$$

$$\theta_{B,SS} = f_{sf,B} \cdot \bar{\rho}_R \cdot \left[\left(\frac{w_1}{w_2} \right)^5 + 2 \cdot \left(\frac{\bar{w}_{SS}}{w_2} \right)^5 + 1 \right] \cdot \frac{L_{hyd,R}}{8 \cdot \rho_2}$$

$$\theta_{B,PS} = f_{sf,B} \cdot \bar{\rho}_R \cdot \left[\left(\frac{w_1}{w_2} \right)^5 + 2 \cdot \left(\frac{\bar{w}_{PS}}{w_2} \right)^5 + 1 \right] \cdot \frac{L_{hyd,R}}{8 \cdot \rho_2}$$

where, in addition to the parameters already defined in the previous sections, the velocities at the midpassage appear. In order to take into account the blade loading effects, Aungier suggested to evaluate the velocity difference between blade suction and pressure sides by means of:

$$\Delta w = \frac{2 \cdot \pi \cdot (D_1 \cdot c_{1u} - D_2 \cdot c_{2u})}{N_{B,R} \cdot L_{hyd,R}}$$

Following this suggestion, the velocities at the midpassage become:

$$\bar{w}_{EW} = \frac{w_1 + w_{2M}}{2}$$

$$\bar{w}_{B,SS} = \bar{w}_{EW} + \frac{\Delta w}{2}$$

$$\bar{w}_{B,PS} = \bar{w}_{EW} - \frac{\Delta w}{2}$$

The passage widths are:

$$b_{B,SS} = b_{B,PS} = O_2$$

$$b_{EW} = b_2$$

Finally, as regards the skin friction coefficients on the end walls ($f_{sf,EW}$) and blade surfaces ($f_{sf,B}$), the procedure previously described is applied by using the previous passage widths in the Reynolds numbers.

Blade loading losses

These losses are given by:

$$Y_{BL} = \frac{1}{24} \cdot \left(\frac{\Delta w}{2 \cdot w_{2M}} \right)^2$$

where Δw has already been defined.

Hub to shroud losses

These losses are given by:

$$Y_{HS} = \frac{1}{6} \cdot \left(\frac{\bar{K}_m \cdot \bar{b} \cdot \bar{w}}{w_{2M}} \right)^2$$

where:

- $\bar{K}_m = \dfrac{\alpha_{C1} - \alpha_{C2}}{L_{m,R}}$

and $L_{m,R}$ is the meridional length, previously defined, and α_C is the flow angle with respect to the axis of rotation. If the fluid enters the rotor approximately radially and exits the rotor approximately axially ($\alpha_{C1} = \pi / 2$ and $\alpha_{C2} = 0$), the previous expression becomes:

$$\bar{K}_m = \frac{\pi}{2 \cdot L_{m,R}}$$

- $\bar{b} = \dfrac{b_1 + b_2}{2}$
- $\bar{w} = \dfrac{w_1 + w_{2M}}{2}$

Blade clearance losses

They can be evaluated in a similar way to radial compressors (Chap. 6) by:

$$Y_{CL} = \frac{2 \cdot m_{CL} \cdot \Delta p_{CL}}{m \cdot \rho_2 \cdot w_{2M}^2}$$

where m is the mass flow rate handled by rotor;

$$\Delta p_{CL} = \frac{m \cdot \psi \cdot u_1^2}{N_{B,R} \cdot L_{hyd,R} \cdot \omega \cdot \left(\frac{D_1+D_{2M}}{4}\right) \cdot \left(\frac{b_1+b_2}{2}\right)}$$

$$u_{CL} = 0.816 \cdot \sqrt{\frac{2 \cdot \Delta p_{CL}}{\rho_1}}$$

$$m_{CL} = \rho_1 \cdot N_{B,R} \cdot \tau_a \cdot L_{hyd,R} \cdot u_{CL}$$

where τ_a is the axial clearance, already defined in Sect. 5.3.

External losses (parasitic loss)

Since these losses are expressed in enthalpy terms, it remains valid what illustrated in Sect. 5.4.1.

5.4.3 *Stator and Stage Efficiency*

The procedure described in Sects. 5.4.1 and 5.4.2 allows evaluating the stage losses, expressed in terms of enthalpy increase compared to isentropic expansion or in terms of pressure loss coefficients.

To assess efficiencies, let us consider the loss model based on enthalpy losses since the procedure for the efficiency calculation through the pressure losses has already been illustrated in Chaps. 3 and 4.

By knowing the isentropic enthalpy at the stator outlet, the adiabatic enthalpy can be calculated by using the stator enthalpy increase due to the losses:

$$h_1 = h_{1,is} + \Delta h_S$$

and, therefore, the stator efficiency (Sect. 1.8.2) can be assessed as:

$$\eta_S = \frac{h_{0t} - h_1}{h_{0t} - h_{1,is}}$$

Similarly, for the stage efficiency, the isentropic enthalpy at the stage outlet can be calculated by using the entropy at the rotor inlet (Fig. 5.2):

$$h_{2',is} = h_{s,p}(s_1, p_2)$$

therefore the adiabatic enthalpy at the stage outlet can be assessed as:

$$h_2 = h_{2',is} + \Delta h_{R,int}$$

Once the thermodynamics of the stage have been calculated on the basis of the stage losses, the various efficiencies can be assessed (Sect. 1.8.2):

$$\eta_{TT} = \frac{h_{0t} - h_{2t}}{h_{0t} - h_{2t,is}}$$

$$\eta_{is} = \frac{h_0 - h_2}{h_0 - h_{2,is}}$$

$$\eta_{TS} = \frac{h_{0t} - h_{2t}}{h_{0t} - h_{2,is}}$$

By following the discussion carried out in Sect. 1.9.6, the isentropic efficiency does not take into account the external losses (for example windage loss) and therefore in the previous expressions only the enthalpy increases generated by internal losses are considered.

Remarks on input parameters

The procedure illustrated in this Sect. 5.4 does not presuppose the introduction of further input parameters, compared to those already identified in Sects. 5.1 and 5.2.

5.5 Input Parameters of the Preliminary Design Procedure

In the previous Sects. 5.1–5.4, the procedure for the calculation of the kinematics, thermodynamics, geometry and efficiency of a radial stage was developed and the input parameters of this procedure were identified.

In this Sect. 5.5 we intend to precisely analyze these input parameters in order to suggest their numerical values (these values, however, can be reviewed in an iterative calculation, see Sect. 5.6).

First of all, let us consider the kinematics calculation procedure (Sect. 5.1).

In general, to determine the velocity triangles at the rotor inlet and outlet, we must set six independent parameters: three for the inlet triangle (station 1 of the stage) and three for the outlet triangle (station 2 of the stage).

Unlike the axial stages, where, being $u_1 = u_2 = u$ and assuming $c_{m1} = c_{m2} = c_m$, the independent parameters are only four, for the radial stages the *independent parameters remain six*.

As clearly illustrated in Sect. 5.1, in fact, the absolute and relative flow angles are:

$$\alpha_1 = f(\varphi, \psi, \xi, \delta_t, \alpha_2)$$
$$\beta_1 = f(\varphi, \psi, \xi, \delta_t, \alpha_2)$$
$$\beta_2 = f(\varphi, \delta_t, \alpha_2)$$

These angles are, therefore, functions of the following five independent parameters:

- the flow coefficient, φ
- the work coefficient, ψ
- the rotor meridional velocity ratio, ξ, or the degree of reaction, R, being (Sect. 5.1):

$$R = f(\varphi, \psi, \xi, \delta_t, \alpha_2)$$

- the rotor tip diameter ratio, δ_t
- the absolute flow angle at the rotor outlet, α_2

As a consequence, by identifying a further independent parameter, for example, the blade speed at the rotor inlet, u_1, the velocity triangles are completely defined.

However, taking into account what emerged in the turbomachinery selection process (Chap. 2), it is advisable to choose the stage efficiency (isentropic or polytropic efficiency) as the additional independent parameter and calculate the blade speed accordingly.

In fact, starting from the data available for the turbomachinery selection process:

- type of fluid
- fluid mass flow rate (kg/s)
- fluid inlet temperature (°C)
- fluid inlet pressure (bar)
- fluid outlet pressure (bar)

the selection process provides ranges of n ($n_{\min} < n < n_{\max}$) and z ($z_{\min} < z < z_{\max}$) compatible with the type of turbomachine chosen. Within these ranges we can choose one or more pairs of these values:

- n: rotational speed (rpm)
- z: number of stages and, consequently, the isentropic enthalpy drop in each stage, $\Delta h_{is,stage}$ (kJ/kg).

For each of these pairs, by using the stage (isoentropic or polytropic) efficiency, set as first iteration guess (Sect. 5.5.3) or calculated by the stage losses (Sect. 5.4), the adiabatic enthalpy drop of each stage can be calculated as well as the blade speed, after choosing the work coefficient.

Indeed, when $c_0 = c_2$ (Sect. 5.1), the blade speed is:

$$u_1 = \sqrt{\frac{\Delta h_{t,ad,stage}}{\psi}} = \sqrt{\frac{\Delta h_{ad,stage}}{\psi}}$$

On the other hand, when $c_0 \neq c_2$ (Sects. 5.1 and 5.3), the work coefficient is expressed as:

$$\psi = \frac{\Delta h_{t,ad,stage}}{u_1^2} = \frac{\Delta h_{ad,stage}}{u_1^2} + \frac{c_0^2 - c_2^2}{2 \cdot u_1^2}$$

and, therefore, expressing the velocity c_2 as a function of the kinematic parameters, we obtain:

$$u_1 = \sqrt{\frac{\Delta h_{ad,stage} + \frac{c_0^2}{2}}{\psi + \frac{1}{2} \cdot \frac{\varphi^2}{\cos^2 \alpha_2}}}$$

After calculating the blade speed, the diameter at the rotor inlet is:

$$D_1 = \frac{60 \cdot u_1}{\pi \cdot n}$$

Alternatively, we can calculate the blade speed, u_1, from the diameter, D_1, obtained, in turn, by the specific diameter, D_s (Chap. 2). In this way, only five independent parameters must be chosen since the work coefficient is calculated by using the stage enthalpy drop, $\Delta h_{ad,stage}$, and the blade speed, u_1. In radial stage design, however, it is preferred to assign the work coefficient, following the procedure illustrated above.

In summary, we must set the following six input parameters in order to design preliminarily the radial turbine stage:

- the flow coefficient, φ
- the work coefficient, ψ
- the degree of reaction, R, or the rotor meridional velocity ratio, ξ
- the rotor tip diameter ratio, δ_t
- the absolute flow angle at the rotor outlet, α_2
- the stage efficiency (isentropic or polytropic efficiency)

Taking into account what is illustrated in Sects. 5.2–5.4, these parameters, with the addition of the stator efficiency, allow to fully calculate the thermodynamics, the geometry and therefore the stage losses. Since these losses allow, in turn, to exactly evaluate the stator and stage efficiencies, the stage efficiency (in the list above) is only a starting value: the calculation will proceed iteratively for each stage until convergence for these efficiencies is achieved (Sect. 5.6).

5.5.1 Flow and Work Coefficient, or Alternative Parameters, Selection

To choose the value of the flow and work coefficients, Smith-type diagrams can be used (as made for the axial stages, Sect. 3.6.1). Among these diagrams, that elaborated by Chen and Baines (1994), and shown in Fig. 5.5 (according to the reworking by

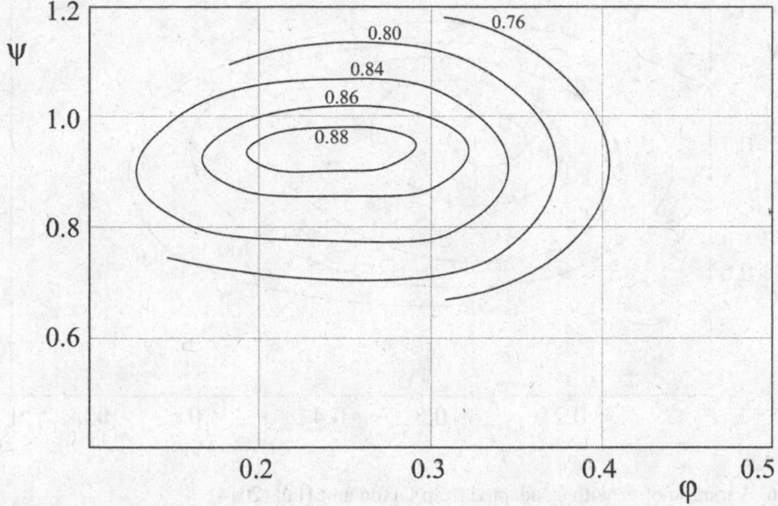

Fig. 5.5 Smith-type diagram. Adapted from Lauriau et al. (2018)

Lauriau et al. 2018), is very effective. This diagram shows the iso-efficiency (total-to-static efficiency) lines of the radial stages as a function of the flow, φ, and work, ψ, coefficients. These lines were elaborated by using experimental data of 40 radial turbines.

The diagram clearly indicates that the maximum efficiency can be achieved in fairly narrow ranges of the flow and work coefficients:

$$\eta_{ts,MAX} \Rightarrow \begin{cases} 0.2 < \varphi < 0.3 \\ 0.9 < \psi < 1.0 \end{cases}$$

Instead of the work coefficient, it is possible to use the *isentropic velocity ratio* defined as:

$$\nu_s = \frac{u_1}{c_{is}} = \frac{u_1}{\sqrt{2 \cdot \Delta h_{is,stage}}}$$

where c_{is} is the *spouting velocity*.

Rodger and Geiser (1987) followed this approach and elaborated the diagram shown in Fig. 5.6 (according to the reworking by Dixon and Hall 2014) by using experimental data of about 30 radial turbines. This diagram shows the iso-efficiency (total-to-static efficiency) lines of the radial stages as a function of the flow coefficient, φ, and the isentropic velocity ratio, ν_s. The diagram clearly indicates that the maximum efficiency can be achieved in ranges of the flow coefficient similar to those of the diagram shown in Fig. 5.5 and for isentropic velocity ratio values close to 0.7:

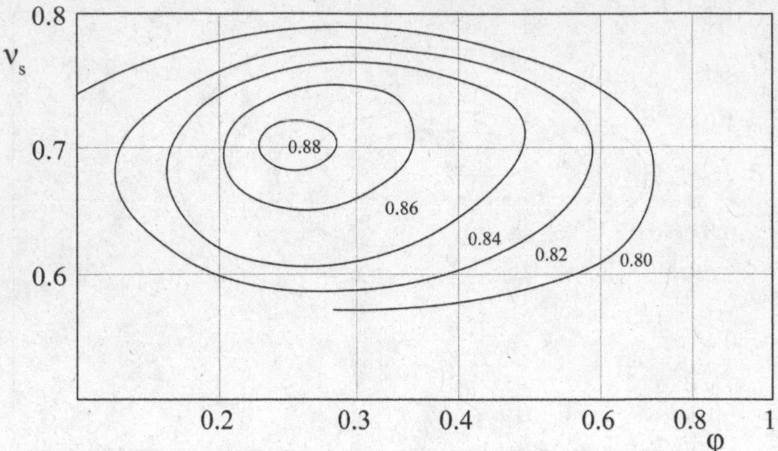

Fig. 5.6 Variation of ν_s with φ. adapted from Dixon and Hall (2014)

$$\eta_{ts,MAX} \Rightarrow \begin{cases} 0.2 < \varphi < 0.3 \\ 0.68 < \upsilon_s < 0.71 \end{cases}$$

Note that the isentropic velocity ratio is related to the isentropic work coefficient through:

$$\psi_{is} = \frac{\Delta h_{is}}{u_1^2} = \frac{1}{2 \cdot \nu_s^2}$$

and, therefore, through the isentropic efficiency (Sect. 5.5.3), to the work coefficient, ψ.

Aungier (2006) provided a correlation between the isentropic velocity ratio and the specific speed of the stage (Chap. 2):

$$\nu_s = 0.737 \cdot \omega_s^{0.2}$$

This correlation can be very useful to establish the isentropic velocity ratio because the specific speed of the radial stage, which, as illustrated in Chap. 2 and shown in Fig. 5.7, falls within a rather narrow range:

$$\eta_{ts,MAX} \Rightarrow 0.4 < \omega_s < 0.8$$

Instead of the flow coefficient, it is possible to use the absolute flow angle at the rotor inlet. In fact, Aungier (2006) provided a correlation between this angle and the specific speed of the stage:

$$\alpha_1 = 79.2 - 14.2 \cdot \omega_s^2$$

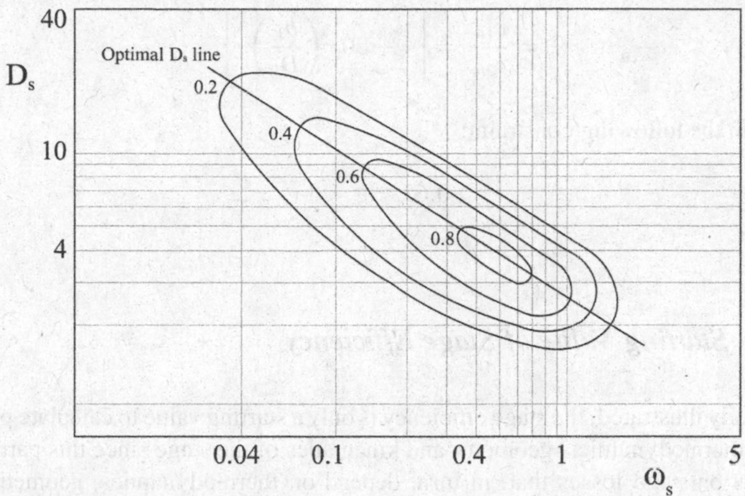

Fig. 5.7 Baljè diagram—radial turbine stages

From the previous considerations, in the radial stages we can choose, as independent parameters, the flow and work coefficients or, alternatively, the isentropic velocity ratio and the inlet absolute flow angle:

$$\varphi, \psi \begin{cases} 0.2 < \varphi < 0.3 \\ 0.9 < \psi < 1.0 \end{cases} \quad \text{or} \quad \nu_s, \alpha_1 = f(\omega_s) \begin{cases} \nu_s = 0.737 \cdot \omega_s^{0.2} \\ \alpha_1 = 79.2 - 14.2 \cdot \omega_s^2 \end{cases}$$

5.5.2 *Degree of Reaction or Rotor Meridional Velocity Ratio Selection*

Aungier (2006) suggested the following range for the degree of reaction:

$$0.45 \leq R \leq 0.65$$

but paying attention to the constraint on the maximum value of the degree of reaction, already discussed in Sect. 5.1.

Alternatively to the degree of reaction, the rotor meridional velocity ratio can be assumed as input parameter. In this regard, always Aungier (2006) proposed to correlate this ratio with the geometric parameters at the rotor inlet, according to the following expression:

$$\xi = \frac{c_{1m}}{c_{2m}} = \left[1 + 20 \cdot \left(\frac{b_1}{D_1}\right)^2\right]^{-1}$$

and with the following constraint:

$$0.65 < \xi < 1$$

5.5.3 Starting Value of Stage Efficiency

As already illustrated, the stage efficiency is only a starting value to calculate preliminarily thermodynamics, geometry and kinematics of the stage since this parameter depends on stage losses that, in turn, depend on thermodynamics, geometry and kinematics of the stage (Sect. 5.4). The efficiency calculation proceeds iteratively for each stage until convergence is achieved (Sect. 5.6).

The starting value of the stage efficiency can be chosen on the diagrams shown in the previous section, after setting the flow and work coefficients, φ and ψ, or the isentropic velocity ratio, ν_s, and the inlet absolute flow angle, α_1.

Also in this case, however, it is very effective to establish the starting value of the stage efficiency (total to static efficiency) as a function of the specific speed, according to the correlation proposed by Aungier (2006):

$$\eta_{ts} = 0.87 - 1.07 \cdot (\omega_s - 0.55)^2 - 0.5 \cdot (\omega_s - 0.55)^3$$

The variation of this efficiency with the specific speed is shown in Fig. 5.8; this diagram clearly indicates that the maximum efficiency can be achieved in a range of the specific speed very similar to that displayed by the Baljé diagram:

$$\eta_{ts,MAX} \quad \Rightarrow \quad 0.4 < \omega_s < 0.8$$

For the thermodynamics evaluation (Sect. 5.2), it is also necessary to choose the starting value of the stator efficiency. It can be set arbitrarily, for example, equal to the stage isentropic efficiency, since also this efficiency will be iteratively calculated by the losses in the stator (Sect. 5.4).

5.5.4 Exit Flow Angle and Rotor Diameter Ratios Selection

Generally, we can assume the exit absolute flow angle as:

$$\alpha_2 = 0$$

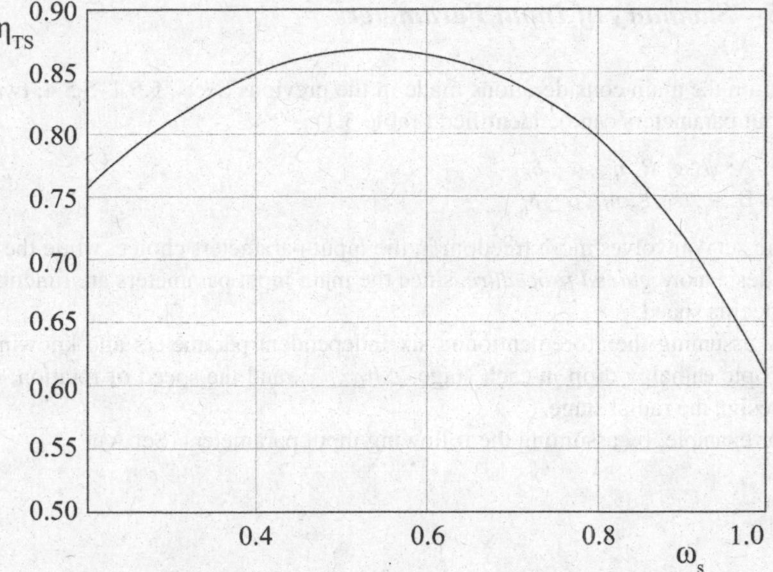

Fig. 5.8 Variation of efficiency with the specific speed

in order to minimize the kinetic energy at the rotor outlet.

For the rotor tip diameter ratio, Rohlik (1968) suggested the following limit:

$$\delta_t = \frac{D_{2t}}{D_1} \leq 0.7$$

and Aungier (2006) proposed the following correlation where this parameter is a function of the specific speed:

$$\delta_t = 1.29 \cdot \omega_s$$

Alternatively, the rotor hub diameter ratio can be chosen as input parameter and Aungier (2006) proposed:

$$\delta_h = \frac{D_{2h}}{D_1} = 0.185$$

Once this parameter has been established, the rotor tip diameter ratio can be calculated by using the continuity equation (Sect. 5.3):

$$\delta_t = \sqrt{\delta_h^2 + \frac{4 \cdot V_2}{\pi \cdot \varphi \cdot u_1 \cdot D_1^2}}$$

with the constraint, $\delta_t \leq 0.7$, previously introduced.

5.5.5 Summary of Input Parameters

Based on the main considerations made in the previous Sects. 5.5.1–5.5.4, two sets of input parameters can be identified (Table 5.1):

- Set A: ψ, φ, R, η_{is}, α_2, δ_h
- Set B: ν_s, α_1, ξ, η_{is}, α_2, δ_h

The set A involves more freedom in the input parameters choice, while the set B provides a more *guided procedure*, since the main input parameters are functions of the specific speed.

By assuming the aforementioned six independent parameters and knowing the isentropic enthalpy drop in each stage, $\Delta h_{is,stage}$, and the speed of rotation, n, we can design the radial stage.

For example, by assuming the following input parameters (Set A):

1. ψ
2. φ
3. R
4. η_{is}
5. α_2
6. δ_h

we can immediately calculate:

- the stage adiabatic enthalpy drop:

$$\Delta h_{ad} = \Delta h_{is} \cdot \eta_{is}$$

- the blade speed at the rotor inlet (for $c_0 \neq c_2$, see Sect. 5.5):

Table 5.1 Input parameters

	Set A	Set B
1	Work coefficient ψ	Isentropic velocity ratio ν_s
2	Flow coefficient φ	Rotor inlet flow angle α_1
3	Degree of reaction R	Rotor meridional velocity ratio ξ
4	Stage efficiency η_{is}	
5	Outlet flow angle α_2	
6	Rotor hub diameter ratio δ_h	

$$u_1 = \sqrt{\frac{\Delta h_{ad}}{\psi}}$$

- the rotor inlet diameter:

$$D_1 = \frac{60 \cdot u_1}{\pi \cdot n}$$

- the rotor tip diameter ratio:

$$\delta_t = \sqrt{\delta_h^2 + \frac{4 \cdot V_2}{\pi \cdot \varphi \cdot u_1 \cdot D_1^2}}$$

- the rotor meridional velocity ratio:

$$\xi = \sqrt{1 - \frac{2 \cdot \psi}{\varphi^2} \cdot \left(R + \frac{\psi}{2} - 1\right) + \tan^2 \alpha_2 \cdot \left(1 - \delta_t^2\right) + \frac{2 \cdot \psi \cdot \delta_t}{\varphi} \cdot \tan \alpha_2}$$

- the flow angles:

$$\tan \alpha_1 = \frac{\psi}{\xi \cdot \varphi} - \frac{\delta_t}{\xi} \cdot \tan \alpha_2$$

$$\tan \beta_1 = \frac{1}{\varphi \cdot \xi} \cdot (\psi - 1) - \frac{\delta_t}{\xi} \cdot \tan \alpha_2$$

$$\tan \beta_2 = \frac{\delta_t}{\varphi} + \tan \alpha_2$$

and therefore we can calculate the kinematics, thermodynamics, geometry and stage losses.

Similarly, by assuming the following input parameters (Set B):

1. ν_s
2. α_1
3. ξ
4. η_{is}
5. α_2
6. δ_h

we can immediately calculate (considering for example $\alpha_2 = 0$):

- the isentropic work coefficient:

$$\psi_{is} = \frac{1}{2 \cdot \nu_s^2}$$

- the blade speed at the rotor inlet:

$$u_1 = \sqrt{\frac{\Delta h_{is}}{\psi_{is}}}$$

- the work coefficient:

$$\psi = \psi_{is} \cdot \eta_{is}$$

- the rotor inlet diameter:

$$D_1 = \frac{60 \cdot u_1}{\pi \cdot n}$$

- the rotor inlet meridional velocity:

$$c_{1m} = u_1 \cdot \frac{\psi}{\tan \alpha_1}$$

- the blade height at the rotor inlet:

$$b_1 = \frac{V_1}{\pi \cdot D_1 \cdot c_{1m}}$$

- the flow coefficient:

$$\varphi = \frac{1}{\xi} \cdot \frac{c_{1m}}{u_1}$$

- the rotor tip diameter ratio:

$$\delta_t = \sqrt{\delta_h^2 + \frac{4 \cdot V_2}{\pi \cdot \varphi \cdot u_1 \cdot D_1^2}}$$

- the degree of reaction:

$$R = 1 - \frac{\psi}{2} + \frac{\varphi^2}{2 \cdot \psi} \cdot \left[\left(1 - \xi^2\right) + \tan^2 \alpha_2 \cdot \left(1 - \delta_t^2\right)\right] + \varphi \cdot \delta_t \cdot \tan \alpha_2$$

- the flow angles:

$$\tan \beta_1 = \frac{1}{\varphi \cdot \xi} \cdot (\psi - 1) - \frac{\delta_t}{\xi} \cdot \tan \alpha_2$$

Table 5.2 Input parameter constraints

Parameter	Range
Specific speed ω_s	0.4–0.8
Flow coefficient φ	0.2–0.3
Work coefficient ψ	0.8–1.0
Degree of reaction R	0.45–0.65
Rotor meridional velocity ratio ξ	0.65–1.0
Rotor tip diameter ratio δ_t	<0.7
δ_h/δ_t	>0.4
Rotor inlet absolute flow angle α_1	66°–78°
Rotor inlet relative flow angle $\beta 1$	20°–40°

$$\tan \beta_2 = \frac{\delta_t}{\varphi} + \tan \alpha_2$$

and therefore we can calculate the kinematics, thermodynamics, geometry and stage losses.

In conclusion, Table 5.2 summarizes the constraints of all the parameters illustrated so far in order to design correctly a radial stage achieving high efficiency.

5.6 The Conceptual Comprehensive Framework of the Proposed Preliminary Design Procedure

To calculate the kinematics, thermodynamics and geometry of a radial stage, it is necessary to assume the stator and stage efficiencies (Sect. 5.5). These efficiencies depend on the losses in the stator and rotor which, in turn, depend on all the parameters listed above (Sect. 5.4). Hence, the preliminary design of a stage is an iterative calculation (Fig. 5.9): first, we calculate the kinematic, thermodynamic and geometric parameters by assuming a starting value of the stage and stator efficiency; then, we can calculate the stator and rotor losses; at this point, the stage and stator efficiencies must be calculated by using these losses; these new efficiencies become the new input parameters of the preliminary design procedure. This iterative procedure ends when the established convergence for the stator and stage efficiencies is achieved.

The iterative procedure shown in Fig. 5.9 could be solved by using different values of the input parameters, as well as of the geometric parameters, in order to optimize the stage performance.

Generally radial inflow turbines are single stage. However, for multistage turbines (Fig. 5.10), the same preliminary design is applied to the stages following the first one. Considering a multistage turbine, it is important to highlight that typically the working fluid enters the turbomachine through an inlet volute and an exhaust diffuser

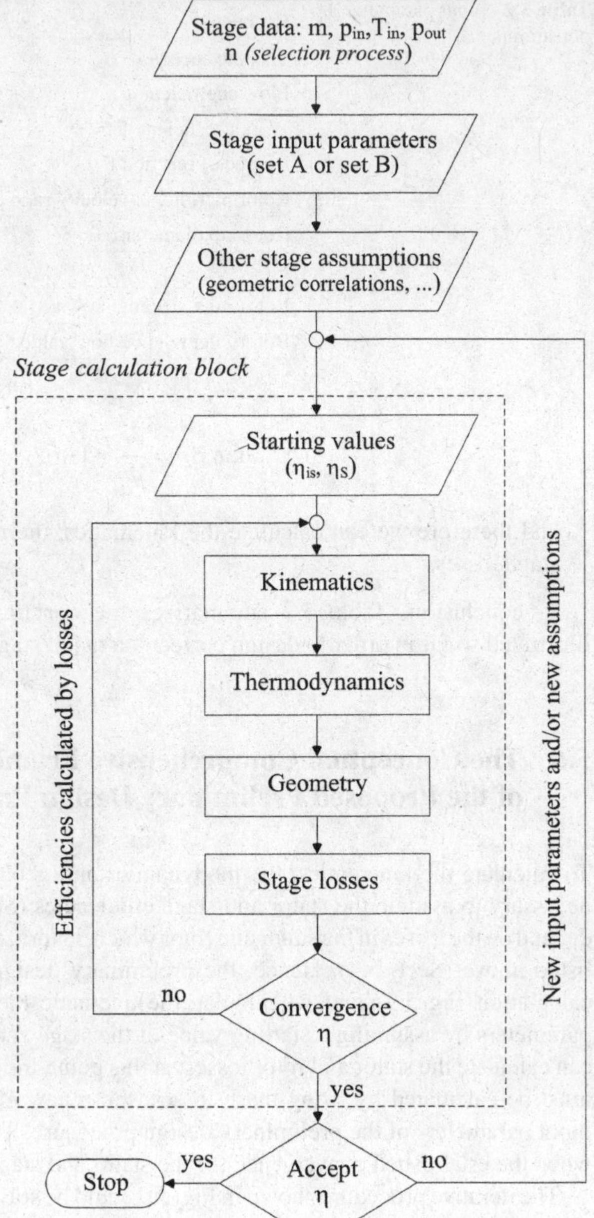

Fig. 5.9 Stage calculation block diagram

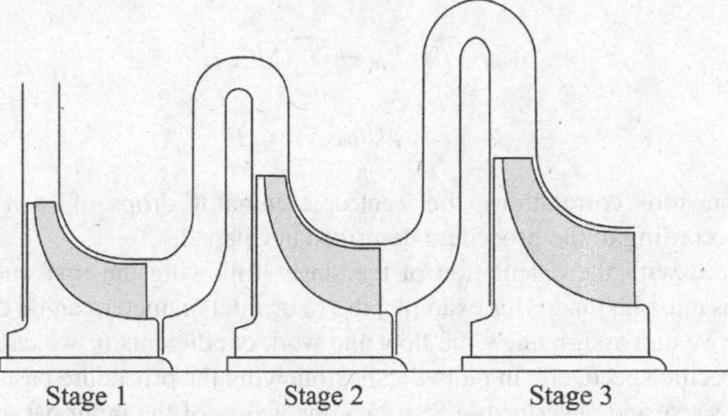

Fig. 5.10 Multistage radial inflow turbine

often follows the last rotor. In these components, the total enthalpy is constant as well as the mass flow rate and, ignoring friction, also the angular momentum is constant. By applying these conservation laws and adopting specific correlations for the geometric design of these components, it is possible to solve completely the turbomachine.

The thermodynamic parameters at the inlet and the pressure at the outlet of stages following the first one (i = 2, z), in absence of heat exchanges (for example reheatings) and neglecting, at least in the preliminary design, the process in the channel linking the rotor outlet station with the stator inlet station of the next stage, are (Fig. 5.11):

$$
\begin{aligned}
p_{in,i} &= p_{out,i-1}\\
T_{in,i} &= T_{out,i-1}\\
h_{in,i} &= h_{out,i-1}\\
s_{in,i} &= s_{out,i-1}\\
\rho_{in,i} &= \rho_{out,i-1}
\end{aligned}
$$

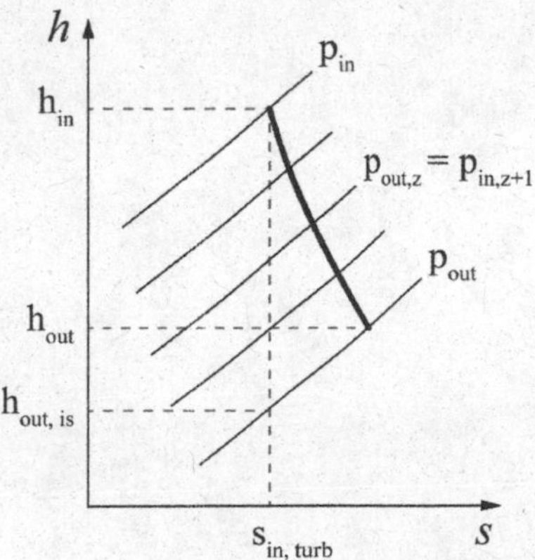

Fig. 5.11 Expansion process in a multistage turbine

$$h_{out,is,i} = h_{in} - \sum_{j=1}^{i} \Delta h_{is,j}$$

$$p_{out,i} = p(h_{out,is,i}, s_{in})$$

In the previous correlations, the isentropic enthalpy drops of each stage are evaluated according to the procedure described in Chap. 2.

To proceed with the calculation of the stages following the first one, specific assumptions must be made: for example, the rotor inlet diameter can be considered constant or we can assign anew the flow and work coefficients or we can calculate anew the specific speed, etc. In more details, following the procedure based on input parameters set A and described in Sect. 5.5, the values of the input parameters (ψ, φ, R, η_{is}, α_2 and δ_h) could be reassigned for the stages following the first one (for example, equal to those of the first stage) while, following the procedure based on input parameters set B, the specific speed must be calculated anew and consequently all the other input parameters that are functions of this specific speed.

Through these input parameters, the kinematics (Sect. 5.1), the thermodynamics (Sect. 5.2) and the stage geometry (Sect. 5.3) are calculated and then, the losses of each stage are assessed (Sect. 5.4).

The turbine design, carried out by assigning a pair of values for the speed of rotation and the number of stages (n, z), provided by the selection process illustrated in Chap. 2, can be accomplished again (Fig. 5.12) by using a different pair of input parameters (n and z), indicated always by the turbomachinery selection process, in order to optimize the preliminary design of the turbine.

Fig. 5.12 Multistage turbine calculation block diagram

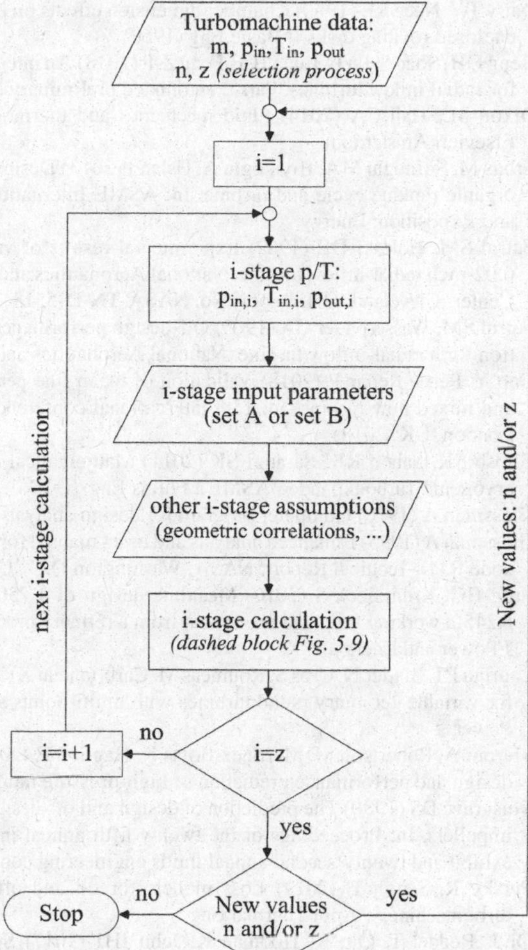

References

Alshammari F, Karvountzis-Kontakiotis A, Pesiridis A, Alatawi I (2020) Design and study of back-swept high pressure ratio radial turbo-expander in automotive organic Rankine cycles. Appl Thermal Eng (2020)

Aungier RH (2006) Turbine aerodynamics. ASME Press, Axial-flow and radial-inflow turbine design and analysis

Baines NC (2003) Radial turbine design. In: Moustapha H (ed) Axial and Radial Turbines. White River Junction, Vt, Concepts NREC

Chen H, Baines NC (1994) The aerodynamic loading of radial and mixed-flow turbines. Int. J. Mech, Sci

Cho SK, Lee J, Lee JI (2018) Comparison of loss models for performance prediction of radial inflow turbine. Int J Fluid Mach Syst (2018)

Daily JW, Nece RE (1960) Chamber dimension effects on induced flow and frictional resistance of enclosed rotating disks. J Basic Eng (1960)
Deng QH, Shao S, Fu L, Luan H-F, Feng Z-P (2018) An integrated design and optimization approach for radial inflow turbines. Part I: automated preliminary design. MDPI Appli Sci (2018)
Dixon SL, Hall CA (2014) Fluid mechanics and thermodynamics of turbomachinery, 7th edn. Elsevier, Amsterdam
Erbas M, Sofuoglu MA, Biyikoglu A, Uslan I (2013) Design and optimization of a low temperature organic rankine cycle and turbine. In: ASME International Mechanical Engineering Congress and Exposition: Energy
Futral SM, Holeski DE (1969) Experimental results of varying the blade-shroud clearance in a 6.02-inch radial-inflow turbine. National Aeronautics and Space Administration, Lewis Research Center; Cleveland, OH. Report No. NASA-TN-D-5513
Futral SM, Wasserbauer CA (1965) Off-design performance prediction with experimental verification for a radial-inflow turbine. National Aeronautics and Space Administration (1965)
Gao Y, Petrie-Repar P (2018) Validation of mean line performance prediction method for radial and mixed flow turbine. In: 13th international conference on turbochargers and turbocharging, London, UK (2018)
Ghosh SK, Sahoo RK, Sarangi SK (2011) Mathematical analysis for off-design performance of cryogenic Turboexpander. ASME J Fluids Eng
Glassman A (1976) Computer program for design analysis of radial-inflow turbines. NASA
Glassman A (1995) Enhanced analysis and users manual for radial-inflow turbine conceptual design code RTD. Techical Report. NASA, Washington D.C., USA
Jung HC, Krumdieck S (2016) Meanline design of a 250 kW radial inflow turbine stage using R245fa working fluid and waste heat from a refinery process. In: Proceedings of IMechE Part A: J Power and Energy
Lauriau PT, Binder N, Cros S, Roumeas M, Carbonneau X (2018) Preliminary design considerations for variable geometry radial turbines with multi-points specifications. Int J Turbomach Propuls Power
Meroni A, Robertson M, Martinez-Botas R, Haglind F. (2018) A methodology for the preliminary design and performance prediction of high-pressure ratio radial-inflow turbines. Energy
Musgrave DS (1980) The prediction of design and off-design efficiency for centrifugal compressor impellers. In: Proceedings of the twenty-fifth annual international gas turbine conference and exhibit and twenty-second annual fluids engineering conference, New Orleans
Persky R, Sauret E (2019) Loss models for on and off-design performance of radial inflow turbomachinery. Appl Thermal Eng
Qi J, Reddel T, Qin K, Hooman K, Jahn IHJ (2017) Supercritical CO_2 radial turbine design performance as a function of turbine size parameters. J Turbomach
Rahbar K, Mahmoud S, Al-Dadah RK (2015) Modelling and optimization of organic Rankine cycle based on a small-scale radial inflow turbine. Energy Convers Manage
Rodgers C (1987) Mainline performance prediction for radial inflow turbines. Technical report on Von Karman Institute for Fluid Dynamics, Lecture Series on Small High Pressure Ratio Turbines
Rodgers C, Geiser R (1987) Performance of a high-efficiency radial/axial turbine. ASME J Turbomach
Rohlik HE (1968) Analytical determination of radial inflow turbine design geometry for maximum efficiency. NASA TN D-4384
Simpson A, Spence S, Watterson J (2008) Numerical and experimental study of the performance effects of varying vaneless space and vane solidity in radial inflow turbine stators. In: Proceedings of ASME turbo expo (2008)
Spence SWT, Artt DW (1998) An experimental assessment of incidence losses in a radial inflow turbine rotor. Proc Instn Mech Engrs
Stanitz JD (1952) Some theoretical aerodynamic investigations of impellers in radial and mixed flow centrifugal compressors. Trans ASME (1952)

Suhrmann JF, Peitsch D, Gugau M, Heuer T, Tomm U (2010) Validation and development of loss models for small size radial turbines. In: Proceedings of ASME turbo expo (2010)

Ventura CAM, Jacobs PA, Rowlands AS, Petrie-Repar P, Sauret E (2012) Preliminary design and performance estimation of radial inflow turbines: an automated approach. ASME J Fluids Eng

Wasserbauer CA, Glassman AJ (1975) Fortran program for predicting off-design performance of radial-inflow turbines. NASA TN D-8063

Whitfield A (1990) The preliminary design of radial inflow turbines. Trans ASME

Wiesner FJ (1967) A review of slip factors for centrifugal compressors. J Eng Power Trans ASME

Chapter 6
Preliminary Design of Centrifugal Compressors

Abstract With reference to a centrifugal compressor stage, a procedure for the calculation of kinematic parameters (Sect. 6.1), thermodynamic parameters (Sect. 6.2), geometric parameters (Sect. 6.3) and stage losses (Sect. 6.4) is provided. Then, Sect. 6.5 discusses the input parameters of this proccdure and suggests their numerical values to be used in the calculations. Finally, Sect. 6.6 illustrates the procedure for extending calculations to multistage compressors. The numerical application of the proposed procedure, aimed at the preliminary design of centrifugal compressors, is instead developed in Chap. 8.

6.1 Kinematic Parameters

In this section, we will deduce the kinematic correlations of a centrifugal stage; they are correlations among kinematic parameters of a stage (degree of reaction R, absolute and relative flow angles, α and β) and dimensionless performance parameters (flow coefficient, φ, and work coefficient, ψ).

Variations of the previously introduced (Chap. 2) dimensionless factors Φ and Ψ are used in practice.

In particular, the ratio of the meridional velocity at the rotor inlet, c_{1m}, to the blade speed at the rotor outlet, u_2, is the *flow coefficient*:

$$\varphi = \frac{c_{1m}}{u_2}$$

Note that there is a proportionality between this flow coefficient and the flow factor, previously introduced (Chap. 2):

$$\Phi = \frac{V}{\omega \cdot D^3} \propto \frac{c_{1m} \cdot D_2^2}{u_2 \cdot D_2^2} = \frac{c_{1m}}{u_2} = \varphi$$

Analogously, the ratio of actual work exchanged between fluid flow and rotor to blade speed squared (at rotor outlet) is the *work coefficient*:

M. Gambini and M. Vellini, *Turbomachinery*, Springer Tracts in Mechanical Engineering, https://doi.org/10.1007/978-3-030-51299-6_6

$$\psi = \frac{W}{u_2^2} = \frac{\Delta h_{t,ad}}{u_2^2}$$

which is proportional to the work factor, Ψ, previously introduced (Chap. 2).

In a radial stage (Fig. 6.1a), the fluid flow is studied (Sect. 1.7) on two stage sections: one perpendicular to the axis of rotation (Fig. 6.1b, right) and the other one containing the axis of rotation (meridional section, Fig. 6.1b left). With reference to these two sections, Fig. 6.1b depicts the velocity triangles of a centrifugal compressor stage.

In the following, the kinematic correlations will be expressed by using $\alpha_1 \neq 0$, even if, in a radial stage, the inlet absolute velocity is generally axial ($\alpha_1 = 0$), unless it is necessary to limit the relative Mach number at the rotor inlet.

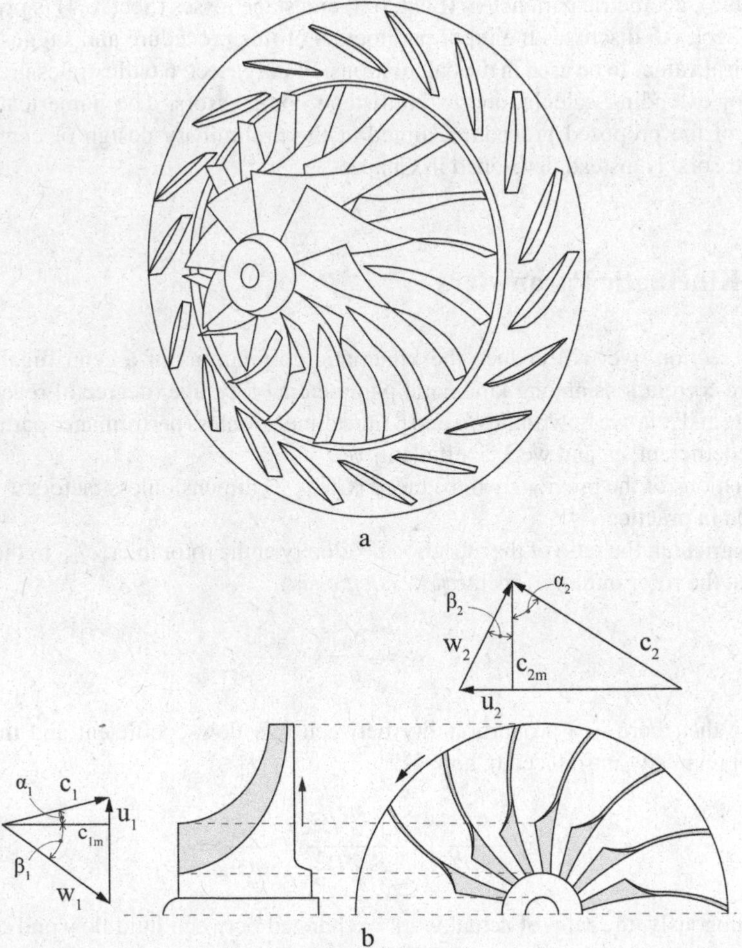

Fig. 6.1 **a** A radial compressor stage, **b** velocity triangles in a radial compressor stage

Two parameters are then introduced: they are the *rotor meridional velocity ratio*, ξ, and the *rotor tip diameter ratio*, δ_t:

$$\xi = \frac{c_{2m}}{c_{1m}}$$

$$\delta_t = \frac{D_{1t}}{D_2}$$

At the stage inlet, in the radial direction, the meridional absolute velocity can be considered constant and the free-vortex flow can be assumed (Sect. 1.7.4):

$$c_{1m(r)} = const$$
$$c_{1u(r)} \cdot r = const$$

and as a consequence:

$$u_{1(r)} \cdot c_{1u(r)} = const$$

Therefore it is possible to express the work transfer by using indifferently the velocity triangles at the tip or the hub at the rotor inlet.

Referring to the tip, the rotor tip diameter ratio, δ_t, can be expressed through the blade speed ratio:

$$\delta_t = \frac{D_{1t}}{D_2} = \frac{u_{1t}}{u_2} = \frac{u_1}{u_2}$$

Next, assuming the angles shown in Fig. 6.1b as positive, the energy equation applied to a compressor stage (Sect. 1.8.3) and Euler's equation (Sect. 1.6.1) provide:

$$W = \Delta h_{t,ad} = h_{3t} - h_{1t} = u_2 \cdot c_{2u} - u_1 \cdot c_{1u}$$

The work coefficient can be therefore expressed as:

$$\psi = \frac{W}{u_2^2} = \frac{c_{2u}}{u_2} - \delta_t \cdot \frac{c_{1u}}{u_2}$$

The angle α_1 is:

$$\tan\alpha_1 = \frac{c_{1u}}{c_{1m}}$$

Since this angle can be assumed as an input parameter (generally, as already mentioned, we can assume $\alpha_1 = 0$), it is convenient to express all the other flow angles as a function of α_1 and of the kinematic parameters introduced above.

Following this approach, the rotor outlet absolute flow angle is expressed as:

$$\tan\alpha_2 = \frac{c_{2u}}{c_{2m}} = \frac{c_{2u}}{u_2}\cdot\frac{u_2}{c_{1m}}\cdot\frac{c_{1m}}{c_{2m}} = \frac{c_{2u}}{u_2}\cdot\frac{1}{\varphi\cdot\xi} = \frac{1}{\varphi\cdot\xi}\cdot(\psi+\varphi\cdot\delta_t\cdot\tan\alpha_1)$$

and thus we obtain:

$$\tan\alpha_2 = \frac{\psi}{\xi\cdot\varphi} + \frac{\delta_t}{\xi}\cdot\tan\alpha_1$$

The rotor relative flow angles are expressed as:

$$\tan\beta_1 = \frac{w_{1u}}{c_{1m}} = \frac{u_1 - c_{1u}}{c_{1m}} = \frac{\delta_t}{\varphi} - \tan\alpha_1$$

$$\tan\beta_2 = \frac{w_{2u}}{c_{2m}} = \frac{u_2 - c_{2u}}{c_{2m}} = \frac{u_2}{c_{1m}}\cdot\frac{c_{1m}}{c_{2m}} - \tan\alpha_2 = \frac{1}{\varphi\cdot\xi} - \tan\alpha_2$$

and thus we obtain:

$$\tan\beta_1 = \frac{\delta_t}{\varphi} - \tan\alpha_1$$

$$\tan\beta_2 = \frac{1}{\varphi\cdot\xi}\cdot(1-\psi) - \frac{\delta_t}{\xi}\cdot\tan\alpha_1$$

The degree of reaction is given by:

$$\begin{aligned} R &= \frac{\Delta h_R}{\Delta h_{t,ad}} = \frac{\Delta h_R}{\psi\cdot u_2^2} = 1 - \frac{\Delta E_{c,R}}{\psi\cdot u_2^2} \\ &= 1 - \frac{c_2^2 - c_1^2}{2\cdot\psi\cdot u_2^2} = 1 - \frac{c_{2u}^2 + c_{2m}^2 - c_{1u}^2 - c_{1m}^2}{2\cdot\psi\cdot u_2^2} \\ &= 1 - \frac{c_{2m}^2\cdot\left(1+\tan^2\alpha_2\right) - c_{1m}^2\cdot\left(1+\tan^2\alpha_1\right)}{2\cdot\psi\cdot u_2^2} \end{aligned}$$

and thus it is:

$$R = 1 - \frac{\psi}{2} + \frac{\varphi^2}{2\cdot\psi}\cdot\left[\left(1-\xi^2\right) + \tan^2\alpha_1\cdot\left(1-\delta_t^2\right)\right] - \varphi\cdot\delta_t\cdot\tan\alpha_1$$

From this equation, we infer that, unlike what happens in axial stages, in radial stages angle α_1 can be assumed as input parameter (for example, $\alpha_1 = 0$) keeping R, φ and ψ (or ξ, φ and ψ) as independent parameters.

The previous equation also allows to express ξ as a function of R:

$$\xi = \sqrt{1 - \frac{2\cdot\psi}{\varphi^2}\cdot\left(R + \frac{\psi}{2} - 1\right) + \tan^2\alpha_1\cdot\left(1-\delta_t^2\right) - \frac{2\cdot\psi\cdot\delta_t}{\varphi}\cdot\tan\alpha_1}$$

Consequently there is a constraint on the maximum value of the degree of reaction. For example, if $\alpha_1 = 0$, it must be:

$$R < 1 + \frac{\varphi^2}{2 \cdot \psi} - \frac{\psi}{2}$$

Remarks on input parameters

From the kinematic correlations reported above, we infer that once the *five independent parameters* φ, ψ, R (or ξ), δ_t and α_1 are set (Sect. 6.5), all flow angles, α_2, β_1, β_2, can be calculated. Once the blade speed, u_2, is also established [in order to determine this speed, it is necessary to know, in addition to the work coefficient ψ, also the stage isentropic efficiency (or the polytropic efficiency, as illustrated in Sect. 6.5)], the kinematics of the rotor can be completely calculated:

$$\begin{aligned}
c_{1m} &= u_2 \cdot \varphi \\
c_{1u} &= u_2 \cdot \varphi \cdot \tan\alpha_1 \\
c_{2m} &= u_2 \cdot \xi \cdot \varphi \\
c_{2u} &= u_2 \cdot \xi \cdot \varphi \cdot \tan\alpha_2 \\
w_{1u} &= u_2 \cdot \varphi \cdot \tan\beta_1 \\
w_{2u} &= u_2 \cdot \xi \cdot \varphi \cdot \tan\beta_2 \\
c_1 &= u_2 \cdot \varphi \cdot \sqrt{1 + \tan^2\alpha_1} \\
c_2 &= u_2 \cdot \xi \cdot \varphi \cdot \sqrt{1 + \tan^2\alpha_2} \\
w_1 &= u_2 \cdot \varphi \cdot \sqrt{1 + \tan^2\beta_1} \\
w_2 &= u_2 \cdot \xi \cdot \varphi \cdot \sqrt{1 + \tan^2\beta_2} \\
u_1 &= u_2 \cdot \delta_t
\end{aligned}$$

In order to complete the kinematics of the stage, the velocity c_3, the absolute velocity at the stator outlet, must also be determined.

Generally we assume:

$$c_3 = c_1$$

If this assumption does not guarantee the respect of limitations and constraints introduced in Sect. 6.3, c_3 must be recalculated. In this case, the blade speed, u_2, can still be calculated as a function of the work coefficient, but now:

$$\Delta h_{t,ad} \neq \Delta h_{ad}$$

so we must remember that:

$$\Delta h_{t,ad} = \Delta h_{ad} + \frac{c_3^2 - c_1^2}{2}$$

6.2 Thermodynamic Parameters

The pressure p_1 and temperature T_1 at the rotor inlet and the pressure p_3 at the stator outlet are assigned based on what is illustrated in Chap.2 (for multistage compressors, these values are calculated by the procedure described in Sect. 6.6).

Now, all the thermodynamic parameters at the rotor inlet and at the stator inlet/outlet can be calculated as follows.

If the working fluid cannot be assimilated to a perfect gas, for the following calculations, we need appropriate state functions, such as those of the NIST libraries. In the following, we will refer to a real working fluid.

Knowing p_1, T_1, we can calculate:

$$h_1 = h_{p,T}(p_1, T_1)$$
$$s_1 = s_{p,T}(p_1, T_1)$$
$$\rho_1 = \rho_{p,T}(p_1, T_1)$$

and (Sect. 1.8.1) the following total quantities (Fig. 6.2):

$$h_{1t} = h_1 + \frac{c_1^2}{2}$$
$$p_{1t} = p_{h,s}(h_{1t}, s_1)$$
$$T_{1t} = T_{h,s}(h_{1t}, s_1)$$
$$\rho_{1t} = \rho_{h,s}(h_{1t}, s_1)$$

and the total relative quantities (Fig. 6.2):

$$h_{1tr} = h_1 + \frac{w_1^2}{2}$$
$$p_{1tr} = p_{h,s}(h_{1tr}, s_1)$$
$$T_{1tr} = T_{h,s}(h_{1tr}, s_1)$$

The rothalpy at the rotor inlet is:

$$I_1 = h_1 + \frac{w_1^2}{2} - \frac{u_1^2}{2} = h_{1tr} - \frac{u_1^2}{2}$$

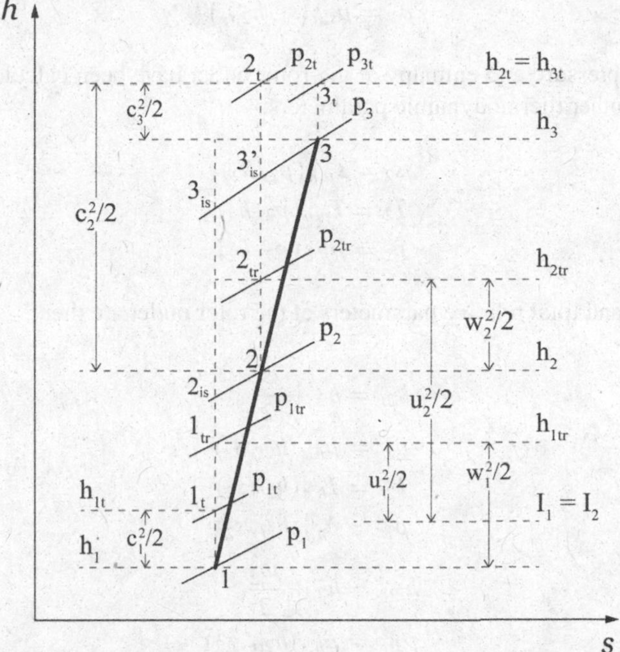

Fig. 6.2 Compression process in a radial compressor stage

To calculate the thermodynamic parameters at the rotor outlet, we must know the rotor efficiency and remember the rothalpy conservation in the rotor (Sect. 1.8.3):

$$I_2 = I_1$$

In fact, the static enthalpy is calculated through the previous equation:

$$h_2 = I_2 - \frac{w_2^2}{2} + \frac{u_2^2}{2}$$

while using the expression of the rotor efficiency, defined as:

$$\bar{\eta}_R = \frac{h_{2,is} - h_1}{h_2 - h_1}$$

isentropic static enthalpy is obtained:

$$h_{2,is} = h_1 + \bar{\eta}_R \cdot \Delta h_R$$

and, therefore, the rotor outlet pressure:

$$p_2 = p_{s,h}(s_1, h_{2,is})$$

Once the pressure and enthalpy at the rotor outlet have been calculated, we can evaluate the other thermodynamic parameters:

$$\begin{aligned} s_2 &= s_{p,h}(p_2, h_2) \\ T_2 &= T_{p,h}(p_2, h_2) \\ \rho_2 &= \rho_{p,h}(p_2, h_2) \end{aligned}$$

The total and total relative parameters at the rotor outlet are then:

$$\begin{aligned} h_{2t} &= h_2 + \frac{c_2^2}{2} \\ p_{2t} &= p_{h,s}(h_{2t}, s_2) \\ T_{2t} &= T_{h,s}(h_{2t}, s_2) \\ \rho_{2t} &= \rho_{h,s}(h_{2t}, s_2) \\ h_{2tr} &= h_2 + \frac{w_2^2}{2} \\ p_{2tr} &= p_{h,s}(h_{2tr}, s_2) \\ T_{2tr} &= T_{h,s}(h_{2tr}, s_2) \end{aligned}$$

Losses in the rotor are related to the entropy increase ($s_2 - s_1$) due to irreversibilities and entail an increase in the final enthalpy ($h_2 - h_{2,is}$) and a reduction in the total pressure expressed by the loss coefficient (Sect. 6.4):

$$Y_R = \frac{p_{2tr,is} - p_{2tr}}{\frac{p_{2tr}}{p_{1tr}} \cdot (p_{1tr} - p_1)}$$

The rothalpy conservation in the rotor allows to calculate the total relative isentropic pressure, necessary to calculate the pressure losses in the stage, according to the previous expression Y_R.

By applying the rothalpy conservation along an ideal (isentropic) and a real (adiabatic) process, we have:

$$dh + wdw = udu$$

$$\left(h_{2,is} + \frac{w_{2,is}^2}{2}\right) - \left(h_1 + \frac{w_1^2}{2}\right) = \frac{u_2^2 - u_1^2}{2}$$

$$\left(h_2 + \frac{w_2^2}{2}\right) - \left(h_1 + \frac{w_1^2}{2}\right) = \frac{u_2^2 - u_1^2}{2}$$

and as a consequence:

$$h_{2tr,is} = h_{2tr}$$

Now it is possible to calculate the total relative isentropic pressure and temperature as:

$$p_{2tr,is} = p_{h,s}\left(h_{2tr,is}, s_1\right)$$
$$T_{2tr,is} = T_{h,s}\left(h_{2tr,is}, s_1\right)$$

At the stator outlet, the pressure is known; therefore the enthalpy, referred to the isentropic compression, can be calculated as:

$$h_{3,is} = h_{p,s}(p_3, s_1)$$

The final enthalpy in the adiabatic compression can be determined by applying the total enthalpy conservation in the stator (Sect. 1.8.3) and knowing c_3 (Sect. 6.3):

$$h_{3t} = h_{2t}$$
$$h_3 = h_{3t} - \frac{c_3^2}{2}$$

or by applying the stage isentropic efficiency (referred to static enthalpy increases, Sect. 1.8.3):

$$\eta_{is} = \frac{h_{3,is} - h_1}{h_3 - h_1}$$

$$h_3 = h_1 + \frac{\left(h_{3,is} - h_1\right)}{\eta_{is}}$$

The calculation of the thermodynamic parameters can, therefore, be completed:

$$s_3 = s_{p,h}(p_3, h_3)$$
$$T_3 = T_{p,h}(p_3, h_3)$$
$$\rho_3 = \rho_{p,h}(p_3, h_3)$$
$$p_{3t} = p_{h,s}(h_{3t}, s_3)$$
$$T_{3t} = T_{h,s}(h_{3t}, s_3)$$
$$\rho_{3t} = \rho_{h,s}(h_{3t}, s_3)$$

Losses in the stator are related to the entropy increase $(s_3 - s_2)$ due to irreversibilities and entail an increase in the final enthalpy $(h_3 - h_{3',is})$ and a reduction in the total relative pressure expressed by the loss coefficient (Sect. 1.9.5):

$$Y_S = \frac{p_{2t} - p_{3t}}{p_{2t} - p_2}$$

From the knowledge of the kinematics and thermodynamics of the stage, the absolute (stator inlet/outlet) and relative (rotor inlet/outlet) Mach numbers can be calculated. From the properties of the working fluid, first, the speeds of sound are calculated:

$$c_{s,1} = c_{s,p,T}(p_1, T_1)$$
$$c_{s2} = c_{s,p,T}(p_2, T_2)$$
$$c_{s3} = c_{s,p,T}(p_3, T_3)$$

and, therefore, the Mach numbers:

$$Ma_1 = \frac{w_1}{c_{s1}}$$
$$Ma_{2,R} = \frac{w_2}{c_{s2}}$$
$$Ma_{2,S} = \frac{c_2}{c_{s2}}$$
$$Ma_3 = \frac{c_3}{c_{s3}}$$

By using the pressure and temperature conditions of the working fluid, we can calculate the fluid viscosities (necessary to assess the Reynolds numbers, Sect. 6.3):

$$\mu_1 = \mu_{p,T}(p_1, T_1)$$
$$\mu_2 = \mu_{p,T}(p_2, T_2)$$
$$\mu_3 = \mu_{p,T}(p_3, T_3)$$

Remarks on input parameters

Based on what is illustrated in this Sect. 6.2, in addition to the parameters already identified in Sect. 6.1 (φ, ψ, R, δ_t, α_1 and η_{is}), the rotor efficiency must be assumed in order to calculate the thermodynamics of the stage.

6.3 Geometric Parameters

Now we will deduce the geometric correlations of a radial stage; they are correlations among geometric (rotor and stator diameters, blade height, chord and blade pitch, etc.), thermodynamic (density, volumetric flow rate) and kinematic parameters

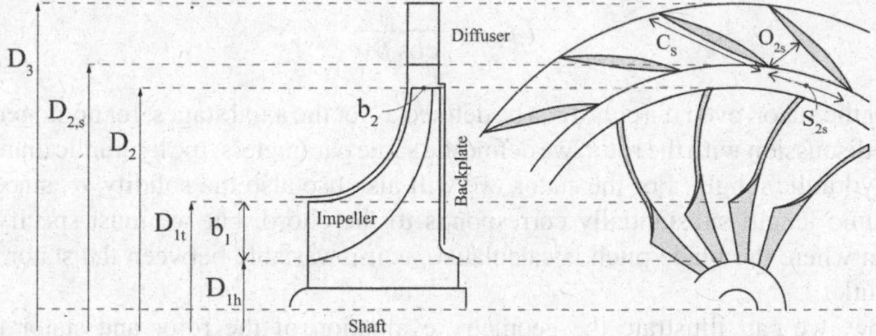

Fig. 6.3 Stage geometry

(meridional velocities, absolute velocity angles α, relative velocity angles β) of the stage.

The rotor outlet diameter (D_2) is determined by the kinematics of the stage (Sect. 6.1), and in particular by the knowledge of the blade speed at the rotor outlet (u_2) and the speed of rotation (n, rpm) (Sect. 6.5):

$$D_2 = \frac{60}{\pi} \cdot \frac{u_2}{n}$$

Before evaluating the rotor and stator geometry, we must highlight the differences between the geometric parameters in the radial and axial stages. In particular, in the axial stages, after determining the blade height through the equation of continuity, the geometry of the rotor and stator channels (Chaps. 3–4) was calculated by using two parameters: the aspect ratio and the solidity. In the radial stages, on the other hand, it is not simple the definition of a rotor chord; moreover, the blade heights and the rotor diameters change, sometimes even a lot, between the rotor inlet and outlet. For these reasons, in order to evaluate the stage losses, an equivalent rotor channel, having a circular section of diameter D_{hyd} (hydraulic diameter) and length L_{hyd} (hydraulic length), is used (Sect. 6.4)

The hydraulic diameter is defined as:

$$D_{hyd} = 4 \cdot \frac{Cross\ sectional\ area}{Wetted\ Perimeter}$$

To define the hydraulic length, first, we must introduce the channel length on the meridional plane, that is, the plane containing the axis of rotation:

$$L_m = Meridional\ length$$

The hydraulic length is correlated with the meridional length through the angle β_M (for the rotor or α_M for the stator) representing the mean inclination of the rotor/stator channel on a plane perpendicular to the axis of rotation; these mean angles are similar to the stagger angle in the axial stages:

$$L_{hyd} = \frac{L_m}{\cos \beta_M}$$

For the stator, even if a chord can be defined as for the axial stages, for homogeneity of the discussion with the rotor, we define the same parameters: the hydraulic diameter and hydraulic length. For the stator, we will also use also the solidity, σ, since the hydraulic length substantially corresponds to the chord, but we must specify the station where the blade pitch is calculated, as it is variable between the stator inlet and outlet.

Now, we can illustrate the geometry evaluation of the rotor and stator for a centrifugal compressor stage.

Rotor geometry

The volumetric flow rates in the rotor (1—rotor inlet, 2—rotor outlet) are functions of the assigned mass flow rate and fluid densities calculated in Sect. 6.2:

$$V_i = \frac{m}{\rho_i} \quad i = 1, 2$$

By using these volumetric flow rates, knowing the meridional velocities c_{1m} and c_{2m} (Sect. 6.1), we can calculate the flow areas:

$$A_i = \frac{V_i}{c_{im}}$$

and therefore, considering a section just downstream of the rotor outlet (where no blade blockage occurs, Sect. 1.8), the blade height at the rotor outlet (Fig. 6.3):

$$b_2 = \frac{A_2}{\pi \cdot D_2}$$

The flow area at the rotor inlet can be correlated to the rotor diameter ratios; in fact we have:

$$A_1 = \frac{\pi}{4} \cdot \left(D_{1t}^2 - D_{1h}^2\right) = \frac{\pi}{4} \cdot D_2^2 \cdot \left(\delta_t^2 - \delta_h^2\right)$$

and, as a consequence, we can obtain the rotor tip diameter ratio:

$$\delta_t = \sqrt{\delta_h^2 + \frac{4 \cdot A_1}{\pi \cdot D_2^2}}$$

since the rotor hub diameter ratio is defined as:

$$\delta_h = \frac{D_{1h}}{D_2}$$

After choosing δ_h, for example Hazby et al. (2017) suggested:

$$\delta_h = 0.35$$

or δ_t (Sect. 6.5), the rotor inlet geometry is determined:

$$D_{1h} = \delta_h \cdot D_2$$

$$D_{1t} = \delta_t \cdot D_2$$

$$b_1 = \frac{D_{1t} - D_{1h}}{2} = \frac{D_2}{2} \cdot (\delta_t - \delta_h)$$

By using these parameters, we can calculate the inlet mean flow angle, that is, the flow angle at the inlet mean diameter:

$$D_{1M} = \frac{D_{1h} + D_{1t}}{2}$$

In fact, remembering that the kinematics at the rotor inlet (Sect. 6.1) is evaluated at the tip diameter, the mean relative flow angle is:

$$\tan \beta_{1M} = \frac{w_{1u,M}}{c_{1m}} = \frac{u_{1M} - c_{1u,M}}{c_{1m}} = \frac{u_{1t}}{u_2} \cdot \frac{D_{1M}}{D_{1t}} \cdot \frac{u_2}{c_{1m}} - \frac{c_{1u,M}}{c_{1m}}$$

$$= \frac{\delta_t}{\varphi} \cdot \frac{D_{1M}}{D_{1t}} - \tan \alpha_{1M}$$

where, the mean absolute flow angle is:

$$\tan \alpha_{1M} = \frac{c_{1u,M}}{c_{1m}} = \frac{c_{1u,t}}{c_{1m}} \cdot \frac{c_{1u,M}}{c_{1u,t}} = \tan \alpha_1 \cdot \frac{D_{1t}}{D_{1M}}$$

and the ratio between the mean and tip diameters is:

$$\frac{D_{1M}}{D_{1t}} = \frac{1}{2} \cdot \left(1 + \frac{\delta_h}{\delta_t}\right)$$

As a consequence, we can write:

$$\tan \beta_{1M} = \frac{1}{2 \cdot \varphi} \cdot (\delta_t + \delta_h) - \frac{2 \cdot \delta_t \cdot \tan \alpha_1}{\delta_t + \delta_h}$$

Therefore, the mean relative velocity is:

$$w_{1M} = \frac{c_{1m}}{\cos \beta_{1M}}$$

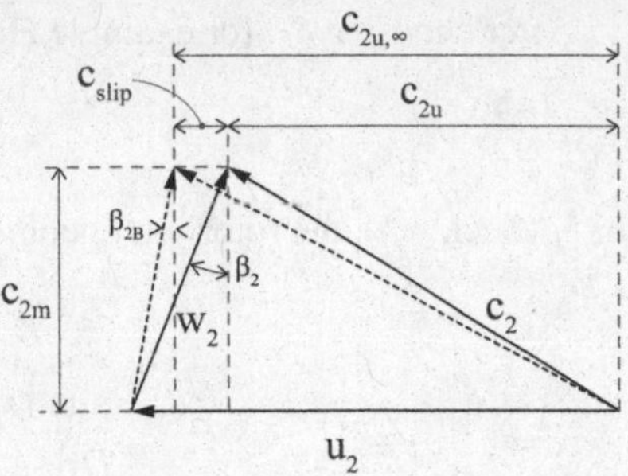

Fig. 6.4 Effect of slip on velocity triangles at the rotor outlet

To complete the rotor geometry, we must consider that the outlet flow angle (β_2) is different from the blade angle (β_{2B}) because the fluid flow will receive less than perfect guidance from the vanes, due to the finite number of blades in the rotor (Fig. 6.4). In order to take into account this phenomenon, also called *slip*, a *slip factor* may be defined as:

$$SF = \frac{c_{2u}}{c_{2u,\infty}} = \frac{c_{2u,\infty} - c_{slip}}{c_{2u,\infty}} = 1 - \frac{c_{slip}}{c_{2u,\infty}}$$

where $c_{2u,\infty}$ is the absolute tangential velocity with an infinite number of infinitely thin blades (the fluid flow is perfectly guided by the blade at the rotor outlet, and hence the flow angle and the blade angle are equal $\beta_{2,\infty} = \beta_{2B}$).

The slip factor can be used to express the blade outlet angle such as to ensure the effective outlet flow angle:

$$\tan\beta_{2B} = \frac{w_{2u,\infty}}{c_{2m}} = \frac{u_2 - c_{2u,\infty}}{c_{2m}} = \frac{u_2}{c_{2m}} - \frac{c_{2u}}{SF \cdot c_{2m}}$$

and, by the kinematic correlations deduced in Sect. 6.1, we obtain:

$$\tan\beta_{2B} = \frac{1}{\xi \cdot \varphi} - \frac{\tan\alpha_2}{SF}$$

In order to calculate the slip factor, and hence the blade outlet angle, we can use the correlation proposed by Wiesner (1967):

$$SF = 1 - \frac{\sqrt{\cos\beta_{2B}}}{N_{B,R}^{0.7}}$$

with the correction suggested by Aungier (2000) when the rotor mean diameter ratio, defined as:

$$\delta_M = \frac{D_{1M}}{D_2} = \frac{\delta_t + \delta_h}{2}$$

exceeds the following limit value:

$$\delta_{M,\lim} = \frac{SF - SF^*}{1 - SF^*}$$

where:

$$SF^* = sin\left[19^\circ + 0.2 \cdot (90^\circ - \beta_2)\right]$$

Hence, when $\delta_M > \delta_{M,\lim}$, Aungier (2000) suggested the following expression of the slip factor, correcting the previous correlation proposed by Wiesner:

$$SF_{cor} = SF \cdot \left[1 - \left(\frac{\delta_M - \delta_{M,\lim}}{1 - \delta_{M,\lim}}\right)^{\sqrt{(90-\beta_2)/10}}\right]$$

To calculate the number of blades, we can use the equation proposed by Osnaghi (2002), according to Eckert and Schnell (1961) and Rodgers (2000):

$$N_{B,R} = \mathrm{int}\left(\frac{2 \cdot \pi \cdot \cos\beta_M}{\zeta \cdot \log(1/\delta_t)}\right)$$

where:

$$\beta_M = \frac{\beta_{1M} + \beta_{2B}}{2}$$
$$\zeta = 0.4$$

More generally, Eckert and Schnell (1961) suggested the optimum values of ζ in the range 0.35–0.45.

Alternatively, the number of blades may be evaluated as a function of the stage pressure ratio by using empirical correlations suggested by Xu (2007) and Xu and Amano (2012).

The previous three equations allow to calculate, iteratively, the three parameters β_{2B}, SF and $N_{B,R}$.

Regarding the number of blades, it is necessary to consider the possibility of adopting *splitter blades*, which are shorter blades, about 50% compared to the full blades, placed at the center line of the blade pitch in the outlet station. These splitter blades, thanks to the reduced length, affect only the radial part of the impeller (Fig. 6.1a). This means that at the inlet station, generally axial, the number of blades is half than that at the outlet station. The splitter blades are generally adopted in highly-pressure ratio stages (greater than 2, Came and Robinson 1998) with high values of the angle β_1. In this case, adopting splitter blades limits the effect of the blade blockage at the inlet and hence the number of Mach; these blades ensure, however, the fluid flow guidance at the rotor outlet, limiting the slip.

Adopting splitter blades, the equivalent number of blades must be used:

$$N_{B,R} = N_{B,FB} + N_{B,SB} \cdot \frac{L_{SB}}{L_{FB}}$$

where the subscript FB indicates the full blades and SB the splitter blades. As already mentioned, the ratio between the length of the splitter blades and the full blades (L_{SB}/L_{FB}) is generally around 0.5.

The blade thickness can be assumed constant along the blade height and equal to:

$$\frac{t_B}{D_2} \cong 0.003 \quad \text{for unshrouded impellers}$$
$$\frac{t_B}{D_2} \cong 0.01 \quad \text{for shrouded impellers}$$

Regarding clearances (necessary for calculating the clearance losses, Sect. 6.5), we can assume that:

$$\frac{\tau}{b_2} = 2.0 \div 5.0\%$$

but the minimum values must be not lower than $\tau = 0.3 \div 0.5$ mm.

In absence of specific data, we can choose:

$$\frac{\tau_a}{b_2} = \frac{\tau_r}{b_2} = \frac{\tau_b}{b_2} = 0.05$$

where (Fig. 6.5):

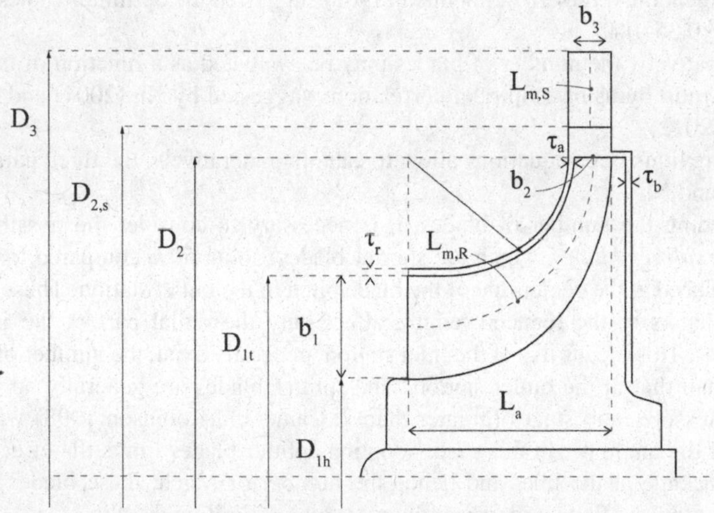

Fig. 6.5 Stage geometry in the meridional section

- τ_a is the axial clearance (blade at rotor outlet)
- τ_r is the radial clearance (blade at rotor inlet)
- τ_b is the backface clearance (on the back of the rotor disc)

The blade pitch at the rotor inlet (considering the mean diameter) is:

$$S_{1,R} = \frac{\pi \cdot D_{1M}}{N_{B,R}} = \frac{\pi \cdot (D_{1t} + D_{1h})}{2 \cdot N_{B,R}}$$

and at the rotor outlet is:

$$S_{2,R} = \frac{\pi \cdot D_2}{N_{B,R}}$$

Now, the rotor geometry must be completed calculating the hydraulic diameter, D_{hyd}, and the hydraulic length, L_{hyd}.

As previously mentioned, the hydraulic diameter is defined as:

$$D_{hyd} = 4 \cdot \frac{Cross\ sectional\ area}{Wetted\ Perimeter}$$

To evaluate the wetted perimeter, and the cross sectional area delimited by this perimeter, it is necessary to calculate the blade distance at the leading edge, O_1, and at the trailing edge, O_2, following what was done for the axial stages.

By assuming as right triangles the triangles having a hypotenuse equal to S, a base equal to O and an angle between them equal to β, we obtain:

$$O_1 = S_{1,R} \cdot \cos \beta_{1M}$$
$$O_2 = S_{2,R} \cdot \cos \beta_{2B}$$

Therefore, the cross sectional areas and the wetted perimeters arc:

$$A_{w1} = O_1 \cdot b_1$$
$$P_{w1} = 2 \cdot (O_1 + b_1)$$
$$A_{w2} = O_2 \cdot b_2$$
$$P_{w2} = 2 \cdot (O_2 + b_2)$$

and the hydraulic diameters at the rotor inlet and outlet become:

$$D_{hyd,1} = 2 \cdot \frac{O_1 \cdot b_1}{O_1 + b_1}$$

$$D_{hyd,2} = 2 \cdot \frac{O_2 \cdot b_2}{O_2 + b_2}$$

The rotor hydraulic diameter is the mean hydraulic diameter between the rotor inlet and outlet:

$$D_{hyd,R} = \frac{D_{hyd,1} + D_{hyd,2}}{2} = \frac{O_1 \cdot b_1}{O_1 + b_1} + \frac{O_2 \cdot b_2}{O_2 + b_2}$$

To calculate the hydraulic length L_{hyd}, as previously illustrated, first, we must assess the channel length, L_m, on the meridional plane. Let us consider the following semi-axes (Fig. 6.5):

$$a = L_a - \frac{b_2}{2}$$

$$b = \frac{D_2 - D_{1M}}{2} = \frac{D_2 - D_{1t} + b_1}{2}$$

The meridional length can be considered equivalent to a quarter circle having an equivalent radius equal to:

$$r_{eq} = \frac{a + b}{2} = \frac{L_a - \frac{b_1}{2} + \frac{D_1 - D_{2t} + b_2}{2}}{2}$$

and, as a consequence, we obtain:

$$L_{m,R} \cong \frac{\pi}{2} \cdot r_{eq} = \frac{\pi}{8} \cdot (2 \cdot L_a - b_2 + D_2 - D_{1t} + b_1)$$

In the previous equations, L_a is the rotor axial length. This parameter, influencing the meridional length, assumes considerable importance in the loss calculations (Sect. 6.4). According to Aungier (2000), the rotor axial length can be evaluated as:

$$L_a = D_2 \cdot \left[0.014 + \frac{0.023}{\delta_h} + 1.58 \cdot \left(\delta_t^2 - \delta_h^2\right) \cdot \varphi\right]$$

Alternatively, on the meridional plane, the shroud can be considered circular and so we obtain (neglecting clearances):

$$L_a = \frac{D_2 - D_{1t}}{2} + b_2$$

After assessing the meridional length, the hydraulic length is:

$$L_{hyd,R} = \frac{L_{m,R}}{\cos \beta_M} = \frac{L_{m,R}}{\cos\left(\frac{\beta_{1M} + \beta_{2B}}{2}\right)}$$

Stator geometry

At the rotor outlet, the fluid enters the diffuser (stator). Generally, the diffuser is composed of a vaneless diffuser followed by a vaned diffuser. By indicating with 2 s and 3 respectively the vaned diffuser inlet and outlet (Fig. 6.5), the vaneless diffuser extends from the rotor outlet (station 2) to the vaned diffuser inlet (2 s).

Aungier (2000) and Kim et al. (2002) suggested the following correlations for the vaneless diffuser:

$$D_{2S} = D_2 \cdot \left(1 + \frac{90^\circ - \alpha_{2S}}{360} + \frac{Ma_2^2}{15}\right)$$

where:

$$Ma_2 = \frac{c_2}{c_{s2}}$$

$$\alpha_{2S} = 72^\circ + \frac{\alpha_2 - 72^\circ}{4} \qquad \text{if} \quad \alpha_2 \geq 72^\circ$$

$$\alpha_{2S} = 72^\circ \qquad \text{if} \quad \alpha_2 < 72^\circ$$

Ignoring friction, the angular momentum conservation (Sect. 1.7.4) in the vaneless diffuser allows calculating the absolute tangential velocity at the vaned diffuser inlet:

$$c_{2Su} = c_{2u} \cdot \frac{D_2}{D_{2S}}$$

and, therefore, the meridional velocity:

$$c_{2Sm} = \frac{c_{2Su}}{\tan \alpha_{2S}}$$

The inlet width of the vaned diffuser is then calculated from the meridional velocity:

$$b_{2S} = \frac{m}{\rho_{2S} \cdot c_{2Sm} \cdot \pi \cdot D_{2S}}$$

If:

$$b_{2S} \leq b_2$$

the value so calculated is accepted; otherwise we set:

$$b_{2S} = b_2$$

and the angle $\alpha_{2,S}$ is recalculated accordingly.

The vaneless diffuser geometry is then completed with the calculation of the hydraulic length ($L_{hyd,vaneless}$), which corresponds to the diffuser radial length:

$$L_{hyd,vaneless} = L_{m,vaneless} = \frac{D_{2S} - D_2}{2}$$

and with the hydraulic diameter ($D_{hyd,vaneless}$):

$$\begin{aligned} D_{hyd,vaneless} &= \frac{D_{hyd,2} + D_{hyd,2S}}{2} \\ &= \frac{1}{2} \cdot \left(\frac{4 \cdot \pi \cdot D_2 \cdot b_2}{2 \cdot \pi \cdot D_2} + \frac{4 \cdot \pi \cdot D_{2S} \cdot b_{2S}}{2 \cdot \pi \cdot D_{2S}} \right) = b_2 + b_{2S} \end{aligned}$$

Still following Aungier's suggestions (2000), the geometry of the vaned diffuser can be defined.

The outlet diameter (Fig. 6.5) can be calculated by:

$$D_3 = D_2 \cdot \left[1.55 + \left(\delta_t^2 - \delta_h^2\right) \cdot \varphi\right]$$

Generally, the vaned diffuser width is assumed constant:

$$b_3 = b_{2S}$$

Consequently, we can calculate:

$$c_{3m} = \frac{V_3}{\pi \cdot D_3 \cdot b_3}$$

If this velocity is lower than the previously assumed c_3 (Sect. 6.1), we can assess:

$$\cos \alpha_3 = \frac{c_{3m}}{c_3}$$

otherwise, c_3 is recalculated as:

$$c_3 = c_{3m} \quad ; \quad \alpha_3 = 0$$

and, consequently, the stage thermodynamics must be calculated anew accordingly to this new velocity (Sect. 6.2).

In any case, the calculation of the stator geometry and kinematics does not end here, since some limits and constraints, illustrated later, must be satisfied.

The number of blades in the vaned diffuser, subjected to checks illustrated below, can be set by:

$$N_{B,S} = N_{B,R} - 1 \quad \text{if} \quad 10 < N_{B,R} < 20$$

otherwise:

$$\left|N_{B,S} - N_{B,R}\right| \geq 8$$

The vaned diffuser geometry is completed by calculating the parameters illustrated below.

Since the stator extends in a completely radial direction, the meridional length, L_m, is:

$$L_{m,vaned} = \frac{D_3 - D_{2S}}{2}$$

and, therefore, the hydraulic length (which in this case corresponds to the chord) is:

$$L_{hyd,vaned} = C_S = \frac{L_{m,vaned}}{\cos \alpha_M} = \frac{D_3 - D_{2S}}{2 \cdot \cos\left(\frac{\alpha_{2S}+\alpha_3}{2}\right)}$$

The blade pitch at the stator inlet is:

$$S_{2S} = \frac{\pi \cdot D_{2S}}{N_{B,S}}$$

and at the stator outlet is:

$$S_3 = \frac{\pi \cdot D_3}{N_{B,S}}$$

The solidity is:

$$\sigma_S = \left(\frac{C}{S}\right)_S = \frac{D_3 - D_{2S}}{2 \cdot S_3 \cdot \cos\left(\frac{\alpha_3+\alpha_{2S}}{2}\right)}$$

To evaluate the hydraulic diameter, it is necessary to calculate the blade distance at the leading edge, O_{2S}, and at the trailing edge, O_3. By assuming as right triangles the triangles having a hypotenuse equal to S, a base equal to O and an angle between them equal to α, we obtain:

$$O_{2S} = S_{2S} \cdot \cos \alpha_{2S}$$
$$O_3 = S_3 \cdot \cos \alpha_3$$

Therefore, the cross sectional areas and the wetted perimeters are:

$$A_{w2S} = O_{2S} \cdot b_{2S}$$
$$P_{w2S} = 2 \cdot (O_{2S} + b_{2S})$$
$$A_{w3} = O_3 \cdot b_3$$

$$P_{w3} = 2 \cdot (O_3 + b_3)$$

and the hydraulic diameters at the stator inlet and outlet become:

$$D_{hyd,2S} = 2 \cdot \frac{O_{2S} \cdot b_{2s}}{O_{2S} + b_{2S}}$$

$$D_{hyd,3} = 2 \cdot \frac{O_3 \cdot b_3}{O_3 + b_3}$$

The stator hydraulic diameter is the mean hydraulic diameter between the stator inlet and outlet:

$$D_{hyd,vaned} = \frac{D_{hyd,2S} + D_{hyd,3}}{2} = \frac{O_{2S} \cdot b_{2S}}{O_{2S} + b_{2S}} + \frac{O_3 \cdot b_3}{O_3 + b_3}$$

To verify the applicability of the kinematics and geometry established so far, Aungier (2000) and Kim et al. (2002) introduced three parameters:

- the divergence angle, θ_C:

$$\tan \theta_C = \frac{\pi \cdot (D_3 \cdot \cos \alpha_3 - D_{2S} \cdot \cos \alpha_{2S})}{2 \cdot N_{B,S} \cdot L_{hyd,vaned}}$$

- the blade loading parameter, BL:

$$BL = \frac{\pi \cdot (D_{2S} \cdot c_{2Su} - D_3 \cdot c_{3u})}{N_{B,S} \cdot L_{hyd,vaned} \cdot (c_{2S} - c_3)}$$

- the area ratio, A_R:

$$A_R = \frac{D_3 \cdot b_3 \cdot \cos \alpha_3}{D_{2S} \cdot b_{2S} \cdot \cos \alpha_{2S}}$$

These parameters must fall within the following limits:

$$7^\circ < 2 \cdot \theta_C < 11^\circ$$

$$0 < BL < 1/3$$

$$1.4 < A_R < 2.4$$

and their optimum values are close to the upper limits:

$$10^\circ < 2 \cdot \theta_{C,opt} < 11^\circ$$

$$0.30 < BL_{opt} < 0.33$$

$$2.2 < A_R < 2.4$$

If these conditions are not satisfied, the assumptions on geometric and kinematic parameters of the stator must be changed in order to meet them.

From the knowledge of the kinematics, thermodynamics and geometry of the stage, the absolute (stator inlet/outlet) and relative (rotor inlet/outlet) Reynolds numbers can be calculated:

$$\mathrm{Re}_1 = \frac{\rho_1 \cdot w_1 \cdot D_{hyd,R}}{\mu_1}$$

$$\mathrm{Re}_2 = \frac{\rho_2 \cdot w_2 \cdot D_{hyd,R}}{\mu_2}$$

$$\mathrm{Re}_{2,S} = \frac{\rho_{2,S} \cdot c_{2,S} \cdot D_{hyd,vaned}}{\mu_{2,S}}$$

$$\mathrm{Re}_3 = \frac{\rho_3 \cdot c_3 \cdot D_{hyd,vaned}}{\mu_3}$$

In these equations, the characteristic length is the hydraulic diameter; if the Reynolds numbers must be calculated by using another characteristic dimension, these new equations will be explicitly expressed in the following section.

Once the Reynolds numbers have been calculated, considerations on surface roughness can be made according to Sect. 1.9.1.

In particular, the *admissible surface roughness*, $k_{s,adm}$, can be assessed; it is the limit roughness below which there are no effects on losses allowing the walls to be considered smooth. This condition occurs when surface roughness is contained within the laminar sub-layer of the boundary layer.

This roughness (Sect. 1.9.1) can be evaluated, for the stator and the rotor, by:

$$\frac{k_{s,adm,R}}{D_{hyd,R}} = \frac{100}{\mathrm{Re}_1}$$

$$\frac{k_{s,adm,S}}{D_{hyd,vaned}} = \frac{100}{\mathrm{Re}_{2,S}}$$

The calculation of $k_{s,adm}$ assumes considerable importance as it allows to establish the finishing level of the channel surfaces in order to contain dissipations.

Taking into account that turbomachinery surface roughness generally falls within the range:

$$5\ \mu\mathrm{m} < k_s < 50\ \mu\mathrm{m}$$

the knowledge of $k_{s,adm}$ allows to evaluate a possible, and economically convenient, surface treatment aimed at respecting this admissible roughness value as well as not resorting to very high finish levels where not required.

In other words, if the admissible surface roughness, $k_{s,adm}$, is lower than the minimum roughness that is technologically and economically achievable, $k_{s,min}$, we assume:

$$k_s = k_{s,\text{min}}$$

otherwise:

$$k_{s,min} \le k_s \le k_{s,adm}$$

In particular, in the latter case, a roughness lower than the admissible one can be assumed in order to take into account that the roughness tends to increase with the turbomachine operation.

Remarks on input parameters

The procedure illustrated in this Sect. 6.3 does not presuppose any further input parameters, with respect to those already identified in Sects. 6.1 and 6.2, taking into account the assumptions, defining the stator and rotor geometry, adopted in this section.

6.4 Stage Losses and Efficiency

The loss sources and the loss models in the different types of turbomachinery are extensively described in Sect. 1.9. In this section we will provide specific correlations for loss models able to quantify the losses introduced in Sect. 1.9.

Losses occurring in the stator and rotor of a radial stage depend on the kinematic (Sect. 6.1), thermodynamic (Sect. 6.2) and geometric (Sect. 6.3) parameters of the stage.

In the literature there are some loss models based on pressure loss coefficients (the model proposed by Aungier 2000, and subsequently adopted by many authors, such as, Gong and Chen 2014; Blanchette et al. 2016) and other loss models based on energy losses (Oh et al. 1997; Sanghera 2013; Khadse et al. 2016; Ameli et al. 2018; Meroni et al. 2018), expressed in terms of enthalpy increase compared to the isentropic compression.

Both loss models will be illustrated below.

6.4.1 Loss Models Based on Pressure Loss Coefficients

In the following, we mainly refer to the Aungier loss model (Aungier 2000).

6.4.1.1 Rotor Losses

Aungier (2000) proposed to calculate the total relative pressure at the rotor outlet by:

$$p_{2tr} = p_{2tr,is} - f_C \cdot (p_{1tr} - p_1) \cdot Y_R$$

where:

- $p_{2tr,is}$, already defined in Sect. 6.2, is the total relative pressure that would be obtained with an isentropic compression in the rotor;
- Y_R is the total relative pressure loss coefficient in the rotor. The definition of this coefficient is different from that introduced for axial compressors, since in the radial turbomachine rotor the total relative enthalpy is not constant and therefore $p_{2tr,is} \neq p_{1tr}$
- f_C is a corrective coefficient introduced by Aungier and defined as:

$$f_C = \frac{\rho_{2tr} \cdot T_{2tr}}{\rho_{1tr} \cdot T_{1tr}}$$

Expressing this coefficient as:

$$f_C = \frac{p_{2tr}}{p_{1tr}}$$

the total relative pressure at the rotor outlet is:

$$p_{2tr} = \frac{p_{2tr,is}}{1 + Y_R \cdot \left(1 - \frac{p_1}{p_{1tr}}\right)}$$

To calculate this pressure, and therefore the real conditions of the fluid exiting the rotor, it is necessary to evaluate the pressure loss coefficient which includes the following main losses:

$$Y_R = Y_{inc} + Y_{sf} + Y_{BL} + Y_{HS} + Y_{mix} + Y_{CL}$$

where:

Y_{inc} represents the incidence losses;
Y_{sf} represent the skin friction losses;
Y_{BL} represents the blade loading losses;
Y_{HS} represents the hub to shroud losses;
Y_{mix} represents the mixing losses;
Y_{CL} represents the clearance losses

Incidence losses

They are composed of two terms:

$$Y_{inc} = 0.8 \cdot \left(1 - \frac{c_{1m}}{w_{1M} \cdot \cos \beta_{1M,B}}\right)^2 + \left(\frac{N_{B,R} \cdot t_B}{\pi \cdot D_{1M} \cdot \cos \beta_{1M,B}}\right)^2$$

The first term is due to the flow incidence at the rotor inlet and is generally zero in design conditions ($\beta_{1M,nom} = \beta_{1M,B}$). The second term is related to the flow area reduction (blade blockage, Sect. 1.8) due to the blade thickness (t_B), defined in Sect. 6.3.

Skin friction losses

In analogy to the passage losses defined for radial turbines, the skin friction losses are:

$$Y_{sf} = 4 \cdot f_{sf} \cdot \frac{L_{hyd,R}}{D_{hyd,R}} \cdot \left(\frac{\bar{w}}{w_{1M}}\right)^2 .$$

where:

- $L_{hyd,R}$ and $D_{hyd,R}$ are the hydraulic length and diameter of the rotor, as defined in Sect. 6.3;
- $\bar{w}$ is the quadratic mean velocity between inlet (at the mean diameter) and outlet relative velocity:

$$\bar{w} = \sqrt{\frac{w_{1M}^2 + w_2^2}{2}}$$

- f_{sf} is the skin friction coefficient and it depends on the flow regime. Let us define the Reynolds number as:

$$\mathrm{Re}_D = \frac{\rho_1 \cdot w_{1M} \cdot D_{hyd,R}}{\mu_1}$$

If the flow is laminar ($\mathrm{Re}_D < 2000$), the skin friction coefficient is:

$$f_{sf} = f_l = \frac{16}{\mathrm{Re}_D}$$

while if $\mathrm{Re}_D > 4000$, the skin friction coefficient is a weighted average between the skin friction coefficient for smooth walls (f_{ts}) and for fully rough walls (f_{tr}):

$$\frac{1}{\sqrt{f_{ts}}} = -4 \cdot \log_{10}\left(\frac{1.255}{\mathrm{Re}_D \cdot \sqrt{f_{ts}}}\right)$$

$$\frac{1}{\sqrt{f_{tr}}} = -4 \cdot \log_{10}\left(\frac{1}{3.71} \cdot \frac{k_s}{D_{hyd,R}}\right)$$

where k_s is the superficial roughness defined in Sect. 6.3

To calculate this weighted average, the following Reynolds number is introduced:

$$\mathrm{Re}_e = (\mathrm{Re}_D - 2000) \cdot \frac{k_s}{D_{hyd,R}}$$

and by using this parameter we can calculate the skin friction coefficient when $Re_D > 4000$. In fact we can apply:

$$f_{sf} = f_t = f_{ts} \quad \text{when} \quad Re_e < 60$$

$$f_{sf} = f_t = f_{ts} + (f_{tr} - f_{ts}) \cdot \left(1 - \frac{60}{Re_e}\right) \quad \text{when} \quad Re_e \geq 60$$

If $2000 < Re_D < 4000$, the boundary layer is in transition from laminar to turbulent flow; the skin friction coefficient in this transition zone is well approximated by a weighted average of the laminar and turbulent values:

$$f = f_{l,(Re_D=2000)} - \left(f_{l,(Re_D=2000)} - f_{t,(Re_D=4000)}\right) \cdot \left(\frac{Re_D}{2000} - 1\right)$$

Blade loading losses

These losses are correlated to the velocity difference between blade suction and pressure sides and they are expressed by:

$$Y_{BL} = \frac{1}{24} \cdot \left(\frac{\Delta w}{w_{1M}}\right)^2$$

where

$$\Delta w = \frac{2 \cdot \pi \cdot D_2 \cdot u_2 \cdot \psi}{N_{B,R} \cdot L_{hyd,R}}$$

Hub to shroud losses

These losses are given by:

$$Y_{HS} = \frac{1}{6} \cdot \left(\frac{\bar{K}_m \cdot \bar{b} \cdot \bar{w}}{w_{1M}}\right)^2$$

where:

- $\bar{K}_m = \frac{\alpha_{C2} - \alpha_{C1}}{L_{m,R}}$
  and $L_{m,R}$ is the meridional length, previously defined, and α_C is the flow angle with respect to the axis of rotation. If the fluid enters the rotor approximately axially and exits the rotor approximately radially ($\alpha_{C1} = 0$ and $\alpha_{C2} = \pi/2$), the previous expression becomes:

$$\bar{K}_m = \frac{\pi}{2 \cdot L_{m,R}}$$

- $\bar{b} = \frac{b_1+b_2}{2}$
- $\bar{w} = \frac{w_{1M}+w_2}{2}$

Mixing losses

They can be expressed through:

$$Y_{mix} = \left(\frac{c_{2m,wake} - c_{2m,mix}}{w_{1M}}\right)^2$$

where:

- $c_{2m,mix} = c_{2m} \cdot \left(1 - \frac{N_{B,R} \cdot t_B}{\pi \cdot D_2}\right)$
- $c_{2m,wake} = \sqrt{w_{sep}^2 - w_{2u}^2}$

To calculate w_{sep} it is necessary to introduce the equivalent diffusion factor:

$$DF_{eq} = \frac{w\text{max}}{w_2} = \frac{w_{1M} + w_2 + \Delta w}{2 \cdot w_2}$$

where Δw is already defined (in blade loading loss). As a consequence we have:

$$w_{sep} = w_2 \qquad \text{if} \quad DF_{eq} \leq 2$$

$$w_{sep} = w_2 \cdot \frac{DF_{eq}}{2} \qquad \text{if} \quad DF_{eq} > 2$$

Clearance losses

They occur only in open impellers and they are expressed as:

$$Y_{CL} = \frac{2 \cdot m_{CL} \cdot \Delta p_{CL}}{m \cdot \rho_1 \cdot w_{1M}^2}$$

where

- m is the mass flow rate of the working fluid handled by the rotor

$$\Delta p_{CL} = \frac{m \cdot \psi \cdot u_2^2}{N_{B,R} \cdot L_{hyd,R} \cdot \omega \cdot \left(\frac{D_{1M}+D_2}{4}\right) \cdot \left(\frac{b_1+b_2}{2}\right)}$$

$$u_{CL} = 0.816 \cdot \sqrt{\frac{2 \cdot \Delta p_{CL}}{\rho_2}}$$

$$m_{CL} = \rho_2 \cdot N_{B,R} \cdot \tau_a \cdot L_{hyd,R} \cdot u_{CL}$$

and τ_a is the axial clearance already defined in Sect. 6.3.

External losses (parasitic losses)

External losses also occur in the rotor (Sect. 1.9): windage and disk friction losses, recirculation flow and leakage. These losses are not included (Sect. 1.9.5) in the stage efficiency (isentropic or polytropic) and are expressed in terms of enthalpy increases contributing to the shaft power.

For the expression of these losses, reference is made to the discussion of Oh et al. (1997).

Disk friction losses

They can be expressed as:

$$\Delta h_{df} = f_{df} \cdot \frac{\bar{\rho}}{2} \cdot \frac{u_2^3 \cdot D_2^2}{8 \cdot m}$$

where:

- m is the mass flow rate of the working fluid handled by the rotor
- $\bar{\rho}$ is a mean average density defined as:

$$\bar{\rho} = \frac{\rho_1 + \rho_2}{2}$$

- the coefficient f_{df} can be expressed by:

$$f_{df} = \frac{2.67}{\mathrm{Re}_{df}^{0.5}} \quad \text{when} \quad \mathrm{Re}_{df} < 3 \cdot 10^5$$

$$f_{df} = \frac{0.0622}{\mathrm{Re}_{df}^{0.5}} \quad \text{when} \quad \mathrm{Re}_{df} \geq 3 \cdot 10^5$$

where

$$\mathrm{Re}_{df} = \frac{\rho_2 \cdot u_2 \cdot (D_2/2)}{\mu_2}$$

Recirculation losses

They can be calculated by:

$$\Delta h_{rc} = 8 \cdot 10^{-5} \cdot \sinh\left(3.5 \cdot \alpha_2^3\right) \cdot DF_f^2 \cdot u_2^2$$

where the diffusion factor is equal to:

$$DF_f = 1 - \frac{w_2}{w_{1t}} + \frac{0.75 \cdot \psi}{\frac{w_{1t}}{w_2} \cdot \left[\frac{N_{B,R}}{\pi} \cdot (1 - \delta_t) + 2 \cdot \delta_t\right]}$$

Leakage losses
They occur in open impellers and are expressed through:

$$\Delta h_{LK} = \frac{m_{CL} \cdot u_{CL} \cdot u_2}{2 \cdot m}$$

where m_{CL} and u_{CL} have already been defined (clearance losses).

6.4.1.2 Stator Losses

The losses in the stator (diffuser) are divided into the losses in the vaneless diffuser and, if present, in the vaned diffuser:

$$Y_S = Y_{vaneless} + Y_{vaned}$$

Losses in the vaneless diffuser

According to the definition already used for axial machines, the total pressure loss coefficient is expressed as:

$$Y_{vaneless} = \frac{p_{2t} - p_{2St}}{p_{2t} - p_2}$$

The losses in the vaneless diffuser include two contributions: friction losses and diffusion losses:

$$Y_{vaneless} = Y_{sf} + Y_{diff}$$

Similarly to the rotor, the skin friction losses are expressed as:

$$Y_{sf} = 4 \cdot f_{sf} \cdot \frac{L_{hyd,vaneless}}{D_{hyd,vaneless}} \cdot \left(\frac{\bar{c}}{c_2}\right)^2$$

where:

- $L_{hyd,vaneless}$ and $D_{hyd,vaneless}$ are the hydraulic length and diameter of the vaneless diffuser, as defined in Sect. 6.3;
- $\bar{c}$ is the quadratic mean velocity between inlet and outlet velocity:

$$\bar{c} = \sqrt{\frac{c_{2S}^2 + c_2^2}{2}}$$

- f_{sf} is the skin friction coefficient and it is calculated in the same way as for the rotor, using the parameters of the vaneless diffuser.

The diffusion losses depend on diffusion efficiency E (Gong and Chen 2014):

$$Y_{diff} = -2 \cdot (1 - E) \cdot \frac{c_{2S,is} - c_2}{c_2}$$

Aungier (2000) suggested a procedure to calculate this diffusion efficiency. The divergence parameter is introduced:

$$D_v = \frac{b_2 \cdot \left(\frac{D_{2S}}{D_2} - 1\right)}{L_{hyd,vaneless}}$$

and the reference parameter, D_{rif}:

$$D_{rif} = 0.4 \cdot \left(\frac{b_2}{L_{hyd,vaneless}}\right)^{0.35}$$

Efficiency is:

$$\begin{aligned}
&E = 1 && \text{when} \quad D_v \leq 0 \\
&E = 1 - 0.2 \cdot \left(\frac{D_v}{D_{rif}}\right)^2 && \text{when} \quad 0 < D_v < D_{rif} \\
&E = 0.8 \cdot \sqrt{\frac{D_{rif}}{D_v}} && \text{when} \quad D_v \geq D_{rif}
\end{aligned}$$

Losses in the vaned diffuser

Similarly to the vaneless diffuser, we have:

$$Y_{vaned} = \frac{p_{2St} - p_{3t}}{p_{2St} - p_{2S}}$$

The losses in the vaned diffuser include three contributions: incidence loss, skin friction loss and mixing loss:

$$Y_{vaned} = Y_{inc} + Y_{sf} + Y_{mix}$$

The formulations of these three losses are completely analogous to those of the rotor.

The incidence losses are made up of two terms:

$$Y_{inc} = 0.8 \cdot \left(1 - \frac{c_{2Sm}}{c_{2S} \cdot \cos \alpha_{2S}}\right)^2 + \left(\frac{N_{B,S} \cdot t_B}{\pi \cdot D_{2S}}\right)^2$$

The skin friction losses, again similarly to the rotor, are expressed as:

$$Y_{sf} = 4 \cdot \frac{f_{sf}}{A} \cdot \frac{L_{hyd,vaned}}{D_{hyd,vaned}} \cdot \left(\frac{\bar{c}}{c_{2S}}\right)^2$$

where:

- $L_{hyd,vaned}$ and $D_{hyd,vaned}$ are the hydraulic length and diameter of the vaned diffuser, as defined in Sect. 6.3;
- $\bar{c}$ is the quadratic mean velocity between inlet and outlet velocity:

$$\bar{c} = \sqrt{\frac{c_{2S}^2 + c_3^2}{2}}$$

- f_{sf} is the skin friction coefficient and it is calculated in the same way as for the rotor, using the parameters of the vaned diffuser;
- A is a correction factor of the skin friction coefficient to take into account the effect of the boundary layer in the diffuser and it is expressed as:

$$A = \left(5.142 \cdot f_{sf} \cdot \frac{L_{hyd,vaned}}{D_{hyd,vaned}}\right)^{0.25}$$

The mixing losses are expressed by:

$$Y_{mix} = \left(\frac{c_{3m,wake} - c_{3m,mix}}{c_{2S}}\right)^2$$

where:

- $c_{3m,mix} = c_{3m} \cdot \left(1 - \frac{N_{B,S} \cdot t_B}{\pi \cdot D_3}\right)$
- $c_{3m,wake} = \sqrt{c_{3sep}^2 - c_{3u}^2}$

To calculate c_{3sep} it is necessary to introduce the equivalent diffusion factor, defined as:

$$DF_{eq} = \frac{c_{2S}}{c_3}$$

and then we can calculate:

$$c_{3sep} = c_3 \quad \text{when } DF_{eq} \leq 2$$
$$c_{3sep} = c_3 \cdot \tfrac{DF_{eq}}{2} \quad \text{when } DF_{eq} > 2$$

6.4.2 Loss Models Based on Energy Losses

In the following, we mainly refer to the Oh loss model (Oh et al. 1997).

6.4.2.1 Rotor Losses

The rotor enthalpy losses, similarly to the losses expressed in terms of total pressure, include the following losses:

$$\Delta h_R = \Delta h_{inc} + \Delta h_{sf} + \Delta h_{BL} + \Delta h_{mix} + \Delta h_{CL}$$

where:

Δh_{inc} represents the incidence losses;
Δh_{sf} represent the skin friction losses;
Δh_{BL} represents the blade loading losses;
Δh_{mix} represents the mixing losses;
Δh_{CL} represents the clearance losses

Incidence losses

They are expressed by:

$$\Delta h_{inc} = f_{inc} \cdot \frac{w_{ui}^2}{2} = f_{inc} \cdot \frac{w_{1M}^2}{2} \cdot (sen\beta_{1M,B} - sen\beta_{1M})^2$$

where:

$$f_{inc} = 0.5 \div 0.7$$

These losses are generally zero in design conditions ($\beta_{1M,nom} = \beta_{1M,B}$).

Skin friction losses

Friction losses in the channel are:

$$\Delta h_{sf} = 2 \cdot f_{sf} \cdot \frac{L_{hyd,R}}{D_{hyd,R}} \cdot \bar{w}^2$$

where $\bar{w}$ is the mean velocity between inlet and outlet velocity:

$$\bar{w} = \frac{w_{1M} + w_2}{2}$$

and all the other parameters are those used in the total pressure loss model, introduced previously.

Blade loading losses

These losses are expressed by:

$$\Delta h_{BL} = 0.05 \cdot DF_f^2 \cdot u_2^2$$

where the diffusion factor, already previously (Sect. 6.4.1) defined, is:

$$DF_f = 1 - \frac{w_2}{w_{1t}} + \frac{0.75 \cdot \psi}{\frac{w_{1t}}{w_2} \cdot \left[\frac{N_{B,R}}{\pi} \cdot (1 - \delta_t) + 2 \cdot \delta_t\right]}$$

Mixing losses

They can be expressed through:

$$\Delta h_{mix} = \frac{c_2^2}{2} \cdot \cos^2 \alpha_2 \cdot \left(1 - \frac{b_{2S}/b_2}{1 - \varepsilon_w}\right)^2$$

where:

- $\varepsilon_w = 1 - \frac{c_{2m,wake}}{c_{2m,mix}}$
- $c_{2m,wake}$, $c_{2m,mix}$ have already been defined in Sect. 6.4.1.

Clearance losses

They can be expressed through:

$$\Delta h_{CL} = 0.6 \cdot \left(\frac{\tau_a}{b_2}\right) \cdot c_{2u} \cdot \sqrt{\frac{2 \cdot \pi}{b_2 \cdot N_{B,R}} \cdot \left[\frac{D_{1t}^2 - D_{1h}^2}{(D_2 - D_{1t}) \cdot (1 + \rho_2/\rho_1)}\right] \cdot c_{1m} \cdot c_{2u}}$$

where all the parameters have been previously defined.

External losses (parasitic losses)

It is completely valid what illustrated in Sect. 6.4.1 where parasitic losses had already been expressed in terms of enthalpy increases.

6.4.2.2 Stator Losses

The losses in the vaneless diffuser and, if present, in the vaned diffuser in enthalpy terms are obtained from those expressed in terms of total pressure by the following correlation:

$$\Delta h_S = Y_{vaneless} \cdot \frac{c_2^2}{2} + Y_{vaned} \cdot \frac{c_{2S}^2}{2}$$

6.4.3 Rotor and Stage Efficiency

The procedure described in Sects. 6.4.1–6.4.2 allows evaluating the stage losses, expressed in terms of pressure loss coefficients or in terms of enthalpy increase compared to isentropic compression.

To assess efficiencies, let us consider the loss model based on the pressure loss coefficients, since the procedure for efficiency calculation through the enthalpy increases has already been illustrated in Chap. 5.

The procedure described in Sect. 6.4.1 allows to evaluate the pressure loss coefficients both in the stator and the rotor:

$$Y_R = \frac{p_{2tr,is} - p_{2tr}}{\frac{p_{2tr}}{p_{1tr}} \cdot (p_{1tr} - p_1)}$$

$$Y_{vaneless} = \frac{p_{2t} - p_{2St}}{p_{2t} - p_2}$$

$$Y_{vaned} = \frac{p_{2St} - p_{3t}}{p_{2St} - p_{2S}}$$

$$Y_S = Y_{vaneless} + Y_{vaned}$$

To assess the rotor efficiency (Sect. 6.2), we must first evaluate the total relative pressure downstream of the rotor (the total relative pressure upstream of the rotor and the isentropic total relative pressure downstream of the rotor were calculated in the thermodynamics of the stage, Sect. 6.2):

$$p_{2tr} = \frac{p_{2tr,is}}{1 + Y_R \cdot \left(1 - \frac{p_1}{p_{1tr}}\right)}$$

Since in the rotor the rothalpy is constant, we know the total relative enthalpy downstream of the rotor:

$$h_{2tr} = h_{1tr} + \frac{u_2^2 - u_1^2}{2}$$

and as a consequence we know the thermodynamic parameters downstream of the rotor (Sect. 6.3):

$$s_2 = s_{p,h}(p_{2tr}, h_{2tr})$$
$$h_2 = h_{p,s}(p_2, s_2)$$

and, therefore, the rotor efficiency can be assessed as:

$$\bar{\eta}_R = \frac{h_{2,is} - h_1}{h_2 - h_1}$$

Similarly, for the stator we have:

$$h_{2t} = h_2 + \frac{c_2^2}{2}$$
$$p_{2t} = p_{s,h}(s_2, h_{2t})$$
$$p_{2St} = p_{2t} - Y_{vaneless} \cdot (p_{2t} - p_2)$$
$$p_{3t} = p_{2St} - Y_{vaned} \cdot (p_{2St} - p_2)$$
$$h_{3t} = h_{2t}$$
$$s_3 = s_{p,h}(p_{3t}, h_{3t})$$
$$h_3 = h_{s,p}(s_3, p_3)$$

Once the thermodynamic parameters of the stage have been calculated on the basis of the stage losses, the various efficiencies can be assessed (Sect. 1.8.3):

$$\eta_{TT} = \frac{h_{3t,is} - h_{1t}}{h_{3t} - h_{1t}}$$

$$\eta_{is} = \frac{h_{3,is} - h_1}{h_3 - h_1}$$

$$\eta_{TS} = \frac{h_{3,is} - h_{1t}}{h_{3t} - h_{1t}}$$

If we consider the enthalpy loss model, since the isentropic efficiency does not take into account the external losses (Sect. 1.9.6), in the previous expressions only the enthalpy increases generated by internal losses are considered. Then, the enthalpy increases generated by external losses contribute to the shaft power to drive the compressor.

Remarks on input parameters

The procedure illustrated in this Sect. 6.4 does not presuppose the introduction of further input parameters, compared to those already identified in Sects. 6.1 and 6.2.

6.5 Input Parameters of the Preliminary Design Procedure

In the previous Sects. 6.1–6.4, the procedure for the calculation of the kinematics, thermodynamics, geometry and efficiency of a centrifugal compressor stage was developed and the input parameters of this procedure were identified.

In this Sect. 6.5, we intend to precisely analyze these input parameters in order to suggest their numerical values (these values, however, can be reviewed in an iterative calculation, see Sect. 6.6).

First of all, let us consider the kinematics calculation procedure (Sect. 6.1).

In general, to determine the velocity triangles at the rotor inlet and outlet, we must set six independent parameters: three for the inlet triangle (station 1 of the stage) and three for the outlet triangle (station 2 of the stage).

Unlike the axial stages, where, being $u_1 = u_2 = u$ and assuming $c_{m1} = c_{m2} = c_m$, the independent parameters are only four, for the radial stages the *independent parameters remain six.*

As clearly illustrated in Sect. 6.1, in fact, the absolute and relative flow angles are:

$$\alpha_2 = f(\varphi, \psi, \xi, \delta_t, \alpha_1)$$
$$\beta_1 = f(\varphi, \delta_t, \alpha_1)$$
$$\beta_2 = f(\varphi, \psi, \xi, \delta_t, \alpha_1)$$

These angles are, therefore, functions of the following five independent parameters:

- the flow coefficient, φ
- the work coefficient, ψ
- the rotor meridional velocity ratio, ξ, or the degree of reaction, R, being (Sect. 6.1):

$$R = f(\varphi, \psi, \xi, \delta_t, \alpha_1)$$

- the rotor tip diameter ratio, δ_t

- the absolute flow angle at the rotor inlet, α_1

As a consequence, by identifying a further independent parameter, for example, the blade speed at the rotor outlet, u_2, the velocity triangles are completely defined.

However, taking into account what emerged in the turbomachinery selection process (Chap. 2), it is advisable to choose the stage efficiency (isentropic or polytropic efficiency) as the additional independent parameter and calculate the blade speed accordingly.

In fact, starting from the data available for the turbomachinery selection process:

- type of fluid
- fluid mass flow rate (kg/s)
- fluid inlet temperature (°C)
- fluid inlet pressure (bar)
- fluid outlet pressure (bar)

the selection process provides ranges of n ($n_{min} < n < n_{max}$) and z ($z_{min} < z < z_{max}$) compatible with the type of turbomachine chosen. Within these ranges we can choose one or more pairs of these values:

- n: rotational speed (rpm)
- z: number of stages and, consequently, the isentropic enthalpy increase in each stage, $\Delta h_{is,stage}$ (kJ/kg)

For each of these pairs, by using the stage (isoentropic or polytropic) efficiency, set as first iteration guess (Sect. 6.5.3) or calculated by the stage losses (Sect. 6.4), the adiabatic enthalpy increase of each stage can be calculated as well as the blade speed, after choosing the work coefficient.

Indeed, when $c_3 = c_1$ (Sect. 6.1), the blade speed is:

$$u_2 = \sqrt{\frac{\Delta h_{t,ad,stage}}{\psi}} = \sqrt{\frac{\Delta h_{ad,stage}}{\psi}}$$

On the other hand, when $c_3 \neq c_1$ (Sects. 6.1 and 6.3), the work coefficient is expressed as:

$$\psi = \frac{\Delta h_{t,ad,stage}}{u_2^2} = \frac{\Delta h_{ad,stage}}{u_2^2} + \frac{c_3^2 - c_1^2}{2 \cdot u_2^2}$$

And, therefore, expressing the velocity c_1 as a function of the kinematic parameters, we obtain:

$$u_2 = \sqrt{\frac{\Delta h_{ad,stage} + \frac{c_3^2}{2}}{\psi + \frac{1}{2} \cdot \frac{\varphi^2}{\cos^2 \alpha_1}}}$$

After calculating the blade speed, the diameter at the rotor outlet is:

$$D_2 = \frac{60 \cdot u_2}{\pi \cdot n}$$

Alternatively, we can calculate the blade speed, u_2, from the diameter, D_2, obtained, in turn, by the specific diameter, D_s (Chap. 2). In this way, only five independent parameters must be chosen since the work coefficient is calculated by using the stage enthalpy increase, $\Delta h_{ad,stage}$, and the blade speed, u_2. In radial stage design, however, it is preferred to assign the work coefficient, following the procedure illustrated above.

In summary, we must set the following six input parameters in order to design preliminarily a centrifugal compressor stage:

- the flow coefficient, φ
- the work coefficient, ψ
- the degree of reaction, R, or the rotor meridional velocity ratio, ξ
- the rotor tip diameter ratio, δ_t
- the absolute flow angle at the rotor inlet, α_1
- the stage efficiency (isentropic or polytropic efficiency)

Taking into account what is illustrated in Sects. 6.2–6.4, these parameters, with the addition of the rotor efficiency, allow to fully calculate the thermodynamics, the geometry and therefore the stage losses. Since these losses allow, in turn, to exactly evaluate the rotor and stage efficiencies, the stage efficiency (in the list above) is only a starting value: the calculation will proceed iteratively for each stage until convergence for these efficiencies is achieved (Sect. 6.6).

6.5.1 Flow and Work Coefficient, or Alternative Parameters, Selection

Since in this section the characteristic stage parameters will be expressed as functions of the specific speed (Chap. 2), first of all, it is necessary to establish the specific speed optimum range for the stage efficiency.

In this regard, Fig. 6.6 (Rusch and Casey 2013) sketches the variation of the isentropic efficiency with the specific speed for different values of the pressure ratio and relative Mach number at the inlet tip diameter.

This diagram highlights the influence of the pressure ratio and Mach number on the stage efficiency as well as identifies the optimum ranges of specific speed, which are much narrower if the pressure ratio increases. From Fig. 6.6 we deduce:

$$\eta_{is,MAX} \quad \Rightarrow \quad 0.4 < \omega_s < 1.0$$

This result is in agreement with the range of the specific speed for radial compressors, established in Chap. 2.

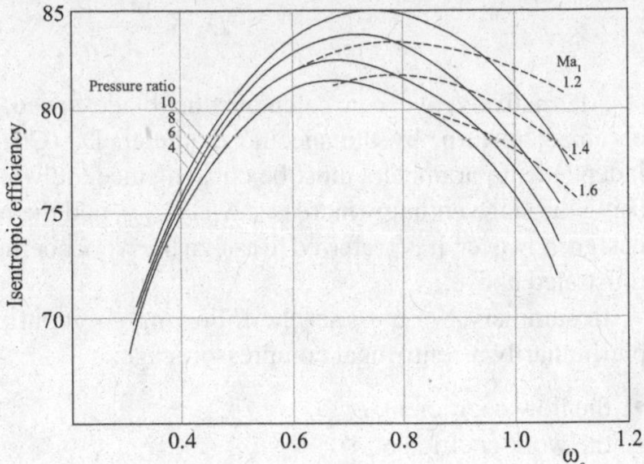

Fig. 6.6 Variation of stage efficiency with the specific speed Adapted from Rusch and Casey (2013)

Figure 6.6 refers to highly pressure ratio stages. When these pressure ratios are lower, the optimum specific speed upper limit for the stage efficiency increases (transition to mixed flow compressors) and, consequently, the correlated flow coefficients increase.

Casey et al. (Casey et al. 2009, 2010; Robinson et al. 2011; Casey and Robinson 2013; Dietmann and Casey 2013) elaborated an in-depth analysis of the characteristic parameters (the work and flow coefficients and efficiency) of radial stages. In particular, Rusch and Casey 2013, starting from the Rodgers' correlations, obtained the stage polytropic efficiency (Fig. 6.7) as a function of the flow coefficient and peripheral Mach number at the rotor outlet (M_{u2}). In this diagram, the flow coefficient is defined as:

$$\varphi' = \frac{V_{in}}{u_2 \cdot D_2^2}$$

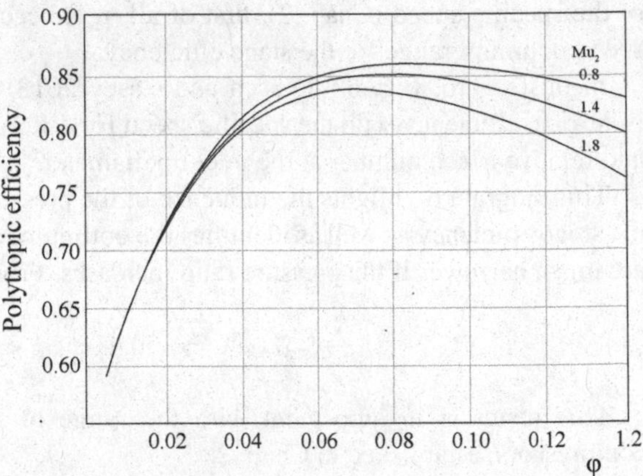

Fig. 6.7 Variation of stage efficiency with the flow coefficient φ′ Adapted from Rusch and Casey (2013)

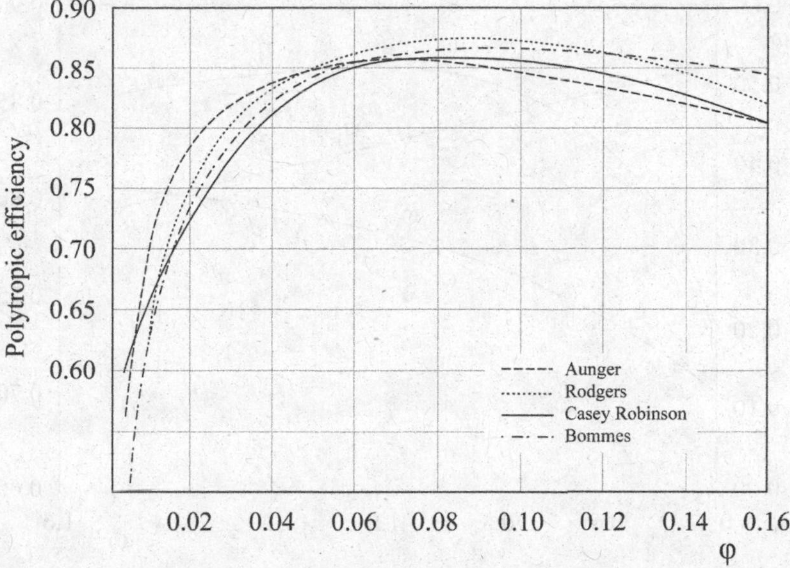

Fig. 6.8 Variation of stage efficiency with the flow coefficient φ′ Adapted from Casey et al. (2010)

and it is correlated with that introduced in Sect. 6.1, through:

$$\varphi' = \frac{V_{in}}{u_2 \cdot D_2^2} = \frac{\pi \cdot b_1 \cdot D_{1M} \cdot c_{1m}}{u_2 \cdot D_2^2} = \frac{\pi}{4} \cdot \left(\delta_t^2 - \delta_h^2\right) \cdot \varphi$$

Similarly, Fig. 6.8 (Casey et al. 2010) shows the polytropic efficiency, calculated by different authors, as a function of the flow coefficient; all these lines substantially correspond.

The diagrams in Figs. 6.7 and 6.8 clearly show that the optimum flow coefficient range for high efficiencies (>80%) is extremely wide:

$$\eta_{p,MAX} \quad \Rightarrow \quad 0.03 < \varphi' < 0.16$$

This situation, not occurring in the turbomachines previously analyzed where the optimum flow coefficient ranges were much narrower, advises against to choose this flow coefficient, φ', as an input parameter of the design procedure.

Unlike this flow coefficient, the optimum work coefficient falls in a much more limited range. In particular, referring to the isentropic work coefficient:

$$\psi_{is} = \frac{\Delta h_{is}}{u_2^2}$$

we find out that (Fig. 6.9) in the specific speed range of radial compressors (0.4–1.0):

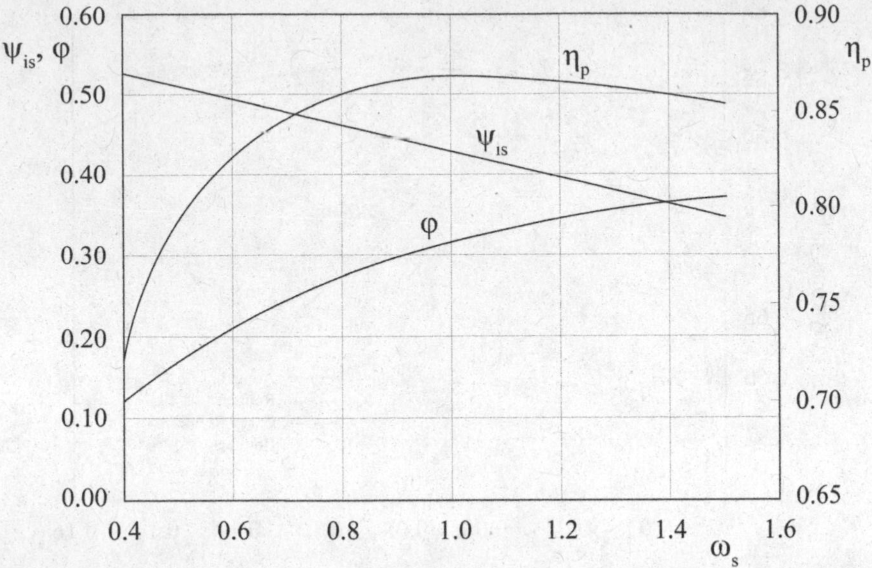

Fig. 6.9 Variation of φ, ψ_{is} and η_p with ω_s

$$\eta_{p,MAX} \quad \Rightarrow \quad 0.4 < \psi_{is} < 0.5$$

Note that the isentropic work coefficient is related to the flow coefficient (Chap. 2) through the specific speed:

$$\omega_s = \omega \cdot \frac{(V_{in})^{1/2}}{(\Delta h_{is})^{3/4}} = \omega \cdot \frac{\left(\varphi' \cdot \omega \cdot D_2^3/2\right)^{1/2}}{\left(\psi_{is} \cdot \omega^2 \cdot D_2^2/4\right)^{3/4}} = 2 \cdot \frac{\varphi'^{1/2}}{\psi_{is}^{3/4}}$$

and, therefore:

$$\varphi' = \frac{\omega_s^2 \cdot \psi_{is}^{3/2}}{4}$$

As a consequence, if the isentropic work coefficient is expressed as a function of the specific speed, also the flow coefficient becomes a function of the specific speed, by using the previous correlation.

Therefore, the procedure is completely analogous to that followed for the radial turbines where the isentropic velocity ratio (directly related to the isentropic work coefficient) was expressed as a function of the stage specific speed.

In this regard, the correlation proposed by Casey et al. (2010) can be effectively used:

$$\psi_{is} = \psi_{medium} \cdot (1 - A) + \psi_{high} \cdot A + (\psi_{medium} - \psi_{low}) \cdot \mathrm{B}$$

where:

$$\psi_{low} = 0.45 \qquad \psi_{medium} = 0.55 \qquad \psi_{high} = 0.02$$

$$A = \frac{1}{1 + e^{-t_1}}$$

$$t_1 = K_1 \cdot \left(K_2 + \log_{10} \omega_s\right) \qquad K_1 = 4 \qquad K_2 = -0.3$$

$$B = e^{-t_2}$$

$$t_2 = K_3 \cdot \left(K_4 + \log_{10} \omega_s\right) \qquad K_3 = 5 \qquad K_4 = 1.0$$

The previous correlations also allow expressing the flow coefficient, φ, defined in the kinematics (Sect. 6.1), as a function of the isentropic work coefficient and specific speed. Indeed, we have:

$$\varphi' = \frac{\pi}{4} \cdot \left(\delta_t^2 - \delta_h^2\right) \cdot \varphi = \frac{\omega_s^2 \cdot \psi_{is}^{3/2}}{4}$$

so, the flow coefficient becomes:

$$\varphi = \frac{\omega_s^2 \cdot \psi_{is}^{3/2}}{\pi \cdot \left(\delta_t^2 - \delta_h^2\right)}$$

The flow coefficient, φ, can be expressed only as a function of the specific speed because also the geometric parameters, δ_t and δ_h, are functions of this speed (see their expressions reported below). By using these expressions, we obtain:

$$\varphi = \frac{\omega_s^2 \cdot \psi_{is}^{3/2}}{\pi \cdot \left(\delta_t^2 - \delta_h^2\right)} = \frac{\omega_s^2 \cdot \psi_{is}^{3/2}}{\pi \cdot \left[\left(0.5 + 1.5 \cdot \frac{\omega_s^2 \cdot \psi_{is}^{3/2}}{\pi}\right)^2 - (0.35)^2\right]} = f(\omega_s)$$

By using also the efficiency correlation provided below, the diagram in Fig. 6.9 shows that, unlike the coefficient φ', the coefficient φ falls within much narrower range, always in the specific speed range of radial compressors (0.4–1.0):

$$\eta_{p,MAX} \quad \Rightarrow \quad 0.2 < \varphi < 0.3$$

Consequently, this flow coefficient can be effectively chosen as an input parameter.

Figure 6.9 also shows, as stated above, a narrow range of the isentropic work coefficient (0.4–0.5). The work coefficient is obtainable combining the isentropic work coefficient with the stage efficiency and its optimum range is:

$$\eta_{p,MAX} \quad \Rightarrow \quad 0.5 < \psi < 0.6$$

Finally, we can choose one of the two rotor diameter ratios, δ_t and δ_h, in place of the flow coefficient, φ, as an independent input parameter; for example, choosing δ_t (δ_h will be set later), we can apply the correlation suggested by Hazby et al. (2017):

$$\delta_t = 0.5 + 1.5 \cdot \frac{\omega_s^2 \cdot \psi_{is}^{3/2}}{\pi}$$

In conclusion, two substantially equivalent sets of independent input parameters are therefore possible:

$$\varphi, \psi \begin{cases} 0.2 < \varphi < 0.3 \\ 0.5 < \psi < 0.6 \end{cases} \quad \text{or} \quad \psi_{is}, \delta_t \begin{cases} \psi_{is} = f(\omega_s) \\ \delta_t = f(\omega_s) \end{cases}$$

6.5.2 Degree of Reaction or Rotor Meridional Velocity Ratio Selection

Dalbert et al. (1999) suggested the following range for the degree of reaction:

$$0.6 \leq R \leq 0.7$$

but we must remember the constraint on the maximum value of the degree of reaction, already discussed in Sect. 6.1.

Alternatively to the degree of reaction, the rotor meridional velocity ratio can be assumed as an input parameter. This parameter, like the degree of reaction, is related to the absolute flow angle at the rotor outlet, α_2. Indeed, remembering the expression of ξ obtained from the kinematics (Sect. 6.1), we have:

$$\xi = \frac{1}{\varphi \cdot \tan \alpha_2} \cdot (\psi + \varphi \cdot \delta_t \cdot \tan \alpha_1)$$

The expression of the angle α_2, suggested by Aungier (2000), as a function of the specific speed can therefore be used:

$$\alpha_2 = 72° - 0.5 \cdot \log\left(\frac{\omega_s^2 \cdot \psi_{is}^{3/2}}{\pi}\right) - 585 \cdot \left(\frac{\omega_s^2 \cdot \psi_{is}^{3/2}}{\pi}\right)^2$$

6.5.3 Starting Value of Stage Efficiency

As already illustrated, the stage efficiency is only a starting value to calculate preliminarily the thermodynamics, geometry and kinematics of the stage, since this parameter depends on stage losses that, in turn, depend on the thermodynamics, geometry and kinematics of the stage (Sect. 6.4). The efficiency calculation proceeds iteratively for each stage until convergence is achieved (Sect. 6.6).

The starting value of the stage efficiency can be calculated by using different correlations, proposed by various authors, as functions of φ and ψ.

Also in this case, however, it is very effective to establish the starting value of the stage efficiency (polytropic efficiency) as a function of the specific speed, according to the correlation proposed by Bommes et al. (2003); Casey et al. (2010) judged valid this correlations also for high specific speed stages:

$$\log_{10}(\eta_p) = -0.097358 - 0.0800538 \cdot \log_{10}\left(\frac{\omega_s}{2.9809}\right) + 0.151771 \cdot \left[\log_{10}\left(\frac{\omega_s}{2.9809}\right)\right]^2 + 0.340467 \cdot \left[\log_{10}\left(\frac{\omega_s}{2.9809}\right)\right]^3$$

The variation of this efficiency with the specific speed is shown in Fig. 6.9. The isentropic efficiency is then calculated from the polytropic one following the procedure illustrated in Sect. 1.4.1.

For the thermodynamics evaluation (Sect. 6.2), it is also necessary to choose the starting value of the rotor efficiency. It can be set arbitrarily, for example equal to the stage isentropic efficiency, since also this efficiency will be iteratively calculated by the losses in the rotor (Sect. 6.4).

6.5.4 Inlet Flow Angle and Rotor Diameter Ratios Selection

Unless the relative Mach number must be reduced, generally, we can assume the inlet absolute flow angle as:

$$\alpha_1 = 0$$

A correlation for the rotor tip diameter ratio, δ_t, has already been provided previously as a function of the specific speed (Sect. 6.5.1).

Finally, for the rotor hub diameter ratio, the value proposed by Hazby et al. (2017) can be used:

$$\delta_h = 0.35$$

Alternatively, Hazby et al. (2017) proposed a correlation for the parameter δ_b defined as the ratio of the blade height to the diameter at the rotor outlet:

$$\delta_b = \frac{b_2}{D_2} = 0.025 + 0.4 \cdot \frac{\omega_s^2 \cdot \psi_{is}^{3/2}}{\pi}$$

6.5.5 Summary of Input Parameters

Based on the main considerations made in the previous Sects. 6.5.1–6.5.4, two sets of input parameters can be identified (Table 6.1):

- Set A: ψ, φ, R, η_{is}, α_1, δ_h
- Set B: ψ_{is}, δ_t,α_2, η_{is}, α_1, δ_h

The set A involves more freedom in the input parameters choice, while the set B provides a more *guided procedure*, since the main input parameters are functions of the specific speed.

By assuming the aforementioned six independent parameters and knowing the isentropic enthalpy increase in each stage, $\Delta h_{is,stage}$, and the speed of rotation, n, we can design centrifugal compressor stages.

For example, by assuming the following input parameters (Set A):

1. ψ
2. φ
3. R
4. η_{is}
5. α_1
6. δ_h

we can immediately calculate:

- the stage adiabatic enthalpy increase:

$$\Delta h_{ad} = \Delta h_{is} \cdot \eta_{is}$$

Table 6.1 Input parameters

	Set A	Set B
1	Work coefficient ψ	Isentropic work coefficient ψ_{is}
2	Flow coefficient φ	Rotor tip diameter ratio δ_t
3	Degree of reaction R	Rotor outlet flow angle α_2
4	Stage efficiency η_{is}	
5	Inlet flow angle α_1	
6	Rotor hub diameter ratio δ_h	

- the blade speed at the rotor outlet (for $c_3 \neq c_1$, see Sect. 6.5):

$$u_2 = \sqrt{\frac{\Delta h_{ad}}{\psi}}$$

- the rotor outlet diameter:

$$D_2 = \frac{60 \cdot u_2}{\pi \cdot n}$$

- the rotor tip diameter ratio:

$$\delta_t = \sqrt{\delta_h^2 + \frac{4 \cdot V_1}{\pi \cdot \varphi \cdot u_2 \cdot D_2^2}}$$

- the rotor meridional velocity ratio:

$$\xi = \sqrt{1 - \frac{2 \cdot \psi}{\varphi^2} \cdot \left(R + \frac{\psi}{2} - 1\right) + \tan^2 \alpha_1 \cdot \left(1 - \delta_t^2\right) - \frac{2 \cdot \psi \cdot \delta_t}{\varphi} \cdot \tan \alpha_1}$$

- the flow angles:

$$\tan \alpha_2 = \frac{\psi}{\xi \cdot \varphi} + \frac{\delta_t}{\xi} \cdot \tan \alpha_1$$

$$\tan \beta_1 = \frac{\delta_t}{\varphi} - \tan \alpha_1$$

$$\tan \beta_2 = \frac{1}{\varphi \cdot \xi} \cdot (1 - \psi) - \frac{\delta_t}{\xi} \cdot \tan \alpha_1$$

and therefore we can calculate the kinematics, thermodynamics, geometry and stage losses.

Similarly, by assuming the following input parameters (Set B):

1. ψ_{is}
2. δ_t
3. α_2
4. η_{is}
5. α_1
6. δ_h

we can immediately calculate:

- the blade speed at the rotor outlet:

$$u_2 = \sqrt{\frac{\Delta h_{is}}{\psi_{is}}}$$

- the work coefficient:

$$\psi = \frac{\psi_{is}}{\eta_{is}}$$

- the rotor outlet diameter:

$$D_2 = \frac{60 \cdot u_2}{\pi \cdot n}$$

- the blade height at the rotor inlet:

$$b_1 = \frac{D_2}{2} \cdot (\delta_t - \delta_h)$$

- the rotor inlet meridional velocity:

$$c_{1m} = 2 \cdot \frac{V_1}{\pi \cdot D_2 \cdot (\delta_t + \delta_h) \cdot b_1}$$

- the flow coefficient:

$$\varphi = \frac{c_{1m}}{u_2}$$

- the rotor meridional velocity ratio:

$$\xi = \frac{1}{\varphi \cdot \tan \alpha_2} \cdot (\psi + \varphi \cdot \delta_t \cdot \tan \alpha_1)$$

- the degree of reaction:

$$R = 1 - \frac{\psi}{2} + \frac{\varphi^2}{2 \cdot \psi} \cdot \left[\left(1 - \xi^2\right) + \tan^2 \alpha_1 \cdot \left(1 - \delta_t^2\right)\right] - \varphi \cdot \delta_t \cdot \tan \alpha_1$$

- the flow angles:

$$\tan \beta_1 = \frac{\delta_t}{\varphi} - \tan \alpha_1$$

$$\tan \beta_2 = \frac{1}{\varphi \cdot \xi} \cdot (1 - \psi) - \frac{\delta_t}{\xi} \cdot \tan \alpha_1$$

and therefore we can calculate the kinematics, thermodynamics, geometry and stage losses.

In conclusion, Table 6.2 summarizes the constraints of all the parameters illustrated so far in order to design correctly a radial stage achieving high efficiency.

Table 6.2 Input parameter constraints

Parameter	Range
Specific speed ω_s	0.4 ÷ 1.0
Flow coefficient φ	0.2 ÷ 0.3
Work coefficient ψ	0.5 ÷ 0.6
Degree of reaction R	0.6 ÷ 0.7
Rotor tip diameter ratio δ_t	0.5 ÷ 0.75
δ_h/δ_t	0.2 ÷ 0.7
Rotor outlet absolute flow angle α_2	60° ÷ 70°
Rotor outlet relative flow angle β_2	0° ÷ 60°
Diffusion factor $DF = w_1/w_2$	0.7 ÷ 0.95

6.6 The Conceptual Comprehensive Framework of the Proposed Preliminary Design Procedure

To calculate the kinematics, thermodynamics and geometry of a radial stage, it is necessary to assume the rotor and stage efficiencies (Sect. 6.5). These efficiencies depend on the losses in the stator and rotor which, in turn, depend on all the parameters listed above (Sect. 6.4). Hence, the preliminary design of a stage is an iterative calculation (Fig. 6.10): first, we calculate the kinematic, thermodynamic and geometric parameters by assuming a starting value of the stage and rotor efficiency; then, we can calculate the stator and rotor losses; at this point, the stage and rotor efficiencies must be calculated by using these losses; these new efficiencies become the new input parameters of the preliminary design procedure. This iterative procedure ends when the established convergence for the rotor and stage efficiencies is achieved.

The iterative procedure shown in Fig. 6.10 could be solved by using different values of the input parameters, as well as of the geometric parameters, in order to optimize the stage performance.

For multistage compressors (Fig. 6.11), the same preliminary design is applied to the stages following the first one. Considering a multistage compressor, it is important to highlight that typically the working fluid exits the turbomachine through an outlet volute. In this component, the total enthalpy is constant as well as the mass flow rate and, ignoring friction, also the angular momentum is constant. By applying these conservation laws and adopting specific correlations for the geometric design of this component, it is possible to solve completely the turbomachine.

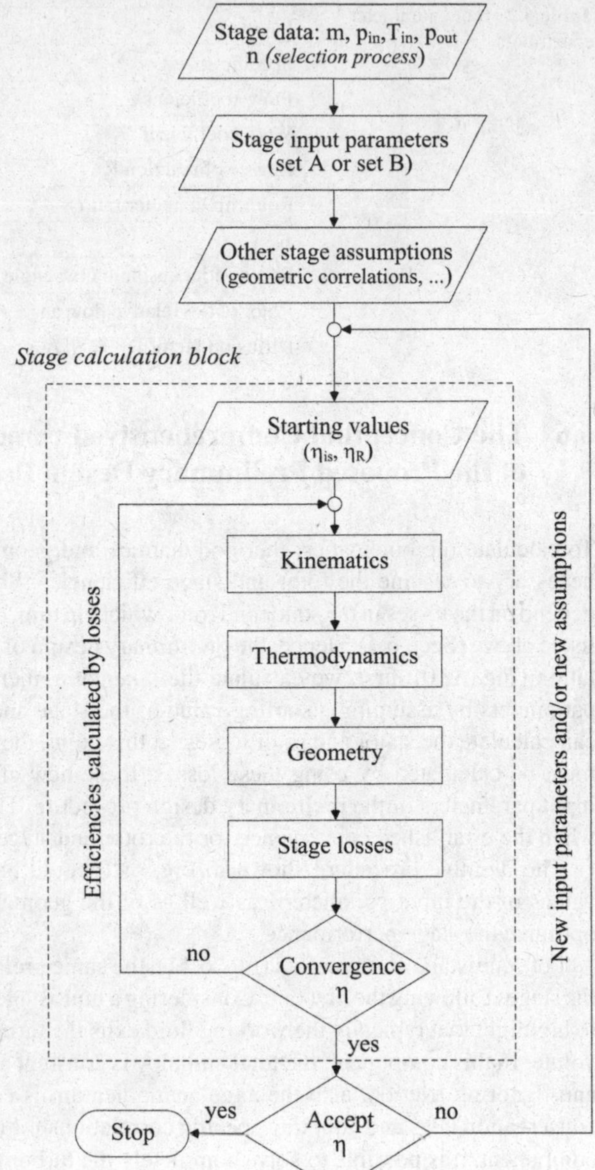

Fig. 6.10 Stage calculation block diagram

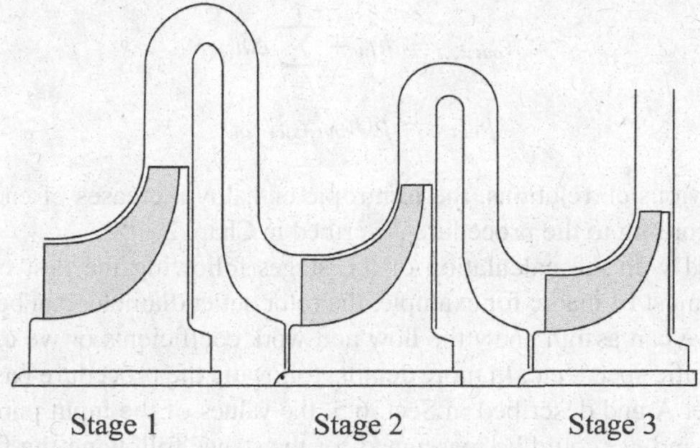

Fig. 6.11 Multistage centrifugal compressor

The thermodynamic parameters at the inlet and the pressure at the outlet of stages following the first one (i = 2, z), in absence of heat exchanges (for example inter-cooling) and neglecting, at least in the preliminary design, the process in the channel linking the stator outlet station with the rotor inlet station of the next stage, are (Fig. 6.12):

$$
\begin{aligned}
p_{in,i} &= p_{out,i-1}\\
T_{in,i} &= T_{out,i-1}\\
h_{in,i} &= h_{out,i-1}\\
s_{in,i} &= s_{out,i-1}\\
\rho_{in,i} &= \rho_{out,i-1}
\end{aligned}
$$

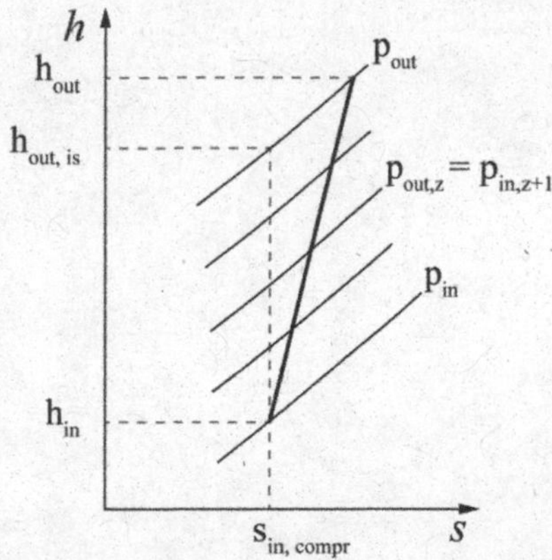

Fig. 6.12 Compression process in a multistage turbomachine

$$h_{out,is,i} = h_{in} + \sum_{j=1}^{i} \Delta h_{is,j}$$

$$p_{out,i} = p(h_{out,is,i}, s_{in})$$

In the previous correlations, the isentropic enthalpy increases of each stage are evaluated according to the procedure described in Chap. 2.

To proceed with the calculation of the stages following the first one, specific assumptions must be made: for example, the rotor outlet diameter can be considered constant or we can assign anew the flow and work coefficients or we can calculate anew the specific speed, etc. In more details, following the procedure based on input parameters set A and described in Sect. 6.5, the values of the input parameters (ψ, φ, R, η_{is}, α_1 and δ_h) could be reassigned for the stages following the first one (for example, equal to those of the first stage) while, following the procedure based on input parameters set B, the specific speed must be calculated anew and consequently all the other input parameters that are functions of this specific speed.

Through these input parameters, the kinematics (Sect. 6.1), the thermodynamics (Sect. 6.2) and the stage geometry (Sect. 6.3) are calculated and then, the losses of each stage are assessed (Sect. 6.4).

The compressor design, carried out by assigning a pair of values for the speed of rotation and the number of stages (n, z), provided by the selection process illustrated in Chap. 2, can be accomplished again (Fig. 6.13) by using a different pair of input parameters (n and z), indicated always by the turbomachinery selection process, in order to optimize the preliminary design of the compressor.

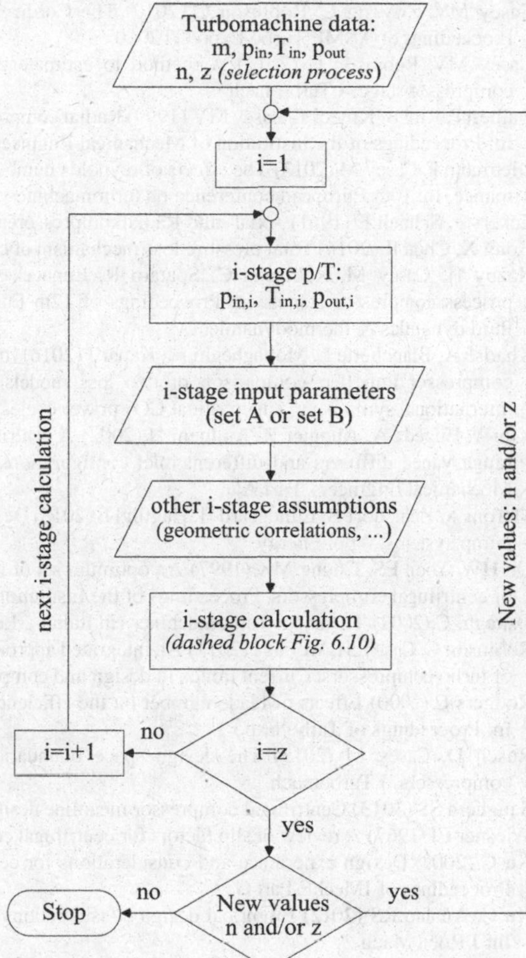

Fig. 6.13 Multistage compressor calculation block diagram

References

Ameli A, Afzalifar A, Turunen-Saaresti T, Backman J (2018) Centrifugal compressor design for near-critical point applications. J Eng Gas Turbin Power

Aungier RH (2000) Centrifugal compressors, A strategy for aerodynamic design and analysis. ASME Press, New York

Blanchette L, Khadse A, Mohagheghi M, Kapat J (2016) Two types of analytical methods for a centrifugal compressor impeller for supercritical CO_2 power cycles. In: 14th international energy conversion engineering conference

Bommes L, Fricke J, Grundmann R (2003) Ventilatoren. Vulkan-Verlag, Essen, Germany

Came PM, Robinson CJ (1998) Centrifugal compressor design. In: Proceedings of the institution of mechanical engineers, Part C: J Mech Eng Sci

Casey MV, Fesich TM (2009) On the efficiency of compressors with diabatic flows. In: Proceedings of ASME Turbo Expo GT 2009

Casey MV, Zwyssig C, Robinson C (2010) The Cordier line for mixed flow compressors. In: Proceedings of ASME Turbo Expo GT2010

Casey MV, Robinson C (2013) A method to estimate the performance map of a centrifugal compressor stage. J Turbomach

Dalbert P, Ribi B, Kmeci T, Casey MV (1999) Radial compressor design for industrial compressors. In: Proceedings of the Institution of Mechanical Engineers, Part C

Dietmann F, Casey M (2013) The effects of reynolds number and roughness on compressor performance. In: 10th European conference on turbomachinery fluid dynamics & thermodynamics

Eckert B, Schnell E (1961) Axial -und Radialkompressoren. Springer

Gong X, Chen R (2014) Total pressure loss mechanism of centrifugal compressors. Mech Eng Res

Hazby H, Casey M, Robinson C, Spataro R, Lunacek O (2017) The design of a family of process compressor stages. In: Proceedings of 12th European conference on turbomachinery fluid dynamics & thermodynamics

Khadse A, Blanchette L, Mohagheghi M, Kapat J (2016) Impact of sCO_2 properties on centrifugal compressor impeller: comparison of two loss models for mean line analyses. In: The 5th international symposium supercritical CO_2 power cycles

Kim Y, Engeda A, Aungier R, Amineni N (2002) A centrifugal compressor stage with wide flow range vaned diffusers and different inlet configurations. In: Proceedings of the Institution of Mechanical Engineers, Part A

Meroni A, Zühlsdorf B, Elmegaard B, Haglind F (2018) Design of centrifugal compressors for heat pump systems. Appl Energy

Oh HW, Yoon ES, Chung MK (1997) An optimum set of loss models for performance prediction of centrifugal compressors. Proceedings of the Institution of Mechanical Engineers, Part A

Osnaghi C (2002) Theory of Turbomachines (in Italian). Leonardo (ed)

Robinson C, Casey M, Woods I (2011) An integrated approach to the aero-mechanical optimisation of turbo compressors. Current trends in design and computation of turbomachinery

Rodgers C (2000) Effects of blade number on the efficiency of centrifugal compressor impellers. In: Proceedings of Turbo Expo

Rusch D, Casey M (2013) The design space boundaries for high flow capacity centrifugal compressors. J Turbomach

Sanghera SS (2013) Centrifugal compressor meanline design using real gas properties

Wiesner FJ (1967) A review of slip factors for centrifugal compressors. J Eng Power, Trans ASME

Xu C (2007) Design experience and considerations for centrifugal compressor development. In: Proceedings of IMechE Part G

Xu C, Amano RS (2012) Empirical design considerations for industrial centrifugal compressors. Int J Rotat Mach

Chapter 7
Preliminary Design of Centrifugal Pumps

Abstract The preliminary design of a centrifugal pump is very similar to that of a centrifugal compressor, even if the fluid incompressibility and the phenomenon of *cavitation*, which is specific for the hydraulic machines, must be taken into account. For these reasons, we devote a specific chapter to pumps. With reference to a centrifugal pump stage, a procedure for the calculation of kinematic parameters (Sect. 7.1), thermodynamic parameters (Sect. 7.2), geometric parameters (Sect. 7.3) and stage losses (Sect. 7.4) is provided. Then, Sect. 7.5 discusses the input parameters of these procedures and suggest their numerical values to be used in the calculations. Finally, Sect. 7.6 illustrates the procedure for extending calculations to multi-stage pumps. The numerical application of the proposed procedure, aimed at the preliminary design of centrifugal pumps, is instead developed in Chap. 8.

7.1 Kinematic Parameters

In this section, we will deduce the kinematic correlations of a centrifugal stage; they are correlations among kinematic parameters of a stage (degree of reaction R, absolute and relative flow angles, α and β) and dimensionless performance parameters (flow coefficient, φ, and work coefficient, ψ).

Regarding the dimensionless coefficients, those already introduced for centrifugal compressors are used (Chap. 6).

In particular, the ratio of the meridional velocity at the rotor inlet, c_{1m}, to the blade speed at the rotor outlet, u_2, is the *flow coefficient*:

$$\varphi = \frac{c_{1m}}{u_2}$$

With regard to the *work coefficient*, we refer to the actual work exchanged between the fluid flow and the rotor (Chap. 1, Sect. 1.5) through the expression:

$$\psi = \frac{W}{u_2^2} = \frac{g \cdot \Delta H}{\eta_{hyd} \cdot u_2^2}$$

M. Gambini and M. Vellini, *Turbomachinery*, Springer Tracts in Mechanical Engineering, https://doi.org/10.1007/978-3-030-51299-6_7

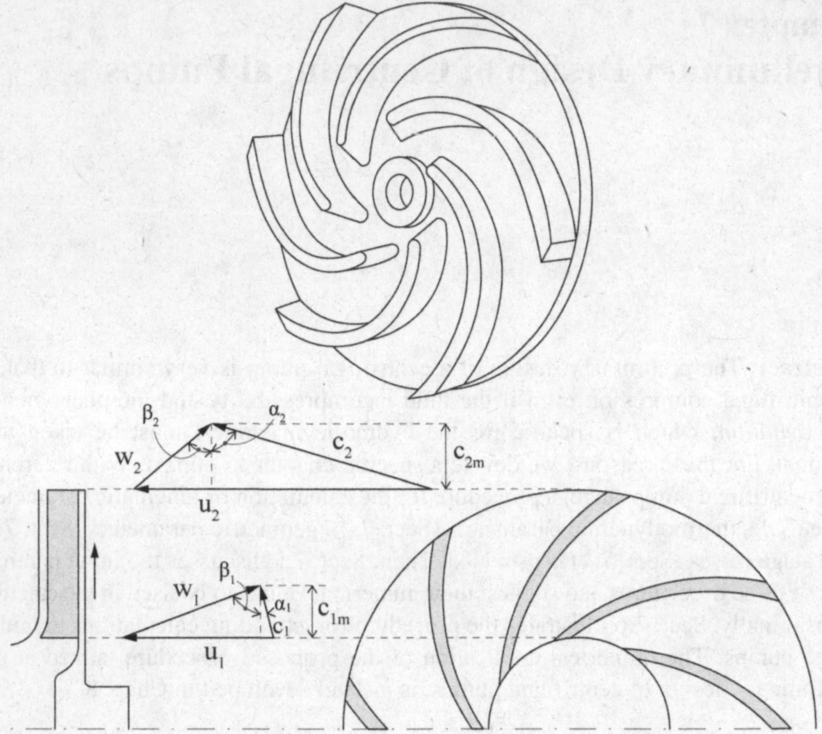

Fig. 7.1 - A centrifugal pump impeller (top) and velocity triangles (bottom)

In a radial stage (Fig. 7.1), the fluid flow is studied (Sect. 1.7) on two stage sections: one perpendicular to the axis of rotation (Fig. 7.1, bottom right) and the other one containing the axis of rotation (meridional section, Fig. 7.1 bottom left). With reference to these two sections, Fig. 7.1 also depicts the velocity triangles of a centrifugal pump stage.

In the following, the kinematic correlations will be expressed by using $\alpha_1 \neq 0$, even if, in a centrifugal pump stage, the inlet absolute velocity is generally perpendicular to the blade speed ($\alpha_1 = 0$), unless it is necessary to adopt $\alpha_1 > 0$ for avoiding cavitation problems.

Two parameters are then introduced: they are the *rotor meridional velocity ratio*, ξ, and the *rotor tip diameter ratio*, δ_t:

$$\xi = \frac{c_{2m}}{c_{1m}}$$

$$\delta_t = \frac{D_{1t}}{D_2}$$

The fluid flow enters the rotor radially (Fig. 7.2-left) or, more generally (Fig. 7.2-right), in a direction intermediate between axial and radial. Referring to the latter

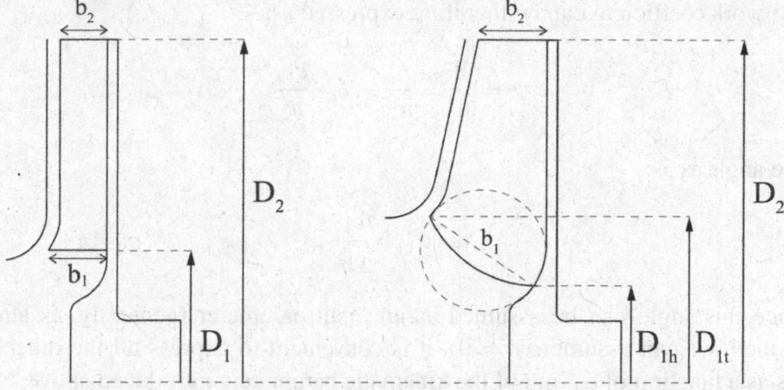

Fig. 7.2 Impeller inlet geometry: radial (left) and mixed (right)

case, the same rotor diameter ratios, δ_t and δ_h, introduced for the compressors, can be used. In this case, to evaluate b_1 it is also necessary to set the blade inclination. But, if a constant channel height between axial and radial direction at the impeller inlet is assumed, b_1 can be expressed as in centrifugal compressors:

$$b_1 = \frac{D_{1t} - D_{1h}}{2} = D_2 \cdot \frac{\delta_t - \delta_h}{2}$$

where, however, D_{1h} is the hub diameter along the vertical from the blade tip (see Fig. 6.5, Sect. 6.3). On the other hand, if the fluid flow enters the rotor radially, we can assume:

$$\delta_t = \frac{D_{1t}}{D_2} = \frac{D_1}{D_2}$$

and we can choose as independent geometric parameter the blade height b_1 (or the dimensionless parameter b_1/D_2) in place of D_{1h} (or δ_h).

In order to the express the work transfer, referring to the velocity triangle at the inlet tip station, the rotor tip diameter ratio, δ_t, can be expressed through the blade speed ratio:

$$\delta_t = \frac{D_{1t}}{D_2} = \frac{u_{1t}}{u_2} = \frac{u_1}{u_2}$$

Next, assuming the angles shown in Fig. 7.1 as positive, Euler's equation (Sect. 1.6.1) provides:

$$W = u_2 \cdot c_{2u} - u_1 \cdot c_{1u}$$

The work coefficient can be therefore expressed as:

$$\psi = \frac{W}{u_2^2} = \frac{c_{2u}}{u_2} - \delta_t \cdot \frac{c_{1u}}{u_2}$$

The angle α_1 is:

$$\tan\alpha_1 = \frac{c_{1u}}{c_{1m}}$$

Since this angle can be assumed as an input parameter (generally, as already mentioned, we can assume $\alpha_1 = 0$), it is convenient to express all the other flow angles as a function of α_1 and of the kinematic parameters introduced above.

Following this approach, the rotor outlet absolute flow angle is expressed as:

$$\tan\alpha_2 = \frac{c_{2u}}{c_{2m}} = \frac{c_{2u}}{u_2} \cdot \frac{u_2}{c_{1m}} \cdot \frac{c_{1m}}{c_{2m}} = \frac{c_{2u}}{u_2} \cdot \frac{1}{\varphi \cdot \xi} = \frac{1}{\varphi \cdot \xi} \cdot (\psi + \varphi \cdot \delta_t \cdot \tan\alpha_1)$$

and thus we obtain:

$$\tan\alpha_2 = \frac{\psi}{\xi \cdot \varphi} + \frac{\delta_t}{\xi} \cdot \tan\alpha_1$$

The rotor relative flow angles are expressed as:

$$\tan\beta_1 = \frac{w_{1u}}{c_{1m}} = \frac{u_1 - c_{1u}}{c_{1m}} = \frac{\delta_t}{\varphi} - \tan\alpha_1$$

$$\tan\beta_2 = \frac{w_{2u}}{c_{2m}} = \frac{u_2 - c_{2u}}{c_{2m}} = \frac{u_2}{c_{1m}} \cdot \frac{c_{1m}}{c_{2m}} - \tan\alpha_2 = \frac{1}{\varphi \cdot \xi} - \tan\alpha_2$$

and thus we obtain:

$$\tan\beta_1 = \frac{\delta_t}{\varphi} - \tan\alpha_1$$

$$\tan\beta_2 = \frac{1}{\varphi \cdot \xi} \cdot (1 - \psi) - \frac{\delta_t}{\xi} \cdot \tan\alpha_1$$

The degree of reaction is given by:

$$R = \frac{W - \Delta E_{C,R}}{W} = 1 - \frac{\Delta E_{C,R}}{\psi \cdot u_2^2} = 1 - \frac{c_2^2 - c_1^2}{2 \cdot \psi \cdot u_2^2} = 1 - \frac{c_{2u}^2 + c_{2m}^2 - c_{1u}^2 - c_{1m}^2}{2 \cdot \psi \cdot u_2^2}$$

$$= 1 - \frac{c_{2m}^2 \cdot \left(1 + \tan^2\alpha_2\right) - c_{1m}^2 \cdot \left(1 + \tan^2\alpha_1\right)}{2 \cdot \psi \cdot u_2^2}$$

and thus it is:

$$R = 1 - \frac{\psi}{2} + \frac{\varphi^2}{2 \cdot \psi} \cdot \left[\left(1 - \xi^2\right) + \tan^2 \alpha_1 \cdot \left(1 - \delta_t^2\right)\right] - \varphi \cdot \delta_t \cdot \tan \alpha_1$$

From this equation we infer that, unlike what happens in axial stages, in radial stages angle α_1 can be assumed as input parameter (for example, $\alpha_1 = 0$) keeping R, φ and ψ (or ξ, φ and ψ) as independent parameters.

The previous equation allows also to express ξ as a function of R:

$$\xi = \sqrt{1 - \frac{2 \cdot \psi}{\varphi^2} \cdot \left(R + \frac{\psi}{2} - 1\right) + \tan^2 \alpha_1 \cdot \left(1 - \delta_t^2\right) - \frac{2 \cdot \psi \cdot \delta_t}{\varphi} \cdot \tan \alpha_1}$$

Consequently there is a constraint on the maximum value of the degree of reaction. For example, if $\alpha_1 = 0$, it must be:

$$R < 1 + \frac{\varphi^2}{2 \cdot \psi} - \frac{\psi}{2}$$

Remarks on input parameters

From the kinematic correlations reported above, we infer that once the *five independent parameters* φ, ψ, R (or ξ), δ_t and α_1 are set (Sect. 7.5), all flow angles, α_2, β_1, β_2, can be calculated. Once the blade speed, u_2, is also established (in order to determine this speed, it is necessary to know, in addition to the work coefficient ψ, also the stage hydraulic efficiency, as illustrated in Sect. 7.5), the kinematics of the rotor can be completely calculated:

$$\begin{aligned}
c_{1m} &= u_2 \cdot \varphi \\
c_{1u} &= u_2 \cdot \varphi \cdot \tan \alpha_1 \\
c_{2m} &= u_2 \cdot \xi \cdot \varphi \\
c_{2u} &= u_2 \cdot \xi \cdot \varphi \cdot \tan \alpha_2 \\
w_{1u} &= u_2 \cdot \varphi \cdot \tan \beta_1 \\
w_{2u} &= u_2 \cdot \xi \cdot \varphi \cdot \tan \beta_2 \\
c_1 &= u_2 \cdot \varphi \cdot \sqrt{1 + \tan^2 \alpha_1} \\
c_2 &= u_2 \cdot \xi \cdot \varphi \cdot \sqrt{1 + \tan^2 \alpha_2} \\
w_1 &= u_2 \cdot \varphi \cdot \sqrt{1 + \tan^2 \beta_1} \\
w_2 &= u_2 \cdot \xi \cdot \varphi \cdot \sqrt{1 + \tan^2 \beta_2} \\
u_1 &= u_2 \cdot \delta_t
\end{aligned}$$

In order to complete the kinematics of the stage, the velocity c_3, the absolute velocity at the stator outlet, must also be determined.

Generally we assume:

$$c_3 = c_1$$

If this assumption does not guarantee the respect of limitations and constraints introduced in the Sect. 7.3, c_3 must be recalculated. In this case, since $c_3 \neq c_1$, the head must be calculated anew.

7.2 Thermodynamic Parameters

The pressure p_1 and temperature T_1 at the rotor inlet and the pressure p_3 at the stator outlet are assigned based on what is illustrated in Chap. 2 (for multistage pumps, these values are calculated by the procedure described in Sect. 7.6).

Now, all the thermodynamic parameters at the rotor inlet and at the stator inlet/outlet can be calculated as follows.

If the working fluid cannot be assimilated to a perfect liquid, for the following calculations, we need appropriate state functions, such as those of the NIST libraries. In this case, the procedure is equal to that followed for centrifugal compressors and therefore it is completely valid what is illustrated in Sect. 6.2. Otherwise, in the case of perfect liquids (Sect. 1.3.2), the density is constant throughout the stage:

$$\rho = const$$

and it is necessary to calculate only the pressure at the rotor outlet and the head of the pump.

To calculate the pressure at the rotor outlet, we must know the rotor efficiency, that can be defined as:

$$\bar{\eta}_{hyd,R} = \frac{\Delta p_R / \rho \cdot g}{\Delta p_R / \rho \cdot g + Z_{H,R}} = \frac{\Delta p_R / \rho}{\frac{u_2^2 - u_1^2}{2} + \frac{w_1^2 - w_2^2}{2}}$$

where $Z_{H,R}$ represents the head losses in the rotor (Sect. 7.5). By using this rotor efficiency, we can calculate the pressure at the rotor outlet as:

$$p_2 = p_1 + \bar{\eta}_{hyd,R} \cdot \frac{\rho}{2} \cdot \left(u_2^2 - u_1^2 + w_1^2 - w_2^2\right)$$

The head (Sect. 1.5), considering $c3 \neq c1$ (Sect. 7.1), is then:

$$\Delta H = \frac{p_3 - p_1}{\rho \cdot g} + \frac{c_3^2 - c_1^2}{2 \cdot g}$$

By using the hydraulic efficiency (Sect. 1.5), the pump work can therefore be calculated; in fact we have:

$$\eta_{hyd} = \frac{g \cdot \Delta H}{W}$$

and thus we obtain:

$$W = u_2 \cdot c_{2u} - u_1 \cdot c_{1u} = g \cdot \left(\Delta H + Z_{H,R} + Z_{H,S}\right) = g \cdot (\Delta H + Z_H)$$

where Z_H represents all the head losses in the stage.

Finally, by using the pressure and temperature conditions of the working fluid, we can calculate the fluid viscosity (necessary to assess the Reynolds number, Sect. 7.4):

$$\mu = \mu_{p,T}(p_1, T_1)$$

Remarks on input parameters

Based on what is illustrated in this Sect. 7.2, in addition to the parameters already identified in Sect. 7.1 (φ, ψ, R, δ_t, α_1 and η_{hyd}), the rotor efficiency, $\bar{\eta}_{hyd,R}$, must be assumed in order to calculate the thermodynamics of the stage.

7.3 Geometric Parameters

Now we will deduce the geometric correlations of a centrifugal pump stage; they are correlations among geometric (rotor and stator diameters, blade height, chord and blade pitch, etc.), thermodynamic (density, volumetric flow rate) and kinematic parameters (meridional velocities, absolute velocity angles α, relative velocity angles β) of the stage.

The rotor outlet diameter (D_2) is determined by the kinematics of the stage (Sect. 7.1), and in particular by the knowledge of the blade speed at the rotor outlet (u_2) and the speed of rotation (n, rpm) (Sect. 7.5):

$$D_2 = \frac{60}{\pi} \cdot \frac{u_2}{n}$$

Before evaluating the rotor and stator geometry, we must highlight the differences between the geometric parameters in the radial and axial stages. In particular, in the axial stages, after determining the blade height through the equation of continuity, the geometry of the rotor and stator channels (Chaps. 3–4) was calculated by using two parameters: the aspect ratio and the solidity. In the radial stages, on the other hand, it is not simple the definition of a rotor chord; moreover, the blade heights (especially in mixed-flow pumps) and the rotor diameters change between the rotor inlet and outlet. For these reasons, in order to evaluate the stage losses, an equivalent rotor channel, having a circular section of diameter D_{hyd} (hydraulic diameter) and length L_{hyd} (hydraulic length), is used (Sect. 7.4).

The hydraulic diameter is defined as:

$$D_{hyd} = 4 \cdot \frac{Cross\ sectional\ area}{Wetted\ Perimeter}$$

To define the hydraulic length, first, we must introduce the channel length on the meridional plane, that is, the plane containing the axis of rotation:

$$L_m = Meridional\ length$$

The hydraulic length is correlated with the meridional length through the angle β_M (for the rotor or α_M for the stator) representing the mean inclination of the rotor/stator channel on a plane perpendicular to the axis of rotation; these mean angles are similar to the stagger angle in the axial stages:

$$L_{hyd} = \frac{L_m}{\cos \beta_M}$$

For the stator, even if a chord can be defined as for the axial stages, for homogeneity of the discussion with the rotor, we define the same parameters: the hydraulic diameter and hydraulic length. For the stator, we will use also the solidity, σ, since the hydraulic length substantially corresponds to the chord, but we must specify the station where the blade pitch is calculated, as it is variable between the stator inlet and outlet.

Now, we can illustrate the geometry evaluation of the rotor and stator for a centrifugal pump stage.

Rotor geometry

The volumetric flow rate in the stage is constant:

$$V = \frac{m}{\rho}$$

By using this volumetric flow rate, knowing the meridional velocities c_{1m} and c_{2m} (Sect. 7.1), we can calculate the flow areas:

$$A_{1,2} = \frac{V}{c_{1,2,m}}$$

and, therefore, considering a section just downstream of the rotor outlet (where no blade blockage occurs, Sect. 1.8), the blade height at the rotor outlet (Fig. 7.3):

$$b_2 = \frac{A_2}{\pi \cdot D_2}$$

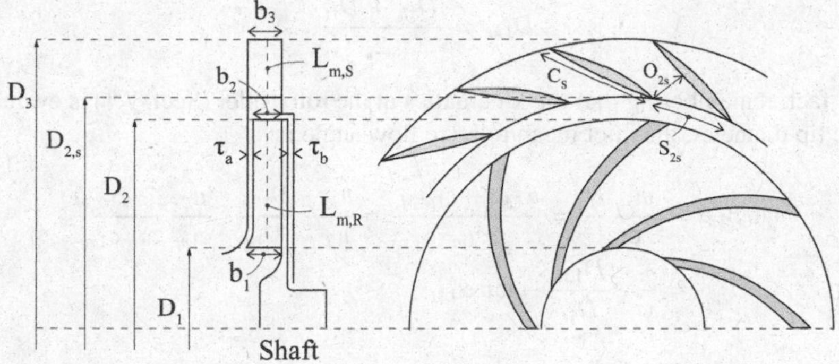

Fig. 7.3 Stage geometry

The flow area at the rotor inlet can be correlated to the rotor diameter ratios (Sect. 7.1); in fact we have:

$$A_1 = \frac{\pi}{4} \cdot \left(D_{1t}^2 - D_{1h}^2\right) = \frac{\pi}{4} \cdot D_2^2 \cdot \left(\delta_t^2 - \delta_h^2\right)$$

and, as a consequence, we can obtain the rotor tip diameter ratio:

$$\delta_t = \sqrt{\delta_h^2 + \frac{4 \cdot A_1}{\pi \cdot D_2^2}}$$

since the rotor hub diameter ratio is defined as:

$$\delta_h = \frac{D_{1h}}{D_2}$$

After choosing δ_h, for example, the following value is generally suggested:

$$\delta_h = 0.3$$

or δ_t (Sect. 7.5), the rotor inlet geometry is determined:

$$D_{1h} = \delta_h \cdot D_2$$

$$D_{1t} = \delta_t \cdot D_2$$

$$b_1 = \frac{D_{1t} - D_{1h}}{2} = \frac{D_2}{2} \cdot (\delta_t - \delta_h)$$

By using these parameters, we can calculate the inlet mean flow angle, that is, the flow angle at the inlet mean diameter:

$$D_{1M} = \frac{D_{1h} + D_{1t}}{2}$$

In fact, remembering that the kinematics at the rotor inlet (Sect. 7.1) is evaluated at the tip diameter, the inlet mean relative flow angle is:

$$\tan\beta_{1M} = \frac{w_{1u,M}}{c_{1m}} = \frac{u_{1M} - c_{1u,M}}{c_{1m}} = \frac{u_{1t}}{u_2} \cdot \frac{D_{1M}}{D_{1t}} \cdot \frac{u_2}{c_{1m}} - \frac{c_{1u,M}}{c_{1m}}$$
$$= \frac{\delta_t}{\varphi} \cdot \frac{D_{1M}}{D_{1t}} - \tan\alpha_{1M}$$

where, the mean absolute flow angle is:

$$\tan\alpha_{1M} = \frac{c_{1u,M}}{c_{1m}} = \frac{c_{1u,t}}{c_{1m}} \cdot \frac{c_{1u,M}}{c_{1u,t}} = \tan\alpha_1 \cdot \frac{D_{1t}}{D_{1M}}$$

and the ratio between the mean and tip diameters is:

$$\frac{D_{1M}}{D_{1t}} = \frac{1}{2} \cdot \left(1 + \frac{\delta_h}{\delta_t}\right)$$

As a consequence, we can write:

$$\tan\beta_{1M} = \frac{1}{2 \cdot \varphi} \cdot (\delta_t + \delta_h) - \frac{2 \cdot \delta_t \cdot \tan\alpha_1}{\delta_t + \delta_h}$$

Therefore, the mean relative velocity is:

$$w_{1M} = \frac{c_{1m}}{\cos\beta_{1M}}$$

If the fluid flow enters the rotor purely radially (Fig. 7.2 left), it is:

$$\beta_{1M} = \beta_1$$
$$w_{1M} = w_1$$
$$D_{1t} = D_{1h} = D_{1M} = D_1$$

To complete the rotor geometry, we must consider that the outlet flow angle (β_2) is different from the blade angle (β_{2B}) because the fluid flow will receive less than perfect guidance from the vanes, due to the finite number of blades in the rotor (Fig. 7.4). In order to take into account this phenomenon, also called *slip*, a *slip factor* may be defined as:

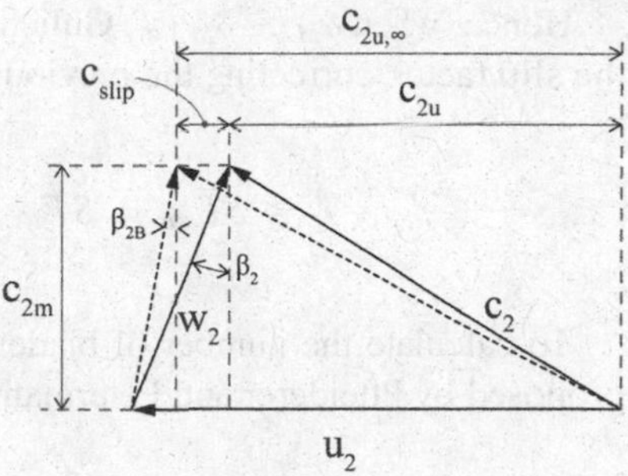

Fig. 7.4 Effect of slip on velocity triangles at the rotor outlet

$$SF = \frac{c_{2u}}{c_{2u,\infty}} = \frac{c_{2u,\infty} - c_{slip}}{c_{2u,\infty}} = 1 - \frac{c_{slip}}{c_{2u,\infty}}$$

where $c_{2u,\infty}$ is the absolute tangential velocity with an infinite number of infinitely thin blades (the fluid flow is perfectly guided by the blade at the rotor outlet and hence the flow angle and the blade angle are equal $\beta_{2,\infty} = \beta_{2B}$).

The slip factor can be used to express the blade outlet angle such as to ensure the effective outlet flow angle:

$$\tan\beta_{2B} = \frac{w_{2u,\infty}}{c_{2m}} = \frac{u_2 - c_{2u,\infty}}{c_{2m}} = \frac{u_2}{c_{2m}} - \frac{c_{2u}}{SF \cdot c_{2m}}$$

and, by the kinematic correlations deduced in Sect. 7.1, we obtain:

$$\tan\beta_{2B} = \frac{1}{\xi \cdot \varphi} - \frac{\tan\alpha_2}{SF}$$

In order to calculate the slip factor and hence the blade outlet angle, we can use the correlation proposed by Wiesner (1967):

$$SF = 1 - \frac{\sqrt{\cos\beta_{2B}}}{N_{B,R}^{0.7}}$$

with the correction suggested by Gulich (2010) when the rotor mean diameter ratio, defined as:

$$\delta_M = \frac{D_{1M}}{D_2} = \frac{\delta_t + \delta_h}{2}$$

exceeds the following limit value:

$$\delta_{M,\lim} = \exp\left(-\frac{8.16 \cdot \cos\beta_{2B}}{N_{B,R}}\right)$$

Hence, when $\delta_M > \delta_{M,\lim}$, Gulich (2010) suggested the following expression of the slip factor, correcting the previous correlation proposed by Wiesner:

$$SF_{cor} = SF \cdot \left[1 - \left(\frac{\delta_M - \delta_{M,\lim}}{1 - \delta_{M,\lim}}\right)^3\right]$$

To calculate the number of blades, some authors suggested to use the equation proposed by Pfleiderer and Petermann (1986):

$$N_{B,R} = \text{int}\left[6.5 \cdot \left(\frac{1/\delta_t + 1}{1/\delta_t - 1}\right) \cdot \cos\beta_M\right]$$

where:

$$\beta_M = \frac{\beta_{1M} + \beta_{2B}}{2}$$

or the equation proposed by Stepanoff (1959):

$$N_{B,R} = \text{int}\left[\frac{90 - \beta_{2B}}{3}\right]$$

Gulich (2010) stated that these correlations are not suitable for calculating the number of rotor blades and proposed to adopt the following values:

$$N_{B,R} = 5 \div 7 \quad \text{when} \quad 0.2 < \omega_s < 2$$

To establish the exact number of blades Gulich (2010) then recommended checking the blade loading:

$$BL = \frac{2 \cdot \pi \cdot \psi}{N_{B,R} \cdot \frac{L_{hyd,R}}{D_{hyd,R}} \cdot \left(\frac{w_{1M} + w_2}{u_2}\right)}$$

This BL value must not exceed the following allowable value:

$$BL_{allowable} = \left(\frac{0.756}{\omega_s}\right)^{0.77}$$

More precisely, it must be imposed:

$$BL \leq 0.9 \cdot BL_{allowable}$$

To calculate the blade loading parameter, BL, however, it is necessary to assess the hydraulic length and the hydraulic diameter.

To this end, first, let us consider the blade pitch at the rotor inlet (considering the mean diameter):

$$S_{1,R} = \frac{\pi \cdot D_{1M}}{N_{B,R}} = \frac{\pi \cdot (D_{1t} + D_{1h})}{2 \cdot N_{B,R}}$$

and the blade pitch at the rotor outlet:

$$S_{2,R} = \frac{\pi \cdot D_2}{N_{B,R}}$$

As previously mentioned, the hydraulic diameter is defined as:

$$D_{hyd} = 4 \cdot \frac{Cross\ sectional\ area}{Wetted\ Perimeter}$$

To evaluate the wetted perimeter, and the cross sectional area delimited by this perimeter, it is necessary to calculate the blade distance at the leading edge, O_1, and at the trailing edge, O_2, following what was done for the axial stages.

By assuming as right triangles the triangles having a hypotenuse equal to S, a base equal to O and an angle between them equal to β, we obtain:

$$O_1 = S_{1,R} \cdot \cos\beta_{1M}$$
$$O_2 = S_{2,R} \cdot \cos\beta_{2B}$$

Therefore, the cross sectional areas and the wetted perimeters are:

$$A_{w1} = O_1 \cdot b_1$$
$$P_{w1} = 2 \cdot (O_1 + b_1)$$
$$A_{w2} = O_2 \cdot b_2$$
$$P_{w2} = 2 \cdot (O_2 + b_2)$$

and the hydraulic diameters at the rotor inlet and outlet become:

$$D_{hyd,1} = 2 \cdot \frac{O_1 \cdot b_1}{O_1 + b_1}$$

$$D_{hyd,2} = 2 \cdot \frac{O_2 \cdot b_2}{O_2 + b_2}$$

The rotor hydraulic diameter is the mean hydraulic diameter between the rotor inlet and outlet:

$$D_{hyd,R} = \frac{D_{hyd,1} + D_{hyd,2}}{2} = \frac{O_1 \cdot b_1}{O_1 + b_1} + \frac{O_2 \cdot b_2}{O_2 + b_2}$$

To calculate the hydraulic length L_{hyd}, as previously illustrated, first, we must assess the channel length, L_m, on the meridional plane (Fig. 7.3).

For a centrifugal pump, this meridional length does not depend on the channel axial length and can be simply calculated by:

$$L_{m,R} = \frac{D_2 - D_{1M}}{2} = \frac{1}{2} \cdot \left(D_2 - \frac{D_{1t} + D_{1h}}{2} \right)$$

After assessing the meridional length, the hydraulic length is:

$$L_{hyd,R} = \frac{L_{m,R}}{\cos \beta_M} = \frac{L_{m,R}}{\cos\left(\frac{\beta_{1M}+\beta_{2B}}{2}\right)}$$

The previous three equations allow to calculate, iteratively, the three parameters β_{2B}, SF and $N_{B,R}$.

Finally, the blade thickness can be assumed constant along the blade height and equal to:

$$\frac{t_B}{D_2} \cong 0.02$$

Regarding clearances, we can assume:

$$\frac{\tau}{b_2} = 2.0 \div 5.0\%$$

but the minimum values must be not lower than $\tau = 0.3$ mm $\div$ 0.5 mm.

In absence of specific data, we can choose:

$$\frac{\tau_a}{b_2} = \frac{\tau_r}{b_2} = \frac{\tau_b}{b_2} = 0.05$$

where (Fig. 7.3):

- τ_a is the axial clearance (blade at rotor outlet)
- τ_r is the radial clearance (blade at rotor inlet)
- τ_b is the backface clearance (on the back of the rotor disc)

Stator geometry

At the rotor outlet, the fluid enters the diffuser (stator). Generally the diffuser is composed of a vaneless diffuser followed by a vaned diffuser. By indicating with 2S and 3 (Fig. 7.3) respectively the vaned diffuser inlet and outlet, the vaneless diffuser extends from the rotor outlet (station 2) to the vaned diffuser inlet (2S).

To limit pressure pulsations, Gulich (2010) suggested to calculate the diameter at the station 2S through the following correlations:

$$D_{2S} = D_2 \cdot \left[1.015 + 0.08 \cdot \left(\frac{\rho \cdot \Delta H}{10^6} - 0.1\right)^{0.8}\right] \quad \text{when} \quad \omega_s \leq 0.75$$

$$D_{2S} = D_2 \cdot [1.04 + 0.053 \cdot (\omega_s - 0.75)] \quad \text{when} \quad \omega_s > 0.75$$

where ΔH is the stage head.

In any case, D_{2S} must be:

$$D_{2S} \geq 1.015 \cdot D_2$$

Ignoring friction, the angular momentum conservation (Sect. 1.7.4) in the vaneless diffuser allows calculating the absolute tangential velocity at the vaned diffuser inlet:

$$c_{2Su} = c_{2u} \cdot \frac{D_2}{D_{2S}}$$

and, therefore, by assuming:

$$\frac{b_{2S}}{b_2} = 1.1$$

or more generally:

$$\frac{b_{2S}}{b_2} = 1.05 \div 1.3$$

we can calculate the meridional velocity at the vaneless diffuser outlet:

$$c_{2Sm} = \frac{V}{\pi \cdot D_{2S} \cdot b_{2S}}$$

As a consequence, the flow angle at the vaned diffuser inlet is:

$$\tan \alpha_{2S} = \frac{c_{2Su}}{c_{2Sm}}$$

The vaneless diffuser geometry is then completed with the calculation of the hydraulic length ($L_{hyd,vaneless}$), which corresponds to the diffuser radial length:

$$L_{hyd,vaneless} = L_{m,vaneless} = \frac{D_{2S} - D_2}{2}$$

and with the hydraulic diameter ($D_{hyd,vaneless}$):

$$D_{hyd,vaneless} = \frac{D_{hyd,2} + D_{hyd,2S}}{2} = \frac{1}{2} \cdot \left(\frac{4 \cdot \pi \cdot D_2 \cdot b_2}{2 \cdot \pi \cdot D_2} + \frac{4 \cdot \pi \cdot D_{2S} \cdot b_{2S}}{2 \cdot \pi \cdot D_{2S}}\right)$$

$$= b_2 + b_{2S}$$

Still following Gulich's suggestions (2010), the geometry of the vaned diffuser can be defined.

The outlet diameter (Fig. 7.3) can be calculated by:

$$D_3 = D_2 \cdot (1.1 + 0.53 \cdot \omega_s)$$

Generally, the vaned diffuser width is assumed constant:

$$b_3 = b_{2S}$$

Consequently, we can calculate:

$$c_{3m} = \frac{V}{\pi \cdot D_3 \cdot b_3}$$

If this velocity is lower than the previously assumed c_3 (Sect. 7.1), we can assess:

$$\cos \alpha_3 = \frac{c_{3m}}{c_3}$$

otherwise, c_3 is recalculated as:

$$c_3 = c_{3m} \quad ; \quad \alpha_3 = 0$$

and, consequently, the stage thermodynamics must be calculated anew accordingly to this new velocity (Sect. 7.2).

In any case, the calculation of the stator geometry and kinematics does not end here, since some limits and constraints, illustrated later, must be satisfied.

The number of diffuser blades can be chosen by using (Table 7.1) correlating number of rotor and stator blades in order to limit pressure pulsations and vibrations.

In particular, the following values of the stator blades (printed bold in Table 7.1) are considered best with respect to low vibrations:

$$N_{B,R} = 5 \Rightarrow N_{B,S} = 8$$
$$N_{B,R} = 6 \Rightarrow N_{B,S} = 10$$
$$N_{B,R} = 7 \Rightarrow N_{B,S} = 10 \div 11$$

Table 7.1 Impeller-diffuser vane combination (Gulich 2010)

$N_{B,R}$	5			6	7				
$N_{B,S}$	7	**8**	12	**10**	9	**10**	**11**	12	(15)

The vaned diffuser geometry is completed by calculating the parameters illustrated below.

Since the stator extends in a completely radial direction, the meridional length, L_m, is:

$$L_{m,vaned} = \frac{D_3 - D_{2S}}{2}$$

Therefore, the hydraulic length (which, in this case, corresponds to the chord) is:

$$L_{hyd,vaned} = C_S = \frac{L_{m,vaned}}{\cos \alpha_M} = \frac{D_3 - D_{2S}}{2 \cdot \cos\left(\frac{\alpha_{2S}+\alpha_3}{2}\right)}$$

The blade pitch at the stator inlet is:

$$S_{2S} = \frac{\pi \cdot D_{2S}}{N_{B,S}}$$

and at the stator outlet is:

$$S_3 = \frac{\pi \cdot D_3}{N_{B,S}}$$

The solidity is:

$$\sigma_S = \left(\frac{C}{S}\right)_S = \frac{D_3 - D_{2S}}{2 \cdot S_3 \cdot \cos\left(\frac{\alpha_3+\alpha_{2S}}{2}\right)}$$

To evaluate the hydraulic diameter, it is necessary to calculate the blade distance at the leading edge, O_{2S}, and at the trailing edge, O_3. By assuming as right triangles the triangles having a hypotenuse equal to S, a base equal to O and an angle between them equal to α, we obtain:

$$O_{2S} = S_{2S} \cdot \cos \alpha_{2S}$$
$$O_3 = S_3 \cdot \cos \alpha_3$$

Therefore, the cross sectional areas and the wetted perimeters are:

$$A_{w2S} = O_{2S} \cdot b_{2S}$$
$$P_{w2S} = 2 \cdot (O_{2S} + b_{2S})$$
$$A_{w3} = O_3 \cdot b_3$$
$$P_{w3} = 2 \cdot (O_3 + b_3)$$

and the hydraulic diameters at the stator inlet and outlet become:

$$D_{hyd,2S} = 2 \cdot \frac{O_{2S} \cdot b_{2s}}{O_{2S} + b_{2S}}$$

$$D_{hyd,3} = 2 \cdot \frac{O_3 \cdot b_3}{O_3 + b_3}$$

The stator hydraulic diameter is the mean hydraulic diameter between the stator inlet and outlet:

$$D_{hyd,vaned} = \frac{D_{hyd,2S} + D_{hyd,3}}{2} = \frac{O_{2S} \cdot b_{2S}}{O_{2S} + b_{2S}} + \frac{O_3 \cdot b_3}{O_3 + b_3}$$

To verify the applicability of the kinematics and geometry established so far, Gulich (2010) introduced two parameters and their optimum values:

- the stator outlet velocity:

$$0.02 < \frac{c_3^2}{2 \cdot g \cdot \Delta H} < 0.04$$

$$0.85 < \frac{c_3}{c_{1m}} < 1.25$$

- the area ratio, A_R:

$$A_R = \frac{D_3 \cdot b_3 \cdot \cos\alpha_3}{D_{2S} \cdot b_{2S} \cdot \cos\alpha_{2S}}$$

The optimum area ratio is:

$$A_{R,opt} = 1.05 + 0.184 \cdot \frac{L_{hyd,vaned}}{r_{eq}}$$

where the equivalent radius, r_{eq}, is:

$$r_{eq} = \sqrt{\frac{D_{2S} \cdot \cos\alpha_{2S}}{N_{B,S}} \cdot b_{2S}}$$

If these conditions are not satisfied, the assumptions on the geometric and kinematic parameters of the stator must be changed in order to meet them.

From the knowledge of the kinematics, thermodynamics and geometry of the stage, the absolute (stator inlet/outlet) and relative (rotor inlet/outlet) Reynolds numbers can be calculated:

$$\mathrm{Re}_1 = \frac{\rho \cdot w_1 \cdot D_{hyd,R}}{\mu}$$

$$\mathrm{Re}_2 = \frac{\rho \cdot w_2 \cdot D_{hyd,R}}{\mu}$$

$$\mathrm{Re}_{2,S} = \frac{\rho \cdot c_{2,S} \cdot D_{hyd,vaned}}{\mu}$$

$$\mathrm{Re}_3 = \frac{\rho \cdot c_3 \cdot D_{hyd,vaned}}{\mu}$$

In these equations, the characteristic length is the hydraulic diameter; if the Reynolds numbers must be calculated by using another characteristic dimension, these new equations will be explicitly expressed in the following section.

Once the Reynolds numbers have been calculated, considerations on surface roughness can be made according to Sect. 1.9.1.

In particular, the *admissible surface roughness*, $k_{s,adm}$, can be assessed; it is the limit roughness below which there are no effects on losses allowing the walls to be considered smooth. This condition occurs when the surface roughness is contained within the laminar sub-layer of the boundary layer.

This roughness (Sect. 1.9.1) can be evaluated, for the stator and the rotor, by:

$$\frac{k_{s,adm,R}}{D_{hyd,R}} = \frac{100}{\mathrm{Re}_1}$$

$$\frac{k_{s,adm,S}}{D_{hyd,vaned}} = \frac{100}{\mathrm{Re}_{2,S}}$$

The calculation of $k_{s,adm}$ assumes considerable importance as it allows to establish the finishing level of the channel surfaces in order to contain dissipations.

Taking into account that turbomachinery surface roughness generally falls within the range:

$$5\ \mu\mathrm{m} < k_s < 50\ \mu\mathrm{m}$$

the knowledge of $k_{s,adm}$ allows to evaluate a possible, and economically convenient, surface treatment aimed at respecting this admissible roughness value as well as not resorting to very high finish levels where not required.

In other words, if the admissible surface roughness, $k_{s,adm}$, is lower than the minimum roughness that is technologically and economically achievable, $k_{s,min}$, we assume:

$$k_s = k_{s,\mathrm{min}}$$

otherwise:

$$k_{s,min} \le k_s \le k_{s,adm}$$

In particular, in the latter case, a roughness lower than the admissible one can be assumed in order to take into account that the roughness tends to increase with the turbomachine operation.

Remarks on input parameters

The procedure illustrated in this Sect. 7.3 does not presuppose any further input parameters, with respect to those already identified in Sects. 7.1 and 7.2, taking into account the assumptions, defining the stator and rotor geometry, adopted in this section.

7.4 Stage Losses and Efficiency

The loss sources and the loss models in the different types of turbomachinery are extensively described in Sect. 1.9. In this section we will provide specific correlations for loss models able to quantify the losses introduced in Sect. 1.9.

Losses occurring in the stator and rotor of a radial stage depend on the kinematic (Sect. 7.1), thermodynamic (Sect. 7.2) and geometric (Sect. 7.3) parameters of the stage.

In the literature there are two loss models based on head losses (Oh and Chung 1999, Oh and Kim 2001, and Gulich 2010). Since the model by Oh and Chung-Oh and Kim is completely analogous to that already illustrated for the compressors, the loss model by Gulich (2010) will be illustrated below.

7.4.1 Rotor Losses ($Z_{H,R}$)

Losses in the rotor include:

$$Z_{H,R} = Z_{H,shock} + Z_{H,sf}$$

where:

- $Z_{H,shock}$ represents the shock losses at the pump inlet;
- $Z_{H,sf}$ represents the skin friction losses

Shock losses

They are expressed through:

$$Z_{H,shock} = 0.3 \cdot \left(\frac{w_{1M} - w_{1q}}{u_2} \right)^2 \cdot \frac{u_2^2}{2 \cdot g}$$

where w_{1q} is the relative velocity at the inlet cross sectional area $(O_1 \cdot b_1)$, defined as:

$$w_{1q} = \frac{V}{N_{B,R} \cdot O_1 \cdot b_1}$$

and w_{1M} is the velocity at the impeller inlet, calculated at the mean diameter (Sect. 7.3).

This loss correlation is valid if:

$$\frac{w_{1q}}{w_{1M}} > 0.65$$

Skin friction losses

They include also mixing losses and they are expressed through:

$$Z_{H,sf} = 4 \cdot f_{sf} \cdot \frac{L_{hyd,R}}{D_{hyd,R}} \cdot \left(\frac{\bar{w}}{u_2}\right)^2 \cdot \frac{u_2^2}{2 \cdot g}$$

where:

- $L_{hyd,R}$ and $D_{hyd,R}$ are the hydraulic length and the hydraulic diameter of the rotor, as defined in Sect. 7.3;
- $\bar{w}$ is the mean velocity defined as:

$$\bar{w} = \frac{2 \cdot V}{N_{B,R} \cdot (O_1 \cdot b_1 + O_2 \cdot b_2)}$$

- f_{sf} is the skin friction coefficient and it depends on the Reynolds number and the superficial roughness (k_s). The skin friction coefficient can be calculated as:

$$f_{sf} = (f_r + 0.0015) \cdot \left(1.1 + 4 \cdot \frac{b_2}{D_2}\right)$$

where:

$$f_r = \frac{0.136}{\left[-\log_{10}\left(0.2 \cdot \frac{k_s}{L_{hyd,R}} + \frac{12.5}{\mathrm{Re}}\right)\right]^{2.15}}$$

In this equation the Reynolds number is calculated as:

$$\mathrm{Re} = \frac{\rho \cdot \bar{w} \cdot L_{hyd,R}}{\mu}$$

External losses (parasitic losses)

External losses also occur in the rotor (Sect. 1.9): windage and disk friction losses, recirculation flow and leakage. These losses are not included (Sect. 1.9.5) in the stage hydraulic efficiency and are expressed as head losses contributing to the shaft power.

For the expression of these losses, we will refer to the Oh's discussion (1999).

Disk friction losses

They are given by:

$$Z_{H,df} = f_{df} \cdot \frac{u_2^3 \cdot D_2^2}{16 \cdot g \cdot V}$$

where the coefficient f_{df} can be expressed by:

$$f_{df} = \frac{2.67}{\mathrm{Re}_{df}^{0.5}} \quad \text{when} \quad \mathrm{Re}_{df} < 3 \cdot 10^5$$
$$f_{df} = \frac{0.0622}{\mathrm{Re}_{df}^{0.5}} \quad \text{when} \quad \mathrm{Re}_{df} \geq 3 \cdot 10^5$$

where:

$$\mathrm{Re}_{df} = \frac{\rho \cdot u_2 \cdot (D_2/2)}{\mu}$$

Recirculation losses

They can be calculated by:

$$Z_{H,rc} = 8 \cdot 10^{-5} \cdot \frac{\sinh(3.5 \cdot \alpha_2^3) \cdot DF_f^2 \cdot u_2^2}{g}$$

where the diffusion factor is equal to:

$$DF_f = 1 - \frac{w_2}{w_{1t}} + \frac{0.75 \cdot \psi}{\frac{w_{1t}}{w_2} \cdot \left[\frac{N_{B,R}}{\pi} \cdot (1 - \delta_t) + 2 \cdot \delta_t\right]}$$

Leakage losses

They occur in open impellers and, in analogy to centrifugal compressors, are expressed through:

$$Z_{H,LK} = \frac{V_{CL} \cdot u_{CL} \cdot u_2}{2 \cdot g \cdot V}$$

where:

- $u_{CL} = 0.816 \cdot \sqrt{\frac{2 \cdot \Delta p_{CL}}{\rho}}$

- $\frac{\Delta p_{CL}}{\rho} = \frac{V \cdot \psi \cdot u_2^2}{\omega \cdot N_{B,R} \cdot L_{hyd,R} \cdot \left(\frac{D_{1M}+D_2}{4}\right) \cdot \left(\frac{b_1+b_2}{2}\right)}$

- $V_{CL} = N_{B,R} \cdot \tau_a \cdot L_{hyd,R} \cdot u_{CL}$

and τ_a is the axial clearance, defined in Sect. 7.3.

7.4.2 Stator Losses ($Z_{H,S}$)

The losses in the stator (diffuser) are divided into the losses in the vaneless diffuser and, if present, in the vaned diffuser:

$$Z_{H,S} = Z_{H,vaneless} + Z_{H,vaned}$$

Losses in the vaneless diffuser

They are due to friction and, always following Gulich's discussion (2010), they can be evaluated by using two different correlations depending on the presence or not of a vaned diffuser downstream.

If the downstream vaned diffuser is present, they are calculated as:

$$Z_{H,vaneless} = \frac{u_2^2}{2 \cdot g} \cdot f_{sf} \cdot \left(\frac{O_{2S}}{D_2} + \frac{b_{2S}}{D_2}\right) \cdot \frac{\pi^3 \cdot \left(\frac{c_{2m}}{u_2} \cdot \frac{b_2}{D_2}\right)^2}{8 \cdot \left(N_{B,S} \cdot \frac{O_{2S}}{D_2} \cdot \frac{b_{2S}}{D_2}\right)^3} \cdot \left(1 + \frac{c_2}{c_{2Sq}}\right)^3$$

where:

- c_{2Sq} is the velocity at the cross sectional area ($O_{2S} \cdot b_{2S}$), defined as:

$$c_{2Sq} = \frac{V}{N_{B,S} \cdot O_{2S} \cdot b_{2S}}$$

- f_{sf} is the skin friction coefficient and it is calculated as:

$$f_{sf} = f_r + 0.0015$$

where f_r is assessed in the same way as for the rotor, using the parameters of the vaneless diffuser.

If instead there is no vaned diffuser downstream of the vaneless diffuser, they are calculated as the sum of two terms:

$$Z_{H,vaneless} = Z_{H,shock} + Z_{H,sf}$$

The shock losses are:

$$Z_{H,shock} = \frac{u_2^2}{2 \cdot g} \cdot \left[\left(\frac{c_{2m}}{u_2} \right)^2 \cdot \left(\frac{\pi \cdot D_2 \cdot \cos \beta_{2B}}{\pi \cdot D_2 \cdot \cos \beta_{2B} - t_B \cdot N_{B,R}} - \frac{b_2}{b_{2S}} \right)^2 \right]$$

and the skin friction losses are:

$$Z_{H,sf} = \frac{u_2^2}{2 \cdot g} \cdot f_r \cdot \frac{D_2}{b_{2S} \cdot \cos \alpha_{2S} \cdot sen^2 \alpha_{2S}} \cdot \left(\frac{c_{2u}}{u_2} \right)^2 \cdot \left(1 - \frac{D_2}{D_{2S}} \right)$$

where f_r is calculated in the same way as for the rotor, using the parameters of the vaneless diffuser.

Losses in the vaned diffuser

These losses depend on coefficient c_p correlated to the pressure recovery downstream of the impeller:

$$Z_{H,vaned} = \frac{u_2^2}{2 \cdot g} \cdot \left\{ \left(\frac{c_{2Sq}}{u_2} \right)^2 \cdot \left[0.3 \cdot \left(\frac{c_2}{c_{2Sq}} - 1 \right)^2 + \left(c_{p,i} - c_p \right) \right] \right\}$$

In this equation c_{2Sq} is the velocity already defined for the vaneless diffuser.

The pressure recovery coefficient, c_p, is defined as the ratio between the pressure energy increase in the diffuser and the kinetic energy at the vaned diffuser inlet:

$$c_p = \frac{p_3 - p_{2S}}{\frac{1}{2} \cdot \rho \cdot c_{2S}^2}$$

In the ideal case (absence of dissipations), the pressure recovery coefficient is a function of the ratio between the inlet and outlet areas of the diffuser only:

$$c_{p,i} = \frac{p_3 - p_{2S}}{\frac{1}{2} \cdot \rho \cdot c_{2S}^2} = \frac{c_{2S}^2 - c_3^2}{c_{2S}^2} = 1 - \frac{A_{2S}^2}{A_3^2} = 1 - \frac{1}{A_R^2}$$

where:

$$A_R = \frac{O_3 \cdot b_3}{O_{2S} \cdot b_{2S}}$$

In the real case, however, this coefficient is lower than the ideal one. If we assume (Sect. 7.3):

$$A_R = A_{R,opt}$$

the pressure recovery coefficient can be considered equal to the optimum one, provided by Gulich through the following correlation:

$$c_p = c_{p,opt} = 0.36 \cdot \left(\frac{L_{hyd,vaned}}{r_{eq}} \right)^{0.26}$$

where the geometric quantities have been already defined in Sect. 7.3.

7.4.3 *Rotor and Stage Efficiency*

Once the losses have been calculated, it is immediate to assess the rotor and stage efficiencies according to the definitions introduced in Sect. 7.2:

$$\bar{\eta}_{hyd,R} = \frac{\Delta p_R / \rho \cdot g}{\Delta p_R / \rho \cdot g + Z_{H,R}} = 1 - \frac{Z_{H,R}}{\frac{u_2^2 - u_1^2}{2 \cdot g} + \frac{w_1^2 - w_2^2}{2 \cdot g}}$$

$$\eta_{hyd} = \frac{g \cdot \Delta H}{W} = \frac{W - g \cdot (Z_{H,R} + Z_{H,S})}{W} = 1 - \frac{g \cdot Z_H}{W}$$

In calculating the hydraulic efficiency, external losses are not taken into account (Sect. 1.9.6); the head losses generated by external losses contribute to the shaft power to drive the pump.

Remarks on input parameters

The procedure illustrated in this Sect. 7.4 does not presuppose the introduction of further input parameters, compared to those already identified in Sects. 7.1 and 7.2.

7.5 Input Parameters of the Preliminary Design Procedure

In the previous Sects. 7.1–7.4, the procedure for the calculation of the kinematics, thermodynamics, geometry and efficiency of a centrifugal pump stage was developed and the input parameters of this procedure were identified.

In this Sect. 7.5, we intend to precisely analyze these input parameters in order to suggest their numerical values (these values, however, can be reviewed in an iterative calculation, see Sect. 7.6).

First of all, let us consider the kinematics calculation procedure (Sect. 7.1).

In general, to determine the velocity triangles at the rotor inlet and outlet, we must set six independent parameters: three for the inlet triangle (station 1 of the stage) and three for the outlet triangle (station 2 of the stage).

Unlike the axial stages, where, being $u_1 = u_2 = u$ and assuming $c_{m1} = c_{m2} = c_m$, the independent parameters are only four, for the radial stages the *independent parameters remain six.*

As clearly illustrated in Sect. 7.1, in fact, the absolute and relative flow angles are:

$$\alpha_2 = f(\varphi, \psi, \xi, \delta_t, \alpha_1)$$
$$\beta_1 = f(\varphi, \delta_t, \alpha_1)$$
$$\beta_2 = f(\varphi, \psi, \xi, \delta_t, \alpha_1)$$

These angles are, therefore, functions of the following five independent parameters:

- the flow coefficient, φ
- the work coefficient, ψ
- the rotor meridional velocity ratio, ξ, or the degree of reaction, R, being (Sect. 7.1):

$$R = f(\varphi, \psi, \xi, \delta_t, \alpha_1)$$

- the rotor tip diameter ratio, δ_t
- the absolute flow angle at the rotor inlet, α_1

As a consequence, by identifying a further independent parameter, for example, the blade speed at the rotor outlet, u_2, the velocity triangles are completely defined.

However, taking into account what emerged in the turbomachinery selection process (Chap. 2), it is advisable to choose the stage efficiency (hydraulic efficiency) as the additional independent parameter and calculate the blade speed accordingly.

In fact, starting from the data available for the turbomachinery selection process:

- type of fluid
- fluid mass flow rate (kg/s)
- fluid inlet temperature (°C)
- fluid inlet pressure (bar)
- fluid outlet pressure (bar).

the selection process provides ranges of n ($n_{min} < n < n_{max}$) and z ($z_{min} < z < z_{max}$) compatible with the type of turbomachine chosen. Within these ranges we can choose one or more pairs of these values:

- n: rotational speed (rpm)
- z: number of stages and, consequently, the head in each stage, ΔH_{stage} (m).

For each of these pairs, by using the stage hydraulic efficiency, set as first iteration guess (Sect. 7.5.3) or calculated by the stage losses (Sect. 7.4), the Euler's work of each stage can be calculated and, therefore, the blade speed, after choosing the work coefficient:

$$u_2 = \sqrt{\frac{W}{\psi}}$$

After calculating the blade speed, the diameter at the rotor outlet is:

$$D_2 = \frac{60 \cdot u_2}{\pi \cdot n}$$

Alternatively, we can calculate the blade speed, u_2, from the diameter, D_2, obtained, in turn, by the specific diameter, D_s (chap. 2). In this way, only five independent parameters must be chosen since the work coefficient is calculated by using the stage work and the blade speed, u_2. In centrifugal pump stage design, however, it is preferred to assign the work coefficient, following the procedure illustrated above.

In summary, we must set the following six input parameters in order to design preliminarily a centrifugal pump stage:

- the flow coefficient, φ
- the work coefficient, ψ
- the degree of reaction, R, or the rotor meridional velocity ratio, ξ
- the rotor tip diameter ratio, δ_t
- the absolute flow angle at the rotor inlet, α_1
- the stage efficiency (hydraulic efficiency)

Taking into account what is illustrated in Sects. 7.2–7.4, these parameters, with the addition of the rotor efficiency, allow to fully calculate the thermodynamics, the geometry and therefore the stage losses. Since these losses allow, in turn, to exactly evaluate the rotor and stage efficiencies, the stage efficiency (in the list above) is only a starting value: the calculation will proceed iteratively for each stage until convergence for these efficiencies is achieved (Sect. 7.6).

7.5.1 *Flow and Work Coefficient, or Alternative Parameters, Selection*

Since in this section the characteristic stage parameters will be expressed as functions of the specific speed (Chap. 2), first of all, it is necessary to establish the specific speed optimum range for the stage efficiency.

In this regard, Fig. 7.5 (Baljè) sketches the variation of the hydraulic efficiency with the specific speed. From this Fig. 7.5, we deduce for a centrifugal stage:

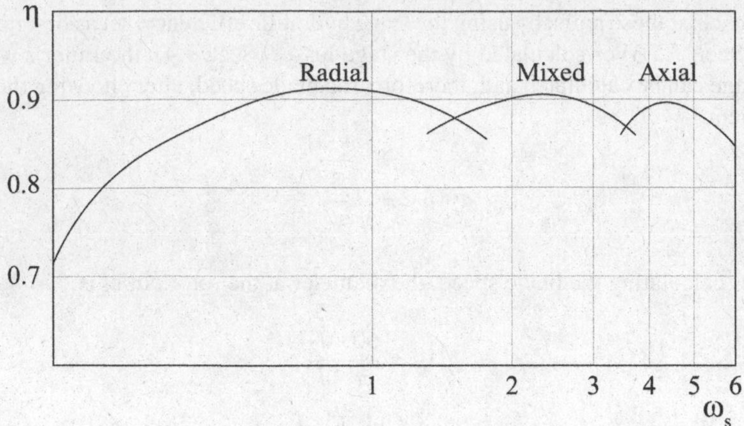

Fig. 7.5 Variation of hydraulic stage efficiency with the specific speed

$$\eta_{hyd,MAX} \Rightarrow 0.5 < \omega_s < 1.3$$

Similarly to the centrifugal compressors, the flow coefficient introduced in Chap. 2 can be analyzed:

$$\varphi' = \frac{V}{u_2 \cdot D_2^2}$$

This flow coefficient, φ', is correlated with that introduced in Sect. 7.1 through:

$$\varphi' = \frac{V}{u_2 \cdot D_2^2} = \frac{\pi \cdot b_1 \cdot D_{1M} \cdot c_{1m}}{u_2 \cdot D_2^2} = \frac{\pi}{4} \cdot \left(\delta_t^2 - \delta_h^2\right) \cdot \varphi$$

In the specific speed range above defined, the optimum flow coefficient range for high efficiencies (>80%) is extremely wide:

$$\eta_{hyd,MAX} \Rightarrow 0.01 < \varphi' < 0.10$$

This situation, similarly to the centrifugal compressors, advises against to choose this flow coefficient, φ', as an input parameter of the design procedure.

Unlike this flow coefficient, the optimum work coefficient falls in a much more limited range. In particular, referring to the ideal work coefficient:

$$\psi_{id} = \frac{g \cdot \Delta H}{u_2^2}$$

we find out that (Fig. 7.6):

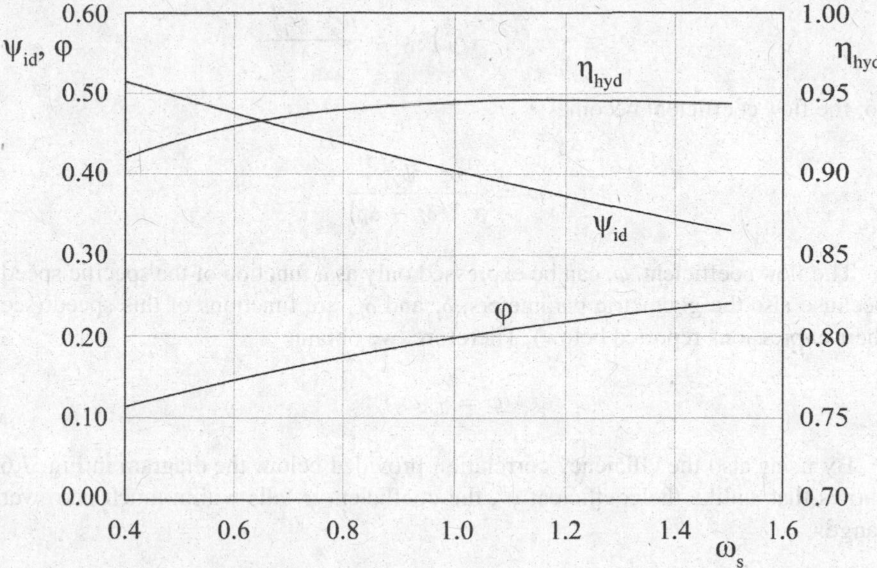

Fig. 7.6 Variation of φ, ψ_{id} and η_{hyd} with ω_s

$$\eta_{hyd,MAX} \Rightarrow 0.35 < \psi_{id} < 0.5$$

Note that the ideal work coefficient is related to the flow coefficient (Chap. 2) through the specific speed:

$$\omega_s = \omega \cdot \frac{(V)^{1/2}}{(g \cdot \Delta H)^{3/4}} = \omega \cdot \frac{\left(\varphi' \cdot \omega \cdot D_2^3/2\right)^{1/2}}{\left(\psi_{id} \cdot \omega^2 \cdot D_2^2/4\right)^{3/4}} = 2 \cdot \frac{\varphi'^{1/2}}{\psi_{id}^{3/4}}$$

and, therefore:

$$\varphi' = \frac{\omega_s^2 \cdot \psi_{id}^{3/2}}{4}$$

As a consequence, if the ideal work coefficient is expressed as a function of the specific speed, also the flow coefficient becomes a function of the specific speed, by using the previous correlation.

In this regard, the correlation proposed by Gulich (2010) can be effectively used:

$$\psi_{id} = 0.605 \cdot e^{-0.408 \cdot \omega_s}$$

The previous correlations also allow expressing the flow coefficient, φ, defined in the kinematics (Sect. 7.1), as a function of the ideal work coefficient and specific speed. Indeed, we have:

$$\varphi' = \frac{\pi}{4} \cdot \left(\delta_t^2 - \delta_h^2\right) \cdot \varphi = \frac{\omega_s^2 \cdot \psi_{id}^{3/2}}{4}$$

so, the flow coefficient becomes:

$$\varphi = \frac{\omega_s^2 \cdot \psi_{id}^{3/2}}{\pi \cdot \left(\delta_t^2 - \delta_h^2\right)}$$

The flow coefficient, φ, can be expressed only as a function of the specific speed because also the geometric parameters, δ_t and δ_h, are functions of this speed (see their expressions reported below). Therefore, we obtain:

$$\varphi = f(\omega_s)$$

By using also the efficiency correlation provided below, the diagram in Fig. 7.6 shows that, unlike the coefficient φ', the coefficient φ falls within much narrower range:

$$\eta_{hyd,MAX} \Rightarrow 0.1 < \varphi < 0.25$$

Consequently, this flow coefficient can be effectively chosen as an input parameter.

Figure 7.6 also shows, as stated above, a narrow range of the ideal work coefficient (0.35–0.5). The work coefficient is obtainable combining the ideal work coefficient with the stage efficiency and its optimum range is:

$$\eta_{hyd,MAX} \Rightarrow 0.4 < \psi < 0.6$$

Finally, we can choose one of the two rotor diameter ratios, δ_t and δ_h, in place of the flow coefficient, φ, as an independent input parameter. In this regard, Gulich (2010) suggested to evaluate δ_t (δ_h will be set later, Sect. 7.5.4) so that the pump design is optimized for the cavitation avoiding. In particular Gulich introduced the following *cavitation coefficients*:

$$\lambda_c = -0.103 \cdot \log \omega_s + 1.1855$$

$$\lambda_w = 0.2144 \cdot e^{0.1745 \cdot \omega_s}$$

These coefficients can be used to calculate the *Net Positive Suction Head required* to avoid cavitation problems:

$$NPSH_R = \lambda_c \cdot \frac{c_{1m}^2}{2 \cdot g} + \lambda_w \cdot \frac{w_{1t}^2}{2 \cdot g}$$

By using these parameters, it is possible to evaluate the optimum rotor tip diameter ratio for the cavitation avoiding:

$$\delta_t = \sqrt{\delta_h^2 + 0.588 \cdot \omega_s^{4/3} \cdot \psi_{id} \cdot \left(\frac{\lambda_c + \lambda_w}{\lambda_w}\right)^{1/3}}$$

In conclusion, two substantially equivalent sets of independent input parameters are therefore possible:

$$\varphi, \psi \begin{cases} 0.1 < \varphi < 0.25 \\ 0.4 < \psi < 0.6 \end{cases} \text{or } \psi_{id}, \delta_t \begin{cases} \psi_{id} = f(\omega_s) \\ \delta_t = f(\omega_s) \end{cases}$$

7.5.2 *Degree of Reaction or Rotor Meridional Velocity Ratio Selection*

Generally, the degree of reaction of a centrifugal pump falls within the following range:

$$0.5 \leq R \leq 0.7$$

but we must remember the constraint on the maximum value of the degree of reaction, already discussed in Sect. 7.1. Gulich (2010) even recommended adopting degrees of reaction of about 0.75 when the specific speed of the stage is medium-low.

As already illustrated, alternatively to the degree of reaction, the rotor meridional velocity ratio can be assumed as an input parameter. In the pumps, where the fluid density can be considered constant, this parameter depends only on geometric parameters:

$$\xi = \frac{c_{2m}}{c_{1m}} = \frac{A_1}{A_2} = \frac{\left(D_{1t}^2 - D_{1t}^2\right)}{4} \cdot \frac{1}{D_2 \cdot b_2} = \frac{\delta_t^2 - \delta_h^2}{4 \cdot \delta_b}$$

where δ_b is defined as:

$$\delta_b = \frac{b_2}{D_2}$$

This last parameter can be assigned as an input since Gulich (2010) suggested to calculate it through the specific speed, according to the following correlation:

$$\delta_b = 0.017 + 0.1386 \cdot \omega_s - 0.0224 \cdot \omega_s^2 + 0.0014 \cdot \omega_s^3$$

7.5.3 Starting Value of Stage Efficiency

As already illustrated, the stage efficiency is only a starting value to calculate preliminarily the thermodynamics, geometry and kinematics of the stage since this parameter depends on stage losses that, in turn, depend on the thermodynamics, geometry and kinematics of the stage (Sect. 7.4). The efficiency calculation proceeds iteratively for each stage until convergence is achieved (Sect. 7.6)

The starting value of the stage efficiency can be calculated by using different correlations, proposed by various authors, as functions of φ and ψ.

Also in this case, however, it is very effective to establish the starting value of the stage efficiency as a function of the specific speed, according to the correlation proposed by Gulich (2010):

$$\eta_{hyd} = 1 - 0.065 \cdot \left(\tfrac{1}{V}\right)^{m} - 0.23 \cdot \left[0.3 - \log_{10}(2.3 \cdot \omega_s)\right]^2 \cdot \left(\tfrac{1}{V}\right)^{0.05} \quad \text{when } \omega_s \leq 1$$
$$\eta_{hyd} = 1 - 0.055 \cdot \left(\tfrac{1}{V}\right)^{m} - 0.09 \cdot \left[\log_{10}(1.18 \cdot \omega_s)\right]^{2.5} \quad \text{when } \omega_s > 1$$

where

$$m = 0.08 \cdot a \cdot \left(\frac{1}{V}\right)^{0.15} \cdot \left(\frac{0.85}{\omega_s}\right)^{0.06}$$

$$a = 1 \quad \text{when} \quad V \leq 1 m^3/s$$
$$a = 0.5 \quad \text{when} \quad V > 1 m^3/s$$

The variation of this efficiency with the specific speed is shown in Fig. 7.6.

For the thermodynamics evaluation (Sect. 7.2), it is also necessary to choose the starting value of the rotor efficiency. It can be set arbitrarily, for example equal to the stage efficiency, since also this efficiency will be iteratively calculated by the losses in the rotor (Sect. 7.4).

7.5.4 Inlet Flow Angle and Rotor Diameter Ratios Selection

Unless arising cavitation problems, we can assume the inlet absolute flow angle as:

$$\alpha_1 = 0$$

A correlation for the rotor tip diameter ratio, δ_t, has already been provided previously as a function of the specific speed.

Finally, for the rotor hub diameter ratio the following empirical value can be used:

$$\delta_h = 0.30$$

Table 7.2 Input parameters

	Set A	Set B
1	Work coefficient ψ	Isentropic work coefficient ψ_{id}
2	Flow coefficient φ	Rotor tip diameter ratio δ_t
3	Degree of reaction R	Rotor outlet width to diameter ratio δ_b
4	Stage efficiency η_{hyd}	
5	Inlet flow angle α_1	
6	Rotor hub diameter ratio δ_h	

7.5.5 Summary of Input Parameters

Based on the main considerations made in the previous Sects. 7.5.1–7.5.4, two sets of input parameters can be identified (Table 7.2):

- Set A: ψ, φ, R, η_{hyd}, α_1, δ_h
- Set B: ψ_{id}, δ_t,δ_b, η_{hyd}, α_1, δ_h

The set A involves more freedom in the input parameters choice, while the set B provides a more *guided procedure*, since the main input parameters are functions of the specific speed.

By assuming the aforementioned six independent parameters and knowing the head in each stage, ΔH_{stage}, and the speed of rotation, n, we can design centrifugal pump stages.

For example, by assuming the following input parameters (Set A):

1. ψ
2. φ
3. R
4. η_{hyd}
5. α_1
6. δ_h

we can immediately calculate:

- the stage work:

$$W = \frac{g \cdot \Delta H}{\eta_{hyd}}$$

- the blade speed at the rotor outlet:

$$u_2 = \sqrt{\frac{W}{\psi}}$$

- the rotor outlet diameter:

$$D_2 = \frac{60 \cdot u_2}{\pi \cdot n}$$

- the rotor tip diameter ratio:

$$\delta_t = \sqrt{\delta_h^2 + \frac{4 \cdot V}{\pi \cdot \varphi \cdot u_2 \cdot D_2^2}}$$

- the rotor meridional velocity ratio:

$$\xi = \sqrt{1 - \frac{2 \cdot \psi}{\varphi^2} \cdot \left(R + \frac{\psi}{2} - 1\right) + \tan^2 \alpha_1 \cdot \left(1 - \delta_t^2\right) - \frac{2 \cdot \psi \cdot \delta_t}{\varphi} \cdot \tan \alpha_1}$$

- the flow angles:

$$\tan \alpha_2 = \frac{\psi}{\xi \cdot \varphi} + \frac{\delta_t}{\xi} \cdot \tan \alpha_1$$

$$\tan \beta_1 = \frac{\delta_t}{\varphi} - \tan \alpha_1$$

$$\tan \beta_2 = \frac{1}{\varphi \cdot \xi} \cdot (1 - \psi) - \frac{\delta_t}{\xi} \cdot \tan \alpha_1$$

and therefore we can calculate the kinematics, thermodynamics, geometry and stage losses.

Similarly, by assuming the following input parameters (Set B):

1. ψ_{id}
2. δ_t
3. δ_b
4. η_{hyd}
5. α_1
6. δ_h

we can immediately calculate:

- the blade speed at the rotor outlet:

$$u_2 = \sqrt{\frac{g \cdot \Delta H}{\psi_{id}}}$$

- the work coefficient:

$$\psi = \frac{\psi_{id}}{\eta_{hyd}}$$

- the rotor outlet diameter:

$$D_2 = \frac{60 \cdot u_2}{\pi \cdot n}$$

- the blade height at the rotor inlet:

$$b_1 = \frac{D_2}{2} \cdot (\delta_t - \delta_h)$$

- the rotor inlet meridional velocity:

$$c_{1m} = 2 \cdot \frac{V}{\pi \cdot D_2 \cdot (\delta_t + \delta_h) \cdot b_1}$$

- the flow coefficient:

$$\varphi = \frac{c_{1m}}{u_2}$$

- the rotor meridional velocity ratio:

$$\xi = \frac{\delta_t^2 - \delta_h^2}{4 \cdot \delta_b}$$

- the degree of reaction:

$$R = 1 - \frac{\psi}{2} + \frac{\varphi^2}{2 \cdot \psi} \cdot \left[(1 - \xi^2) + \tan^2 \alpha_1 \cdot (1 - \delta_t^2)\right] - \varphi \cdot \delta_t \cdot \tan \alpha_1$$

- the flow angles:

$$\tan \alpha_2 = \frac{\psi}{\xi \cdot \varphi} + \frac{\delta_t}{\xi} \cdot \tan \alpha_1$$

$$\tan \beta_1 = \frac{\delta_t}{\varphi} - \tan \alpha_1$$

$$\tan \beta_2 = \frac{1}{\varphi \cdot \xi} \cdot (1 - \psi) - \frac{\delta_t}{\xi} \cdot \tan \alpha_1$$

and therefore we can calculate the kinematics, thermodynamics, geometry and stage losses.

In conclusion, Table 7.3 summarizes the constraints of all the parameters illustrated so far in order to design correctly a radial stage achieving high efficiency.

Table 7.3 Input parameter constraints

Parameter	Range
Specific speed ω_s	$0.5 \div 1.3$
Flow coefficient φ	$0.1 \div 0.25$
Work coefficient ψ	$0.4 \div 0.6$
Degree of reaction R	$0.5 \div 0.75$
Rotor tip diameter ratio δ_t	$0.4 \div 0.8$
δ_h/δ_t	$0.3 \div 0.7$
Rotor outlet width to diameter ratio δ_b	$0.04 \div 0.2$
Rotor inlet relative flow angle β_1	$65° \div 80°$
Rotor outlet blade angle β_{2B}	$60° \div 75°$
Diffusion factor DF $= w_1/w_2$	<1.4
Number of rotor blades	$5 \div 12$

7.6 The Conceptual Comprehensive Framework of the Proposed Preliminary Design Procedure

To calculate the kinematics, thermodynamics and geometry of a radial stage, it is necessary to assume the rotor and stage efficiencies (Sect. 7.5). These efficiencies depend on the losses in the stator and rotor which, in turn, depend on all the parameters listed above (Sect. 7.4). Hence, the preliminary design of a stage is an iterative calculation (Fig. 7.7): first, we calculate the kinematic, thermodynamic and geometric parameters by assuming a starting value of the stage and rotor efficiency; then, we can calculate the stator and rotor losses; at this point, the stage and rotor efficiencies must be calculated by using these losses; these new efficiencies become the new input parameters of the preliminary design procedure. This iterative procedure ends when the established convergence for the rotor and stage efficiencies is achieved.

The iterative procedure shown in Fig. 7.7 could be solved by using different values of the input parameters, as well as of the geometric parameters, in order to optimize the stage performance.

For multistage pumps (Fig. 7.8), the same preliminary design is applied to the stages following the first one. Considering a multistage pump, it is important to highlight that typically the working fluid exits the turbomachine through an outlet volute. In this component, the total enthalpy is constant as well as the mass flow rate and, ignoring friction, also the angular momentum is constant. By applying these conservation laws and adopting specific correlations for the geometric design of this component, it is possible to solve completely the turbomachine.

The thermodynamic parameters at the inlet and outlet of stages following the first one ($i = 2, z$), neglecting, at least in the preliminary design, the process in the channel

Fig. 7.7 Stage calculation block diagram

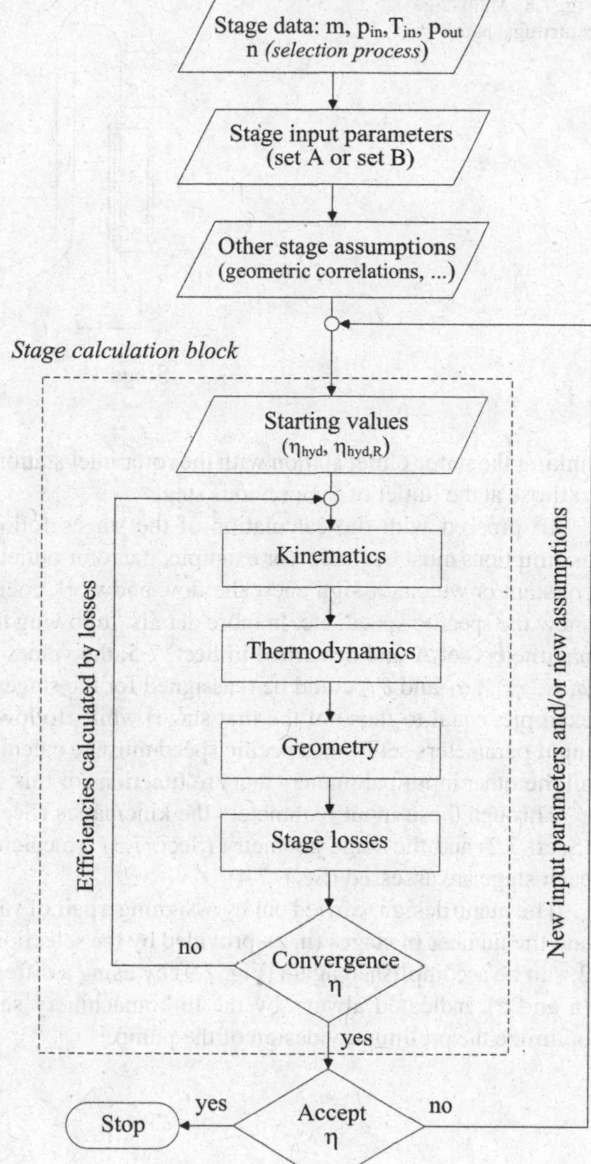

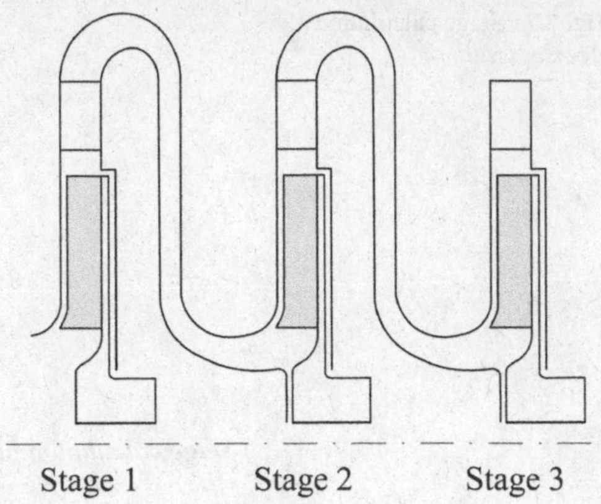

Fig. 7.8 Multistage centrifugal pump

linking the stator outlet station with the rotor inlet station of the next stage, are equal to those at the outlet of the previous stage.

To proceed with the calculation of the stages following the first one, specific assumptions must be made: for example, the rotor outlet diameter can be considered constant or we can assign anew the flow and work coefficients or we can calculate anew the specific speed, etc. In more details, following the procedure based on input parameters set A and described in Sect. 7.5, the values of the input parameters (ψ, φ, R, η_{hyd}, α_1 and δ_h) could be reassigned for the stages following the first one (for example, equal to those of the first stage) while, following the procedure based on input parameters set B, the specific speed must be calculated anew and consequently all the other input parameters that are functions of this specific speed.

Through these input parameters the kinematics (Sect. 7.1), the thermodynamics (Sect. 7.2) and the stage geometry (Sect. 7.3) are calculated and then the losses of each stage are assessed (Sect. 7.4).

The pump design, carried out by assigning a pair of values for the speed of rotation and the number of stages (n, z), provided by the selection process illustrated in chap. 2, can be accomplished again (Fig. 7.9) by using a different pair of input parameters (n and z), indicated always by the turbomachinery selection process, in order to optimize the preliminary design of the pump.

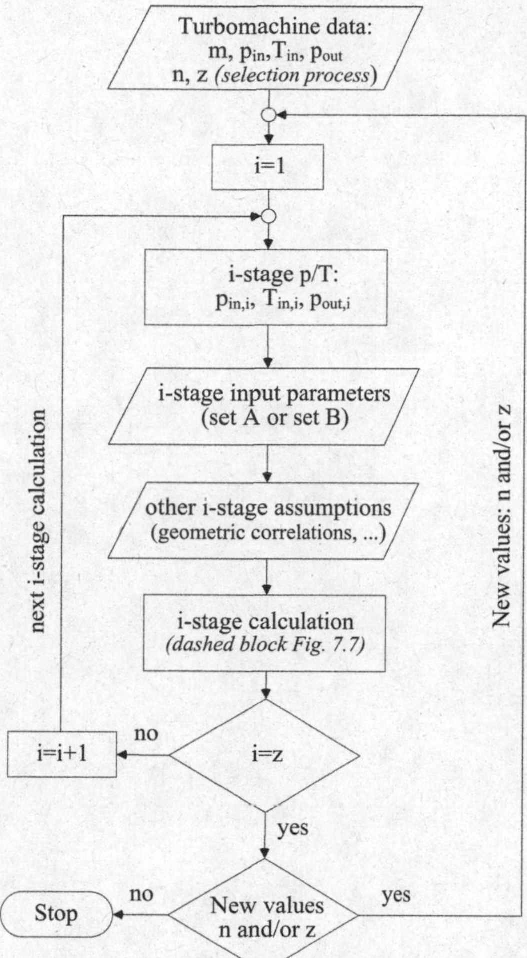

Fig. 7.9 Multistage pump calculation block diagram

References

Gulich JF (2010) Centrifugal pumps, 2nd edn. Springer, Berlin

Oh HW, Chung MK (1999) Optimum values of design variables versus specific speed for centrifugal pumps. Proc Instn Mech Engrs Part A

Oh HW, Kim KY (2001) Conceptual design optimization of mixed-flow pump impellers using mean streamline analysis. Proc Instn Mech Engrs Part A

Pfleiderer C, Petermann H (1986) Strömungsmaschinen. Springer, Berlin

Stepanoff AJ (1959) Radial- und Axialpumpen. Springer, Berlin

Wiesner FJ (1967) A review of slip factors for centrifugal compressors. J Eng Power, Trans ASME

Chapter 8
Case Study: Turbomachines for Concentrating Solar Power Plants

Abstract In this chapter numerical applications of the procedures proposed in the previous chapters are illustrated: from the turbomachinery selection process (Chap. 2) to their preliminary design (Chaps. 3–7). The Concentrating Solar Power (CSP) sector has been chosen in order to provide design examples of turbomachinery operating with unconventional fluids (that is, fluids different from air, water, steam and flue gas used in conventional steam cycles, gas turbines and combined cycles). We will analyze power blocks integrated in a solar field equipped with a central tower; in this application, Brayton closed cycles, operating with Helium, Argon or supercritical CO_2 (sCO_2), are proposed, alone or in combination with Organic Rankine Cycles (ORC) fed by the waste heat of the Brayton cycle. First of all, we will analyze these cycles evaluating their performance and sizing 10 MWe power blocks (Sect. 8.1). Knowing the flow rates handled by turbomachines as well as their inlet and outlet conditions (inlet pressure and temperature, and outlet pressure) we will proceed (Sect. 8.2) to the selection and preliminary design of each turbomachine. It is important to highlight that when defining the thermodynamic cycle of each power block, we assume the turbomachinery efficiency; but this efficiency must necessarily be verified through the loss calculation, which requires knowledge of the kinematics, the thermodynamics and the geometry of the turbomachines (Chaps. 3–7 and this chapter), especially considering the working fluids used in these CSP applications. Otherwise, the cycle performance assessment is not realistic.

8.1 Brayton Cycles for Solar Power Tower Applications

In a solar tower power plant, there are many heliostats (that is, flat mirrors that follow the sun by using two-axis rotation motion) able to focus sunlight onto a *solar receiver* (placed at the top of a tall tower) where a *heat transfer fluid* (HTF) is heated. This heat is profitably employed to feed a thermodynamic cycle able to generate electricity in a dedicated power block.

In these systems, the HTF, which circulates inside the receiver, can reach high operating temperatures, up to 565 °C when using molten salts, up to 800–1000 °C

M. Gambini and M. Vellini, *Turbomachinery*, Springer Tracts in Mechanical Engineering, https://doi.org/10.1007/978-3-030-51299-6_8

when using gas (air, CO_2, Helium, etc.). Consequently, molten salts request to use power blocks based on traditional steam cycles (subcritical or supercritical ones) to ensure suitable conversion efficiency. Vice versa, other HTFs offer the two following advantages: (i) usage of Brayton cycles operating at temperatures higher than those of the steam cycles and, compatibly with the needs of a possible storage system, (ii) exploiting the same HTF as working fluid in the thermodynamic cycle.

Stein and Buck (2017) provided a review of advanced power cycle for concentrating solar power.

Regarding Brayton cycles, attention is increasingly given to fluids other than air (e.g. sCO_2, Helium, Argon); these fluids exhibit thermodynamic properties (such as specific heat, thermal conductivity, density, etc.) leading to possible advantages over usage of air: for example smaller dimensions of turbomachinery and/or heat exchangers, reduction of pressure drops in heat exchangers, etc. Moreover, due to the higher top cycle temperatures, these cycles attain higher efficiencies than those of traditional subcritical and supercritical steam cycles (about 40–45%). However, these fluids necessarily request closed-loop Brayton cycles.

Beserati and Goswami (2017) provided a review of closed-loop Brayton cycles for power generation in solar fields equipped with a central tower. These cycles will be analyzed in the following sections with reference to different working fluids; comparisons will be made with the results of Kusterer et al. (2012, 2013) for Helium and Argon cycles and with those of Turchi et al. (2012, 2013) and Beserati and Goswami (2014) for the supercritical CO_2 cycles (sCO_2). Finally we will consider a more complex system, composed by a closed-loop Brayton cycle as topping cycle and an ORC, fed by waste heat of the topping cycle, as bottoming cycle.

8.1.1 Brayton Cycles Operated with Helium and Argon

In closed-loop Brayton cycles, Helium has the advantage of having a high specific heat, high thermal conductivity and no corrosive behavior. However, Helium has a low molecular weight and low density. Conversely, Argon has a lower specific heat but a higher density involving more compact turbomachines.

Kusterer et al. (2012, 2013) first analyzed Helium cycles (2012) and subsequently compared them with Argon cycles (2013).

With reference to Helium, three closed Brayton cycle configurations were analyzed, all equipped with regeneration: in the first one there are one compressor and one turbine, in the second one there is a compression with intercooling (two compressors and one turbine) and in the last configuration there are the same compression process and an expansion process with sequential heating (two compressors and two turbines). The best compromise between performance and power plant complexity was found in the cycle with intercooled compression. The corresponding flow scheme is shown in Fig. 8.1, where IC stands for intercooler, FC for final cooler, R for recuperator and SR for solar receiver.

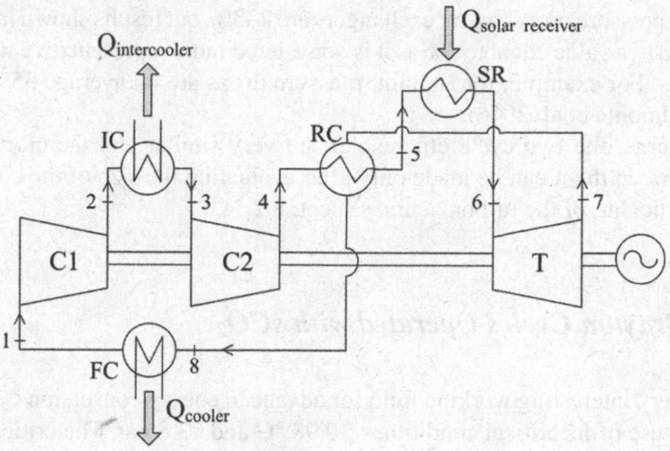

Fig. 8.1 Closed Brayton cycle operated with helium or argon

The cycle performance was assessed considering both Helium and Argon as working fluids. The input parameters and efficiency are summarized in Table 8.1.

Thermodynamic efficiency depends, of course, on the input parameters, and in particular on the *turbine inlet temperature*, as well as on pressure drops in heat exchangers. By increasing the pressure drops from 1 to 4%, the efficiency reduces about 9 percentage points for Helium and about 6.5 percentage points for Argon. This means that for pressure drops greater than 2% Argon has a slightly higher efficiency than Helium while for pressure drops lower than 2%, Helium prevails, even if in a very limited way.

Table 8.1 Performance of closed Brayton cycle operated with helium and argon (Kusterer et al. 2012, 2013)

Working fluid		Helium	Argon
Input parameters			
Compressor inlet temperature (C1 and C2)	°C	30	30
Compressor inlet pressure (C1)	bar	1.5	1.2
Pressure ratio (C1 and C2)	–	1.6	1.85
Turbine inlet temperature	°C	900	900
Pressure drops (FC, IC, R, SR)	%	1	1
Compressor polytropic efficiency (C1 and C2)		0.86	0.86
Turbine polytropic efficiency	–	0.89	0.89
Recuperator effectiveness		0.90	0.90
Output parameters			
Cycle efficiency		0.467	0.461

Typical pressure drops in heat exchangers are 2–3%, but results shown in Table 8.1 are referred to a value equal to 1% as it is considered more representative when using these fluids. For example, for Helium, pressure drops are on average 35% of those for air (Belmonte et al. 2016).

In any case, the two cycle efficiencies are very similar and therefore the final choice between them can be made only after evaluating the size of the components and, in particular, of the turbomachines (Sect. 8.2.1).

8.1.2 Brayton Cycles Operated with sCO_2

CO_2 is a very interesting working fluid for advanced energy conversion cycles especially because of its critical conditions: 30.98 °C and 73.8 bar. The critical temperature is in fact very close to the ambient temperature and moreover the density near the critical conditions is very high (at the critical point density is 469 kg/m^3), that is, a density more similar to a liquid than to a gas. This means that the CO_2 compression, starting from thermodynamic conditions close to the critical point, requires extremely limited energy (work) compared to other gaseous working fluids (air, Helium, Argon). This is precisely the main advantage of sCO_2 (supercritical CO_2) compared to other fluids operating in closed Brayton cycles.

The thermal conductivity near the critical point is also high: about 5 times greater than that of air at ambient conditions.

Another important thermodynamic parameter is the specific heat: near the critical point, it is very high and it varies greatly with temperature. This circumstance has a strong influence on the design of the recuperator in regenerative Brayton cycles: indeed, unlike fluids comparable to perfect gases (for example air, Helium, Argon) having temperature profiles roughly parallel during a regenerative heat exchange, sCO_2 has different specific heats in the hot and cold sides of the recuperator and therefore the pinch-point can occur inside the recuperator (Dostal et al. 2006a, b).

In order to improve the recuperator effectiveness, two compressors are generally installed in sCO_2 cycles (Fig. 8.2): a *main compressor* (MC) and a *recompression compressor* (RC). In this configuration (two compressors and one turbine), the hot fluid exiting the low-temperature recuperator (LTR), station 11 in Fig. 8.2, is divided into two streams: the first one, station 12, is sent to the final cooler (FC), then back to the main compressor (MC) and to the LTR, while the second stream, station 4, is directly pressurized in a recompression compressor (RC) and then it is sent downstream of the LTR where the two streams are mixed, station 6, before entering the high-temperature recuperator (HTR).

The main advantage of this configuration in comparison with the simple regenerative cycle (a compressor and a turbine) is a more efficient regenerative heat exchange. In fact, through the aforementioned subdivision of the streams downstream of the LTR, the thermal capacity of the cold fluid (fluid at high pressure) in the LTR becomes similar to that of the hot fluid (fluid at low pressure), thus avoiding the pinch point problems previously illustrated.

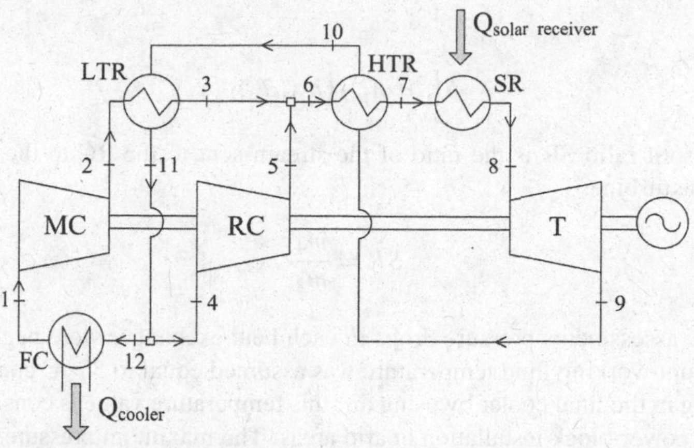

Fig. 8.2 Recompression sCO2 Brayton cycle

Table 8.2 Performance of recompression sCO_2 Brayton cycle (Beserati and Goswami 2014)

Working fluid		sCO_2
Input parameters		
Main compressor inlet temperature (MC)	°C	55
Compressor outlet pressure (MC and RC)	bar	250
Compressor pressure ratio (MC and RC)	–	3.4
Turbine inlet temperature	°C	800
Pressure drops in heat exchangers	%	0
Compressor isentropic efficiency (MC and RC)	–	0.89
Turbine isentropic efficiency	–	0.90
Recuperator effectiveness	–	0.95
Split ratio		0.3
Output parameters		
Cycle efficiency		0.493

Note that the flow fraction sent to the RC becomes a very important design parameter affecting directly the thermodynamic cycle efficiency.

The performance of these sCO_2 cycles was analyzed by various authors (Dostal et al. 2006a, b; Turchi et al. 2012, 2013; Beserati and Goswami 2014).

Referring to the assessments by Beserati and Goswami (2014), Table 8.2 summarizes the input parameters and the efficiency of the sCO_2 cycle shown in Fig. 8.2.

The effectiveness of the recuperator is defined as:

$$\varepsilon = \frac{h_9 - h_{11}}{h_9 - h_{11}^*}$$

where

$$h_{11}^{*} = h_{p,T}(p_{11}, T_2)$$

while the split ratio SR is the ratio of the stream sent to the RC to the total flow entering the turbine:

$$SR = \frac{m_4}{m_8}$$

In these assessments pressure drops in each heat exchanger were neglected and the minimum working fluid temperature was assumed equal to 55 °C considering a dry cooling in the final cooler by using air; this temperature value is consistent with a possible power block installation in arid areas. The maximum pressure was set to 250 bar.

Consequently, considering a turbine inlet temperature of 800 °C, that is, 100 °C lower than that of the previously analyzed Helium and Argon cycles, a thermodynamic efficiency of about 50% is achieved, showing the considerable potential of these sCO_2 cycles in pursuit of high efficiency even with top cycle temperatures much lower than those of traditional open Brayton cycles fueled by natural gas (turbine inlet temperatures of about 1300 °C and thermodynamic efficiency of about 40%).

However, compared to Helium and Argon cycles, much higher operating pressures must be adopted in the sCO2 cycles.

8.1.3 *Combined Cycles*

To increase the cycle efficiency, each of the previously analyzed Brayton cycles (operated with Helium, Argon and sCO_2) can be equipped with a bottoming cycle fed by the topping cycle waste heat. Considering the low temperatures downstream of the recuperator, the best bottoming cycle is an Organic Rankine Cycle (ORC). Adopting the recompression sCO2 cycle as a topping cycle, the combined cycle is shown in Fig. 8.3. As can be seen from this figure, the heat recovery steam generator (HRSG) is placed downstream of the stream split. Therefore, the stream entering the HRSG is the stream sent to the main compressor; if the total flow had entered the HRSG, the stream sent to the recompression compressor would be colder and there would be negative consequences on cycle efficiency.

Beserati and Goswami (2014) analyzed many organic fluids in order to select the most suitable ones in terms of cycle efficiency as well as expansion ratio, defined (Fig. 8.3) as the ratio between volumetric flow rate at the inlet and outlet of the turbine handling the organic fluid:

$$ER = \frac{V_{15}}{V_{14}}$$

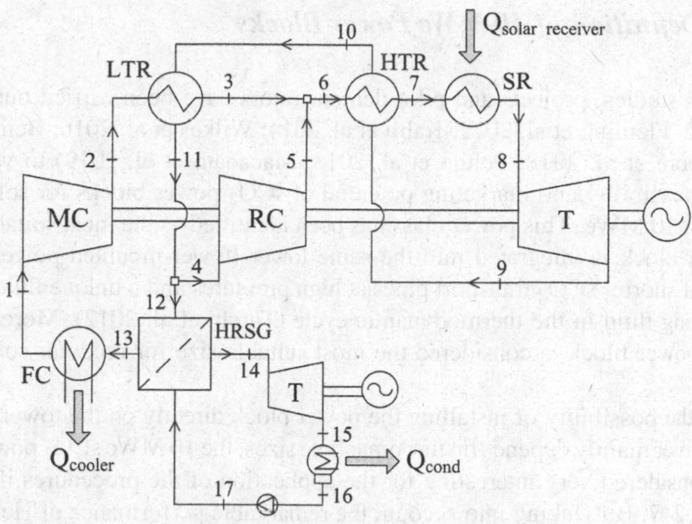

Fig. 8.3 Combined cycle: recompression sCO_2 Brayton cycle and ORC

In fact, extreme expansion ratios are not recommended because of supersonic flow problems, larger turbine size or greater number of stages.

Adopting for the topping cycle the input parameters in Table 8.2, the combined cycle performance evaluation is summarized in Table 8.3.

The sCO2 topping cycle had an efficiency of 49.3%, while the different combined cycles listed in Table 8.3 achieve efficiencies of about 54% and therefore the addition of an ORC cycle, as bottoming cycle, leads to efficiency gains, compared to the sCO2 cycle, of about 5 percentage points.

Table 8.3 Combined cycle (recompression sCO2 Brayton cycle-ORC) performance (Beserati and Goswami, 2014)

Fluids	T_{14} (°C)	p_{14} (bar)	T_{16} (°C)	p_{17}/p_{16}	ER	η_{cc}
R236ea	132.7	29.9	55.0	3.45	8.48	0.5400
R245fa	139.4	27.9	55.0	3.84	9.78	0.5416
Butane	137.4	29.8	55.0	3.84	7.43	0.5398
Butene	126.0	28.5	55.0	3.45	5.42	0.5367
Cis-butene	140.5	29.6	55.0	4.13	7.32	0.5387
Isobutane	120.3	28.5	55.0	3.26	5.03	0.5357

8.1.4 Definition of 10 MWe Power Blocks

Numerous studies, projects and pilot demonstrators have been carried out (Turchi et al. 2012; Fleming et al. 2012; Kalra et al. 2014; Wilkes et al. 2016; Bennet et al. 2017; Moore et al. 2018; Pelton et al. 2018; Macadam et al. 2019) to verify the technical feasibility and marketing potential of sCO_2 power blocks for solar tower systems of 10 MWe. This power class has been identified as the most suitable when the power block is integrated into the same tower (tower-mounted power block), because of shorter sCO_2 transport pipes at high pressures and a unique fluid as HTF and working fluid in the thermodynamic cycle (Turchi et al. 2012). Moreover, the 10MWe power block is considered the most suitable size for modular solar tower systems.

Since the possibility of installing the power block directly on the tower near the solar receiver mainly depends on turbomachine sizes, the 10 MWe sCO_2 power block can be considered very interesting for the application of the procedures illustrated in Chaps. 2–7. But, taking into account the remarkable performance of Helium and Argon cycles for CSP applications in the same power class identified for the sCO_2 power block (Kusterer et al. 2012, 2013), also these cycles have been chosen to apply the turbomachine selection and design procedures illustrated in the Chaps. 2–7. To this end (Sect. 8.2), first, we will establish the operating parameters of the three thermodynamic cycles (sCO_2, Helium, Argon) to calculate the turbomachine inlet and outlet conditions necessary for their selection and design. To define these operating parameters, it is advisable to first compare the performance of the three thermodynamic cycles.

Dostal et al. (2006a, b) assessed the sCO_2 cycle efficiency variation with the turbine inlet temperature and compared these efficiencies with those achieved by other thermodynamic cycles (Helium Brayton cycle, superheated and supercritical steam cycle). These results are shown in Fig. 8.4.

As can be seen from Fig. 8.4, the recompression sCO_2 cycle attains always higher thermal efficiencies than those of Helium Brayton cycles; these efficiencies are also higher than those of steam cycles when the top cycle temperatures are greater than 550 °C. Moreover, the sCO_2 cycle efficiencies are less affected by pressure drops in comparison with Helium cycles because of the higher operating pressures as well as the lower compression work.

Another advantageous feature of the sCO_2 cycles, as will be illustrated in more detail in Sect. 8.2, is the high power density and the extreme compactness of the turbomachines.

Finally, taking into account the high operating pressures and possible CO_2 dissociation phenomena (Kusterer et al. 2012), the best cycle top temperature, considering as compromise between efficiency and power plant requirements, is around 700 °C.

Based on the previous considerations, we have decided to establish the input parameters of these three power blocks so that similar levels of thermodynamic efficiency are achieved; this means that the three power blocks need the same solar field area. Consequently, the turbine inlet temperature is 700 °C for the sCO_2 cycle,

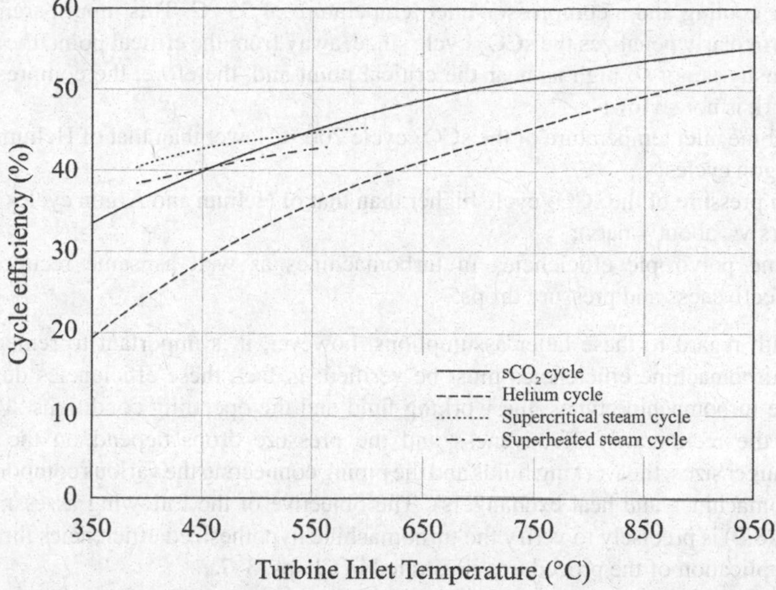

Fig. 8.4 Cycle efficiency variation with turbine inlet temperature (adapted from Dostal et al.)

adopting however the highest operating pressures, and 900 °C for Helium and Argon cycles, adopting much lower operating pressures. For these reasons, we have assessed the thermodynamic efficiencies of the three cycles introduced in Sect. 8.1.1 and 8.1.2. These evaluations have been developed by using the NIST Libraries (Lemmon et al.—REFPROP 10.0) and adopting (Table 8.4) the following assumptions:

Table 8.4 Cycle evaluation assumptions

Working fluid		Helium	Argon	sCO_2
Input parameters				
Compressor inlet temperature (C1, C2, MC)	°C	55.0	55.0	55.0
Compressor pressure ratio (C1, C2, MC, RC)	–	1.6	1.85	2.5
Turbine inlet pressure	bar	3.75	4.0	250
Turbine inlet temperature	°C	900	900	700
Pressure drops (cooler, receiver, recuperator)	%	1	1	1
Compressor polytropic efficiency (C1 and C2)	–	0.89	0.89	0.89
Turbine polytropic efficiency	–	0.90	0.90	0.90
Recuperator effectiveness	–	0.95	0.95	0.95
Split ratio	–	–	–	0.36
Output parameters				
Cycle efficiency		0.482	0.472	0.478

- dry cooling and a compressor inlet temperature of 55 °C. This input parameter particularly penalizes the sCO_2 cycle since, away from the critical point, the fluid density is not so high as near the critical point and, therefore, the compression work is not so low;
- turbine inlet temperature of the sCO_2 cycle 200 °C lower than that of Helium and Argon cycles;
- top pressure of the sCO_2 cycle higher than that of Helium and Argon cycles (250 bars vs. about 4 bars);
- same polytropic efficiencies in turbomachines as well as same recuperator effectiveness and pressure drops.

With regard to these latter assumptions, however, it is important to remember that turbomachine efficiencies must be verified: in fact, these efficiencies depend on the turbomachine types, the working fluid and the operating conditions. Moreover, the recuperator effectiveness and the pressure drops depend on the heat exchanger sizes, the working fluids and the piping connecting the various components (turbomachines and heat exchangers). The objective of the following assessments (Sect. 8.2) is precisely to verify the turbomachine hypothesized efficiencies through the application of the procedures illustrated in Chaps. 3–7.

By using the assumptions listed in Table 8.4, the three-cycle efficiencies are very similar and close to 48%. Also the power plant layouts are similar: two compressors and a turbine in each configuration.

With reference to the 10 MWe power blocks, and assuming in all configurations a mechanical/electric efficiency equal to 0.98, we obtain the turbomachine inlet/outlet conditions shown in Table 8.5.

Then, we have considered a combined cycle composed of a simple sCO_2 regenerative Brayton cycle as topping cycle and an ORC as a bottoming cycle (Fig. 8.5). According to Beserati and Goswami (2014), we have selected the Cis-butene as working fluid in the bottoming ORC. We have chosen this cycle configuration to limit the number of power block components: in this way, in fact, even if the bottoming cycle needs two turbomachines (pump and turbine) and two heat exchangers (HRSG and condenser), the topping cycle is simpler, giving up the recompression compressor and the LTR.

Therefore, the analysis of this combined cycle may be of interest to assess whether the addition of the bottoming ORC can compensate, in terms of efficiency, for the elimination of the recompression compressor and the low-temperature recuperator. These assessments are reported in Table 8.6.

From Table 8.6 it can be seen that, by using the same input parameters of the recompression sCO_2 cycle (Table 8.3) with the exception of the pressure ratio, that is in this case higher than the previous one, the combined cycle attains the same efficiency (47.8%) of the recompression sCO_2 cycle.

Table 8.5 Turbomachine inlet/outlet conditions

Working fluid		Helim	Argon	sCO_2
Compressor (C1-He, C1-Ar, MC-sCO_2)				
Inlet flow rate	kg/s	11.8	97.6	72.4
Inlet pressure	bar	1.5	1.2	100
Inlet temperature	°C	55.0	55	55
Outlet pressure	bar	2.41	2.22	255
Gross power	MW	4.8	5.3	3.0
Compressor (C2-He, C2-Ar, RC-sCO_2)				
Inlet flow rate	kg/s	11.8	97.6	40.7
Inlet pressure	bar	2.38	2.20	101
Inlet temperature	°C	55.0	55.0	142.6
Outlet pressure	bar	3.83	4.1	253.8
Gross power	MW	4.8	5.3	3.2
Turbine				
Inlet flow rate	kg/s	11.8	97.6	113.1
Inlet pressure	bar	3.75	4.0	250
Inlet temperature	°C	900	900	700
Outlet pressure	bar	1.53	1.22	101.5
Gross power	MW	19.8	20.8	16.4
Power block				
Gross power	MW	10.2	10.2	10.2
Mechanical/electric efficiency	–	0.98	0.98	0.98
Net power	MW	10.0	10.0	10.0

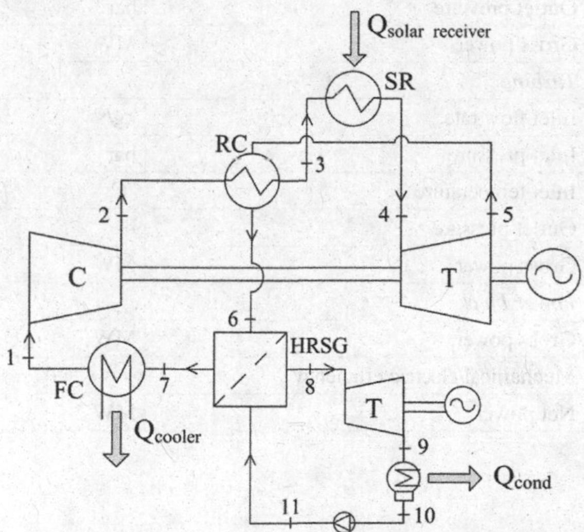

Fig. 8.5 Combined cycle: simple sCO_2 regenerative Brayton and ORC

Table 8.6 Combined cycle: input parameters and performance

Combined cycle		Topping cycle	Bottoming cycle
Working fluid		sCO_2	Cis-butene
Compressor (sCO_2)/Pump (Cis-butene) inlet temperature	°C	55.0	55.0
Pressure ratio	–	5.0	5.8
Turbine inlet pressure	bar	250	29.6
Turbine inlet temperature	°C	700	140.5
Pressure drops (heat exchangers)	%	1	3
Compressor polytropic/pump hydraulic efficiency	–	0.89	0.85
Turbine polytropic efficiency	–	0.90	0.88
Recuperator effectiveness	–	0.95	–
Combined cycle efficiency		0.478	

With reference to a 10 MWe power block, and assuming a mechanical/electric efficiency equal to 0.98, we obtain the turbomachine inlet/outlet conditions shown in Table 8.7.

Table 8.7 Turbomachine inlet/outlet conditions (combined cycle)

Turbomachine inlet/outlet conditions			
Working fluid		sCO_2	Cis-butene
Compressor (sCO_2)—pump (Cis-butene)			
Inlet flow rate	kg/s	66.6	30.6
Inlet pressure	bar	50.0	5.1
Inlet temperature	°C	55.0	55.0
Outlet pressure	bar	255.	31.1
Gross power	MW	7.65	0.152
Turbine			
Inlet flow rate	kg/s	66.6	30.6
Inlet pressure	bar	250.	29.6
Inlet temperature	°C	700	140.5
Outlet pressure	bar	51.0	5.1
Gross power	MW	16.0	1.96
Power block			
Gross power	MW	10.2	
Mechanical/electric efficiency	–	0.98	
Net power	MW	10	

8.2 Turbomachinery Selection and Preliminary Design

The analyses carried out in Sect. 8.1 provide (Tables 8.5 and 8.7) the flow rates and the inlet/outlet conditions (inlet pressure and temperature, outlet pressure) of each turbomachine in the power blocks assessed; these parameters allow the turbomachinery selection and preliminary design.

8.2.1 Turbomachines for Brayton Cycles Operated with Helium

The main results related to the turbomachines in the Brayton cycle operated with Helium and assessed in Sect. 8.1 are synthetically reported in Table 8.8.

On the basis of these parameters, following the procedure illustrated in Chap. 2, Figs. 8.6, 8.7 and 8.8 show the specific speed variation of each turbomachine with the number of stages for different rotational speeds.

Assuming the same rotational speed for the three turbomachines, Figs. 8.6, 8.7 and 8.8 suggest that a good compromise between efficiency and number of stages is:

- rotational speed of 6000 rpm
- axial configurations for both the compressors and the turbine
- 10 stages for C1, 14 for C2 and 10 for the turbine.

For this last turbomachine, a good solution could be an 8-stage configuration, but, in this case, the blade speed becomes very high (400 m/s) so the 10-stage configuration has been chosen. Table 8.9 summarizes the main results of the turbomachinery selection process, highlighting those related to the first and last stage of each turbomachine.

After choosing the turbomachine type, the rotational speed and the number of stages, the turbomachinery preliminary design may start. We have designed the two compressors C1 and C2 by applying the procedure described in Chap. 4. Table 8.10 reports the main results concerning the first and last stage. The input parameters and the kinematic parameters are referred to the mean diameter. The flow coefficient has been assumed to ensure a 20% stall margin (Sect. 4.6.1).

Table 8.8 Turbomachine inlet/outlet conditions (Helium cycle)

		Compressor C1	Compressor C2	Turbine
Inlet flow rate	kg/s	11.8	11.8	11.8
Inlet pressure	bar	1.50	2.38	3.75
Inlet temperature	°C	55.0	55.0	900
Outlet pressure	bar	2.41	3.83	1.53

The compressor C1 has a mean diameter of about 1.2 m; the blade height varies between 7.1 cm (first stage) and 5.6 cm (last stage); the axial length is about 1.2 m, considering an axial clearance, that is, the axial distance between blade rows, equal to about 30% of the axial chord; the maximum blade tip speed reaches 390 m/s; the degree of reaction varies from 0.43 (hub) to 0.56 (tip) and the Mach number is always lower than 0.3. In terms of overall performance, the compressor transfers a power of 4.9 MW from the blades to the fluid with a polytropic efficiency of 87%. Such efficiency is, therefore, two percentage points lower than the assumption made in Sect. 8.1.4.

The compressor C2 has a mean diameter of about 1 m; the blade height varies between 6.3 cm (first stage) and 4.9 cm (last stage); the axial length is about 1.5 m, considering an axial clearance equal to about 30% of the axial chord; the maximum blade tip speed reaches 332 m/s; the degree of reaction varies from 0.43 (hub) to 0.56 (tip) and the Mach number is always lower than 0.25. In terms of overall performance, the compressor transfers a power of 4.9 MW from the blades to the fluid with a polytropic efficiency of 87%. Such efficiency is, therefore, two percentage points lower than the assumption made in Sect. 8.1.4.

We have designed the turbine by applying the procedure described in Chap. 3. Table 8.11 reports the main results concerning the first and last stage. The input parameters and the kinematic parameters are referred to the mean diameter. The work coefficient has been assumed equal to 1.4 to limit the blade speed to about 400 m/s.

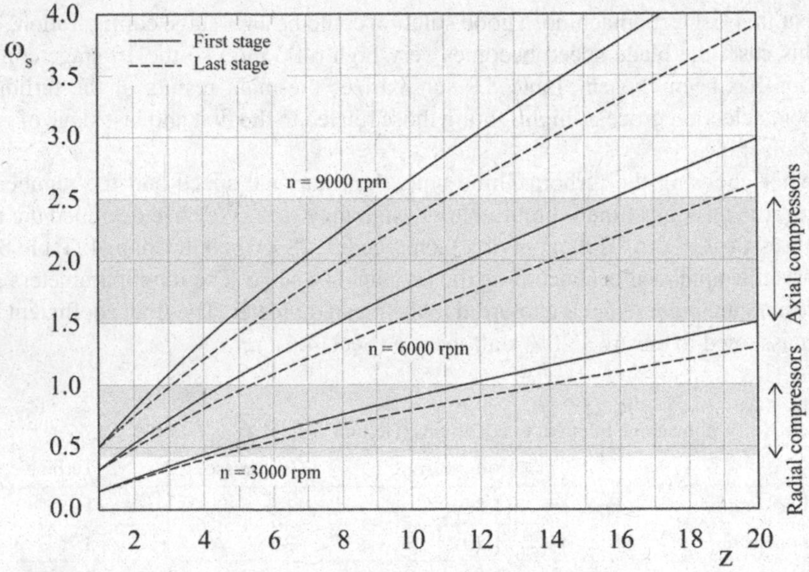

Fig. 8.6 Helium Compressor, C1

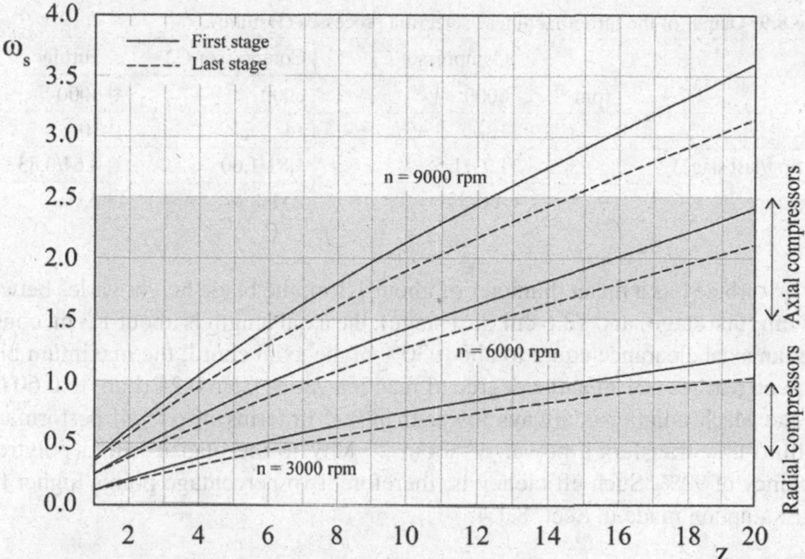

Fig. 8.7 Helium compressor, C2

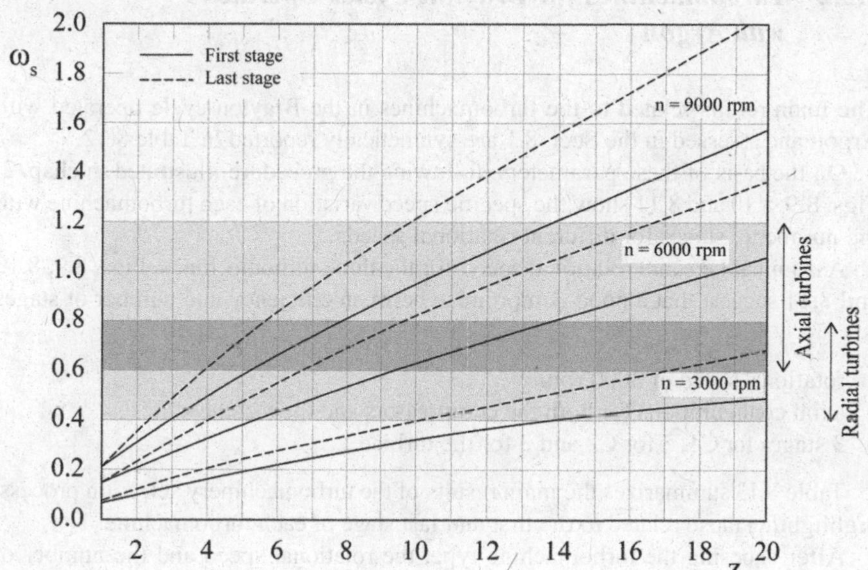

Fig. 8.8 Helium turbine

Table 8.9 Output of the turbomachinery selection processes (Helium)

		Compressor C1	Compressor C2	Turbine
n	rpm	6000	6000	6000
z	–	10	14	10
ω_s (first/last stage)		1.77/1.58	1.81/1.60	0.64/0.83
Stage type	–	Axial	Axial	Axial

The turbine has a mean diameter of about 1.1 m; the blade height varies between 10.8 cm (first stage) and 18.6 cm (last stage); the axial length is about 1.3 m, considering an axial clearance equal to about 30% of the axial chord; the maximum blade tip speed reaches 401 m/s; the degree of reaction varies from 0.24 (hub) to 0.60 (tip) and the Mach number is always lower than 0.3. In terms of overall performance, the fluid flow transfers a power of about 20 MW to the blades with a polytropic efficiency of 92%. Such efficiency is, therefore, two percentage points higher than the assumption made in Sect. 8.1.4.

8.2.2 *Turbomachines for Brayton Cycles Operated with Argon*

The main results related to the turbomachines in the Brayton cycle operated with Argon and assessed in the Sect. 8.1 are synthetically reported in Table 8.12.

On the basis of these parameters, following the procedure illustrated in chap. 2, Figs. 8.9, 8.10 and 8.11 show the specific speed variation of each turbomachine with the number of stages for different rotational speeds.

Assuming the same rotational speed for the three turbomachines, Figs. 8.9, 8.10 and 8.11 suggest that a good compromise between efficiency and number of stages is:

- rotational speed of 3000 rpm
- axial configurations for both the compressors and the turbine
- 3 stages for C1, 5 for C2 and 3 for the turbine.

Table 8.13 summarizes the main results of the turbomachinery selection process, highlighting those related to the first and last stage of each turbomachine.

After choosing the turbomachine type, the rotational speed and the number of stages, the turbomachinery preliminary design may start. We have designed the two compressors C1 and C2 by applying the procedure described in Chap. 4. Table 8.14 reports the main results concerning the first and last stage. The input parameters and the kinematic parameters are referred to the mean diameter. The flow coefficient has been assumed to ensure a 20% stall margin (Sect. 4.6.1).

The compressor C1 has a mean diameter of about 1.5 m; the blade height varies between 8.4 cm (first stage) and 6.3 cm (last stage); the axial length is about 40 cm,

Table 8.10 Compressor design: main output (Helium)

Compressors		Compressor C1		Compressor C2	
		First stage	Last stage	First stage	Last stage
Input parameters					
ψ	–	0.300	0.312	0.300	0.311
φ	–	0.550	0.550	0.550	0.550
R	–	0.500	0.494	0.500	0.494
Output parameters					
Kinematics					
α_1	°	32.49	32.49	32.49	32.49
α_2	°	49.79	50.31	49.79	50.27
β_1	°	49.79	49.79	49.79	49.79
β_2	°	32.49	31.58	32.49	31.65
u	m/s	368.4	368.4	311.8	311.8
Thermodynamics					
p_2	bar	1.540	2.358	2.425	3.770
T_2	°C	58.92	130.7	57.80	132.0
p_3	bar	1.580	2.410	2.470	3.830
T_3	°C	62.83	134.9	60.61	135.0
Geometry					
D_M	cm	117.3	117.3	99.3	99.3
$h_{B,R}$	cm	7.14	5.66	6.30	4.95
$h_{B,S}$	cm	7.04	5.60	6.23	4.91
C_R	cm	6.49	5.15	5.72	4.50
C_S	cm	6.40	5.09	5.67	4.47
$C_{a,R}$	cm	5.11	4.08	4.51	3.56
$C_{a,S}$	cm	5.04	3.99	4.46	3.50
$N_{B,R}$	–	56	74	54	72
$N_{B,S}$	–	57	75	55	72
Dimensionless numbers					
Ma_1	–	0.294	0.266	0.249	0.225
$Ma_{2,S}$	–	0.292	0.268	0.248	0.226
Re_1		2.11×10^5	1.85×10^5	2.50×10^5	2.17×10^5
$Re_{2,S}$		2.09×10^5	1.85×10^5	2.49×10^5	2.19×10^5
Performances					
P_W	MW	4.90		4.91	
η_{TT}	%	85.61		85.88	
η_p	%	87.06		87.26	

Table 8.11 Turbine design: main output (Helium)

		Turbine	
		First stage	Last stage
Input parameters			
ψ	–	1.400	1.467
φ	–	0.600	0.600
R	–	0.500	0.466
Output parameters			
Kinematics			
α_1	°	63.43	64.66
α_2	°	18.43	18.43
β_1	°	18.43	24.00
β_2	°	63.43	63.43
u	m/s	345.8	345.8
Thermodynamics			
p_1	bar	3.609	1.608
T_1	°C	883.9	586.1
p_2	bar	3.474	1.530
T_2	°C	867.8	570.4
Geometry			
D_M	cm	110.1	110.1
$h_{B,S}$	cm	10.83	17.95
$h_{B,R}$	cm	11.09	18.55
C_R	cm	6.80	7.36
C_S	cm	8.04	8.78
$C_{a,R}$	cm	4.57	4.96
$C_{a,S}$	cm	5.51	5.74
$N_{B,R}$	–	64	60
$N_{B,S}$	–	54	50
Dimensionless numbers			
$Ma_{1,S}$		0.232	0.281
Ma_2		0.233	0.271
$Re_{1,S}$		1.09×10^5	0.92×10^5
Re_2		0.91×10^5	0.73×10^5
Performances			
P_W	MW	20.21	
η_{TT}	%	93.22	
η_p	%	92.00	

Table 8.12 Turbomachine inlet/outlet conditions (Argon cycle)

		Compressor C1	Compressor C2	Turbine
Inlet flow rate	kg/s	97.6	97.6	97.6
Inlet pressure	bar	1.20	2.20	4.0
Inlet temperature	°C	55	55.0	900
Outlet pressure	bar	2.22	4.08	1.22

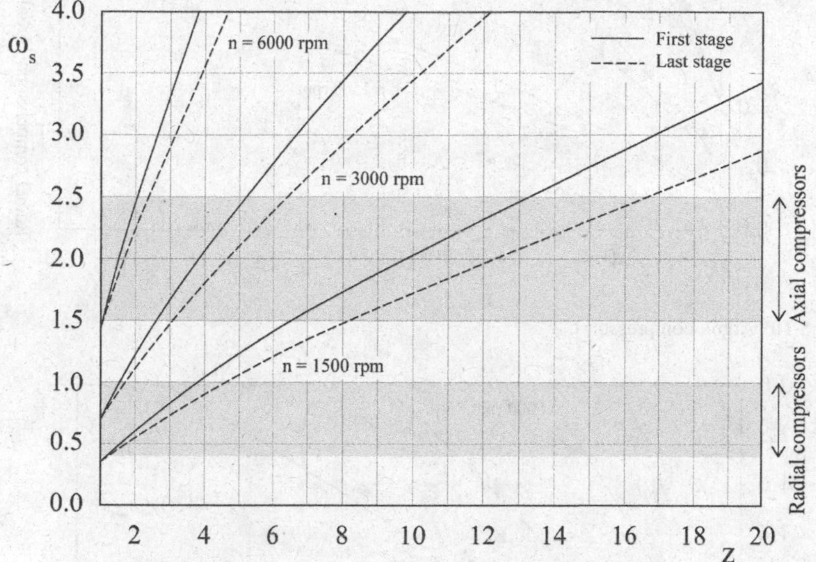

Fig. 8.9 Argon compressor, C1

considering an axial clearance equal to about 30% of the axial chord; the maximum blade tip speed reaches 255 m/s; the degree of reaction varies from 0.44 (hub) to 0.55 (tip) and the Mach number reaches, at most, values of about 0.6. In terms of overall performance, the compressor transfers a power of about 5.2 MW from the blades to the fluid with a polytropic efficiency of almost 91%. Such efficiency is, therefore, two percentage points higher than the assumption made in Sect. 8.1.4.

The compressor C2 has a mean diameter of about 1.2 m; the blade height varies between 7.6 cm (first stage) and 5.6 cm (last stage); the axial length is about 60 cm, considering an axial clearance equal to about 30% of the axial chord; the maximum blade tip speed reaches 200 m/s; the degree of reaction varies from 0.43 (hub) to 0.56 (tip) and the Mach number is always lower than 0.5. In terms of overall performance, the compressor transfers a power of 5.3 MW from the blades to the fluid with a polytropic efficiency of almost 90%. Such efficiency is, therefore, one percentage point higher than the assumption made in Sect. 8.1.4.

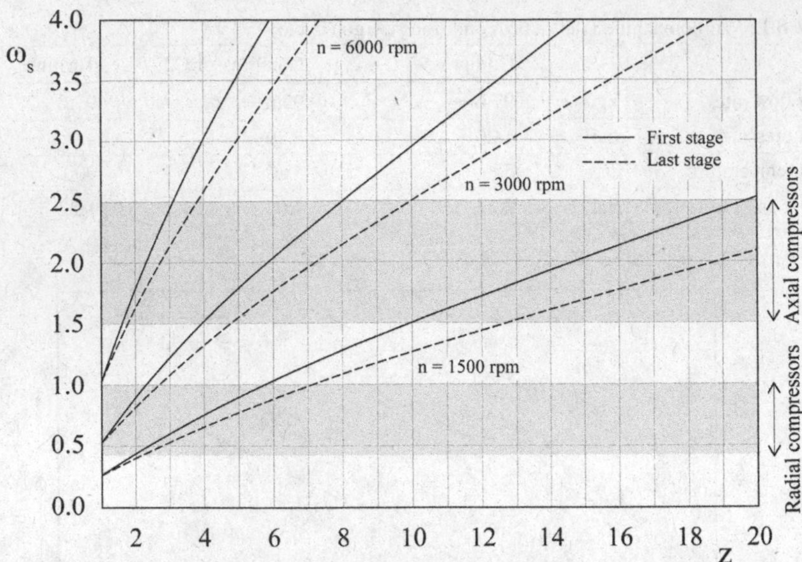

Fig. 8.10 Argon compressor, C2

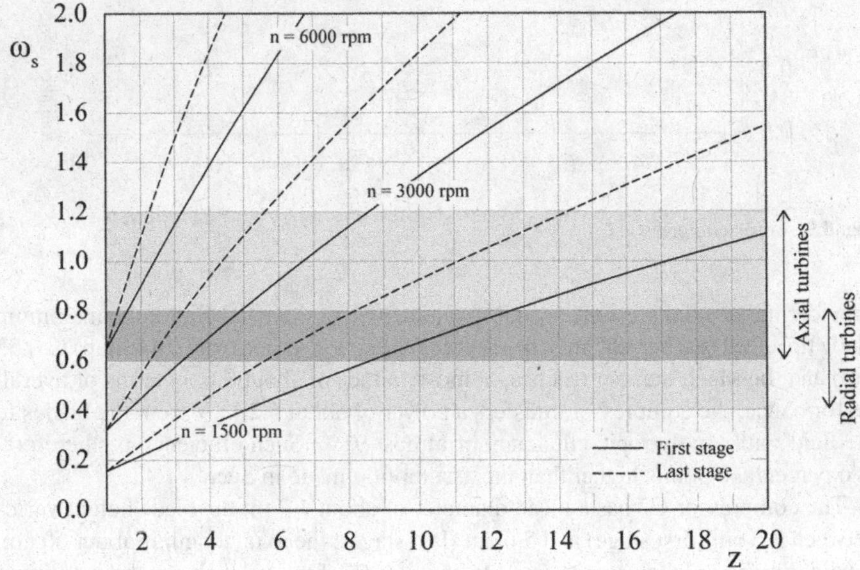

Fig. 8.11 Argon turbine

Table 8.13 Output of the turbomachinery selection processes (Argon)

		Compressor C1	Compressor C2	Turbine
n	rpm	3000	3000	3000
z	–	3	5	3
ω_s (first/last stage)		1.65/1.47	1.79/1.55	0.58/0.76
Stage type	–	Axial	Axial	Axial

We have designed the turbine by applying the procedure described in Chap. 3. Table 8.15 reports the main results concerning the first and last stage. The input parameters and the kinematic parameters are referred to the mean diameter.

The turbine has a mean diameter of about 1.4 m; the blade height varies between 10.4 cm (first stage) and 19.4 cm (last stage); the axial length is about 50 cm, considering an axial clearance equal to about 30% of the axial chord; the maximum blade tip speed reaches 250 m/s; the degree of reaction varies from 0.33 (hub) to 0.57 (tip) and the Mach number is always lower than 0.6. In terms of overall performance, the fluid flow transfers a power of about 21 MW to the blades with a polytropic efficiency of almost 93%. Such efficiency is, therefore, 3 percentage points higher than the assumption made in Sect. 8.1.4.

Comparing these results with those reported in Sect. 8.2.1, the Argon turbomachines have mean diameters slightly higher than those of the Helium turbomachines. The first ones are much more compact in terms of number of stages (about 1/3 of those of Helium) and attain higher polytropic efficiencies (up to 3 percentage points for the compressors and 1 percentage point for the turbine).

8.2.3 *Turbomachines for Brayton Cycles Operated with sCO$_2$*

The main results related to the turbomachines in the Brayton cycle operated with sCO_2 and assessed in the Sect. 8.1 are synthetically reported in Table 8.16.

On the basis of these parameters, following the procedure illustrated in Chap. 2, Figs. 8.12, 8.13 and 8.14 show the specific speed variation of each turbomachine with the number of stages for different rotational speeds.

Assuming the same rotational speed for the three turbomachines, Figs. 8.12, 8.13 and 8.14 suggest that a good compromise between efficiency and number of stages is:

- rotational speed of 24,000 rpm
- centrifugal configurations for the two compressors and axial configuration for the turbine
- 2 stages for MC, 3 for RC and 4 for the turbine.

For this last turbomachine, a good solution could be also an axial 2-stage configuration, but, in this case, the blade heights become very small and so, to have a minimum blade height of 3 cm, the axial 4-stage configuration has been chosen.

Table 8.14 Compressor design: main output (Argon)

Compressors		Compressor C1		Compressor C2	
		First stage	Last stage	First stage	Last stage
Input parameters					
ψ	–	0.300	0.307	0.300	0.309
φ	–	0.550	0.550	0.550	0.550
R	–	0.500	0.497	0.500	0.495
Output parameters					
Kinematics					
α_1	°	32.49	32.49	32.49	32.49
α_2	°	49.79	50.09	49.79	50.19
β_1	°	49.79	49.79	49.79	49.79
β_2	°	32.49	31.97	32.49	31.80
u	m/s	241.9	241.9	188.3	188.3
Thermodynamics					
p_2	bar	1.345	2.023	2.358	3.859
T_2	°C	71.88	140.0	65.22	148.2
p_3	bar	1.499	2.220	2.521	4.080
T_3	°C	88.75	157.4	75.45	158.8
Geometry					
D_M	cm	154.1	154.1	119.9	119.9
$h_{B,R}$	cm	8.36	6.62	7.62	5.78
$h_{B,S}$	cm	7.85	6.28	7.33	5.60
C_R	cm	7.60	6.02	6.92	5.25
C_S	cm	7.14	5.71	6.67	5.09
$C_{a,R}$	cm	5.98	4.75	5.45	4.15
$C_{a,S}$	cm	5.62	4.49	5.25	4.00
$N_{B,R}$	–	63	82	54	74
$N_{B,S}$	–	67	86	56	76
Dimensionless numbers					
Ma_1	–	0.610	0.555	0.475	0.424
$Ma_{2,S}$	–	0.595	0.547	0.468	0.422
Re_1		1.12×10^6	0.97×10^6	1.46×10^6	1.22×10^6
$Re_{2,S}$		1.08×10^6	0.94×10^6	1.43×10^6	1.21×10^6
Performances					
P_W	MW	5.20		5.27	
η_{TT}	%	88.78		88.20	
η_p	%	90.57		89.88	

Table 8.15 Turbine design: main output (Argon)

		Turbine	
		First stage	Last stage
Input parameters			
ψ	–	1.40	1.461
φ	–	0.60	0.60
R	–	0.50	0.469
Output parameters			
Kinematics			
α_1	°	63.43	64.56
α_2	°	18.43	18.43
β_1	°	18.43	23.52
β_2	°	63.43	63.43
u	m/s	224.9	224.9
Thermodynamics			
p_1	bar	3.394	1.525
T_1	°C	832.0	548.8
p_2	bar	2.856	1.220
T_2	°C	764.0	482.1
Geometry			
D_M	cm	143.2	143.2
$h_{B,S}$	cm	10.36	16.78
$h_{B,R}$	cm	11.53	19.37
C_R	cm	7.92	8.32
C_S	cm	9.77	10.23
$C_{a,R}$	cm	5.75	6.22
$C_{a,S}$	cm	7.10	7.24
$N_{B,R}$	–	72	69
$N_{B,S}$	–	58	56
Dimensionless numbers			
$Ma_{1,S}$		0.487	0.588
Ma_2		0.503	0.589
$Re_{1,S}$		7.30×10^5	5.88×10^5
Re_2		5.54×10^5	4.24×10^5
Performances			
P_W	MW	21.24	
η_{TT}	%	94.07	
η_p	%	92.71	

Table 8.16 Turbomachine inlet/outlet conditions (sCO_2- cycle)

		Compressor MC	Compressor RC	Turbine
Inlet flow rate	kg/s	72.4	40.7	113.1
Inlet pressure	bar	100	101	250
Inlet temperature	°C	55	142.6	700
Outlet pressure	bar	255	253.8	101.5

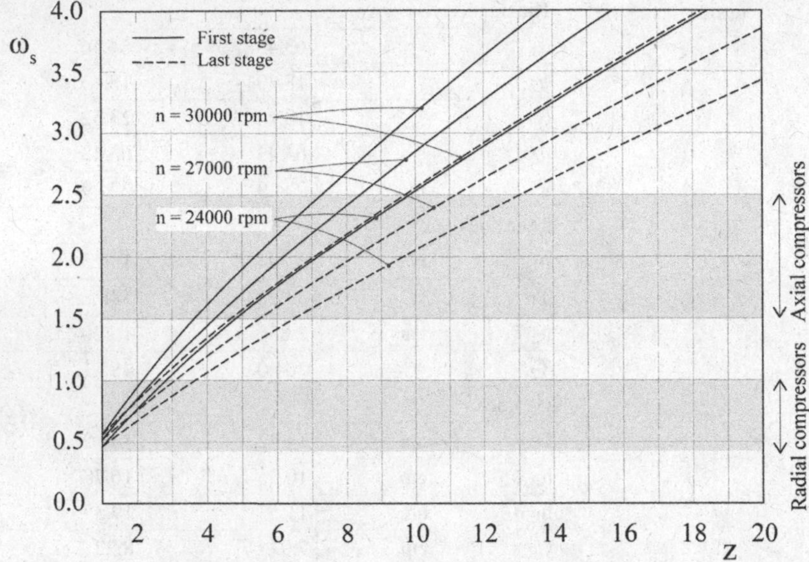

Fig. 8.12 sCO_2 compressor, MC

Again for the turbine, a 1-stage radial solution can also be considered. This solution will be analyzed below and compared with the axial solution.

Regarding the compressors, instead, the axial solution is inadvisable because of the high number of stages and the very low blade heights leading to substantial tip clearance losses.

Table 8.17 summarizes the main results of the turbomachinery selection process, highlighting those related to the first and last stage of each turbomachine; as already mentioned, both the multistage axial and the single-stage radial solutions will be analyzed for the turbine.

After choosing the turbomachine type, the rotational speed and the number of stages, the turbomachinery preliminary design may start. We have designed the two compressors MC and RC by applying the procedure described in Chap. 6 (procedure using Set B as input parameters). Table 8.18 reports the main results concerning the first and last stage (for MC, therefore, the results concern the whole turbomachine).

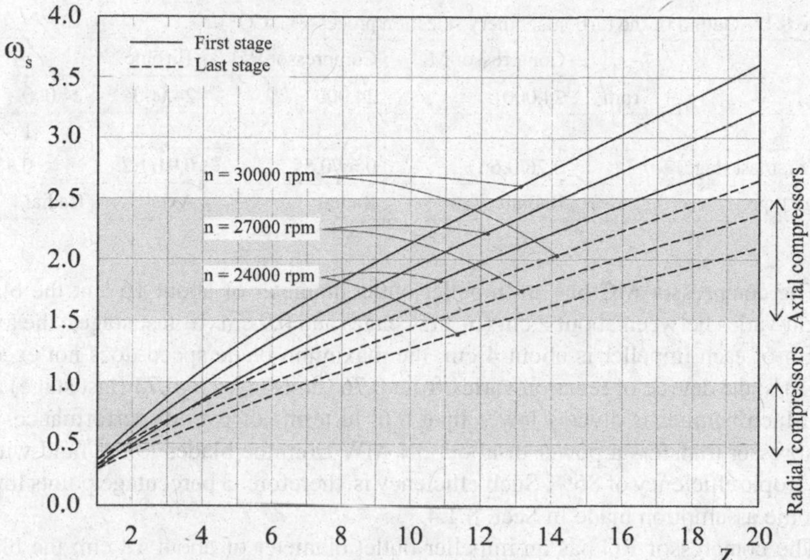

Fig. 8.13 sCO_2 compressor, RC

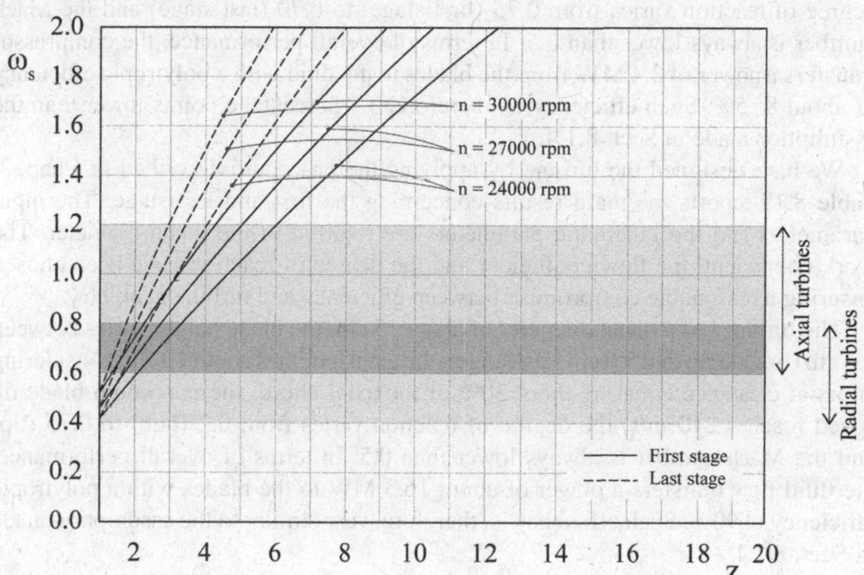

Fig. 8.14 sCO_2 turbine

Table 8.17 Output of the turbomachinery selection processes (sCO_2)

		Compressor MC	Compressor RC	Turbine	
n	rpm	24,000	24,000	24,000	24,000
z	–	2	3	4	1
ω_s (first/last stage)		0.76/0.66	0.69/0.55	0.91/1.2	0.42
Stage type	–	Radial	Radial	Axial	Radial

The compressor MC has an impeller outlet diameter of about 16 cm; the blade height varies between about 2 cm (b_1 first stage) and 0.9 cm (b_2 last stage); the axial length of each impeller is about 4 cm; the maximum blade speed does not exceed 200 m/s; the degree of reaction varies from 0.76 (first stage) to 0.73 (last stage) and the Mach number is always lower than 0.6. In terms of overall performance, the compressor transfers a power of about 3.1 MW from the blades to the fluid with a polytropic efficiency of 86%. Such efficiency is, therefore, 3 percentage points lower than the assumption made in Sect. 8.1.4.

The compressor RC has an impeller outlet diameter of about 18 cm; the blade height varies between 2 cm (b_1 first stage) and 0.7 cm (b_2 last stage); the axial length of each impeller is about 4 cm; the maximum blade speed is nearly 220 m/s; the degree of reaction varies from 0.75 (first stage) to 0.70 (last stage) and the Mach number is always lower than 0.5. In terms of overall performance, the compressor transfers a power of 3.4 MW from the blades to the fluid with a polytropic efficiency of about 85.5%. Such efficiency is, therefore, 3.5 percentage points lower than the assumption made in Sect. 8.1.4.

We have designed the turbine by applying the procedure described in Chap. 3. Table 8.19 reports the main results concerning the first and last stage. The input parameters and the kinematic parameters are referred to the mean diameter. The work coefficient, the flow coefficient and the degree of reaction have been chosen ensuring a reasonable compromise between efficiency and turbine geometry.

The turbine has a mean diameter of about 15 cm; the blade height varies between 2.6 cm (first stage) and 5.0 cm (last stage); the axial length is about 10 cm, considering an axial clearance equal to about 30% of the axial chord; the maximum blade tip speed reaches 240 m/s; the degree of reaction varies from 0.2 (hub) to 0.75 (tip) and the Mach number is always lower than 0.5. In terms of overall performance, the fluid flow transfers a power of about 16.3 MW to the blades with a polytropic efficiency of 90%. Such efficiency is, therefore, very similar to the assumption made in Sect. 8.1.4.

As already illustrated above, a single-stage radial turbine solution can also be considered. By applying the procedure illustrated in Chap. 5 (procedure using Set B as input parameters), we have designed a radial turbine and the main output parameters are reported in Table 8.20.

As can be seen from Table 8.20, the radial turbine is feasible, the blade speed is admissible and the Mach number is lower than 1. The penalty in terms of efficiency,

Table 8.18 Compressor design: main output (sCO_2)

Compressors		Compressor MC		Compressor RC	
		First stage	Last stage	First stage	Last stage
Input [$f(\omega_s)$]					
ψ_{is}	–	0.468	0.484	0.479	0.502
δ_t	–	0.588	0.570	0.576	0.550
α_2	–	71.39	72.28	72.01	73.03
Assumptions					
α_{1M}		0.00	0.00	0.00	0.00
δ_h		0.350	0.350	0.350	0.350
Output parameters					
Dimensionless coefficients					
ψ		0.544	0.577	0.559	0.608
φ		0.263	0.230	0.241	0.186
Kinematics					
ξ	–	0.708	0.802	0.752	0.994
R	–	0.754	0.728	0.743	0.697
β_{1M}	°	60.70	63.45	62.48	67.51
β_2	°	67.33	66.47	67.60	64.72
u_2	m/s	196.7	194.6	220.2	217.4
Thermodynamics					
p_2	bar	151.2	231.5	130.6	235.8
T_2	°C	84.47	115.6	169.2	235.9
p_3	bar	168.9	255.0	141.2	253.8
T_3	°C	92.57	123.0	177.9	245.4
Geometry					
D_{1t}	cm	9.21	8.82	10.09	9.52
D_2	cm	15.65	15.49	17.52	17.30
D_3	cm	25.18	24.73	28.05	27.40
b_1	cm	1.86	1.70	1.98	1.73
b_2	cm	0.99	0.85	1.01	0.69
SF	–	0.896	0.887	0.890	0.874
$D_{hyd,R}$	cm	1.16	1.10	1.27	1.07
$L_{hyd,R}$	cm	13.37	13.73	15.62	15.67
$D_{hyd,vaned}$	cm	1.16	1.18	1.37	0.89
$L_{hyd,vaned}$	cm	7.44	10.35	12.40	9.48
$N_{B,R}$	–	14	13	13	12
$N_{B,S}$	–	18	12	12	21

(continued)

Table 8.18 (continued)

Compressors		Compressor MC		Compressor RC	
		First stage	Last stage	First stage	Last stage
Dimensionless numbers					
Ma_1	–	0.567	0.418	0.464	0.369
$Ma_{2,S}$	–	0.425	0.350	0.415	0.380
Re_1		1.61×10^7	1.44×10^7	9.69×10^6	9.37×10^6
$Re_{2,S}$		1.38×10^7	1.68×10^7	1.23×10^7	1.02×10^7
Performances					
P_W	MW	3.13		3.40	
η_{TT}	%	83.44		83.15	
η_p	%	86.10		85.40	

compared to the axial solution (over 5 penalty points), however, makes this latter solution preferable.

As can be seen from comparing the results reported in Sects. 8.2.1and 8.2.2, the sCO_2 allows an extremely more compact power block: in fact, in comparison with the 10-MWe power block operated with Helium and Argon, the 10-MWe sCO_2 power block has turbomachine dimensions of about ten times smaller both in terms of diameters and axial lengths. This result is very important for CSP applications: these dimensions, in fact, allow to mount the power block directly on the central tower and thus the correlated piping can be extremely short. On the other hand, however, mainly because of adopting centrifugal compressors, the turbomachine polytropic efficiencies are lower than those of the other power blocks.

8.2.4 Turbomachines for the Bottoming ORC

Since the selection and design of sCO2 turbomachines have just been illustrated (Sect. 8.2.3), in this section we will only refer to the turbomachines for the bottoming ORC, as reported in Table 8.21.

On the basis of these parameters, following the procedure illustrated in Chap. 2, Figs. 8.15 and 8.16 show the specific speed variation of each turbomachine with the number of stages for different rotational speeds.

Figure 8.15 suggests a single-stage radial turbine at 15,000 rpm as rotational speed. But also a 3-stage axial solution can be considered: this configuration is preferable to the 2-stage axial one in terms of efficiency and minimum blade height. This 3-stage axial solution will be analyzed below and compared with the radial solution.

Figure 8.16 suggests a centrifugal 6-stage pump at 3,000 rpm to avoid cavitation problems ($NPSH_R$ less than 10 m.)

Table 8.22 summarizes the main results of the turbomachinery selection process,

Table 8.19 Axial turbine design: main output (sCO_2)

		Turbine	
		First stage	Last stage
Input parameters			
ψ	–	1.00	1.049
φ	–	0.40	0.40
R	–	0.60	0.575
Output parameters			
Kinematics			
α_1	°	66.04	67.15
α_2	°	14.04	14.04
β_1	°	−14.04	−7.21
β_2	°	70.02	70.02
u	m/s	187.6	187.6
Thermodynamics			
p_1	bar	230.0	116.7
T_1	°C	688.4	596.6
p_2	bar	202.7	101.5
T_2	°C	670.9	578.3
Geometry			
D_M	cm	14.93	14.93
$h_{B,S}$	cm	2.58	4.43
$h_{B,R}$	cm	2.81	4.89
C_R	cm	1.35	1.48
C_S	cm	1.18	1.33
$C_{a,R}$	cm	0.69	0.78
$C_{a,S}$	cm	0.78	0.86
$N_{B,R}$	–	47	43
$N_{B,S}$	–	52	46
Dimensionless numbers			
$Ma_{1,S}$		0.372	0.423
Ma_2		0.450	0.488
$Re_{1,S}$		6.27×10^6	4.69×10^6
Re_2		7.86×10^6	5.38×10^6
Performances			
P_W	MW	16.33	
η_{TT}	%	90.63	
η_p	%	89.96	

Table 8.20 Radial turbine design: main output (sCO_2)

		Turbine
Input [$f(\omega_s)$]		
ψ_{is}	–	1.30
α_1	°	76.67
ξ	–	0.944
Assumptions		
α_{2M}		0.00
δ_h		0.185
Output parameters		
Dimensionless coefficients		
ψ		1.117
φ		0.280
Kinematics		
δ_t	–	0.584
R	–	0.445
β_1	°	23.91
β_{2M}	°	53.91
u_1	m/s	350.0
Thermodynamics		
p_1	bar	163.7
T_1	°C	637.0
p_2	bar	101.5
T_2	°C	584.4
Geometry		
D_0	cm	36.6
D_1	cm	27.9
D_{2t}	cm	16.3
b_1	cm	1.51
b_2	cm	5.56
$D_{hyd,R}$	cm	2.25
$L_{hyd,R}$	cm	12.3
$D_{hyd,S}$	cm	1.22
$L_{hyd,S}$	cm	4.66
$N_{B,R}$	–	15
$N_{B,S}$	–	25
Dimensionless numbers		
$Ma_{1,S}$	–	0.849
Ma_2	–	0.503

(continued)

Table 8.20 (continued)

		Turbine
$Re_{1,S}$		1.13×10^7
Re_2		6.13×10^6
Performances		
P_W	MW	15.5
η_{TT}	%	85.97
η_p	%	84.99

Table 8.21 Turbomachine inlet/outlet conditions (ORC - Cis-butene)

		Turbine	Pump
Inlet flow rate	kg/s	30.6	30.6
Inlet pressure	bar	29.6	5.1
Inlet temperature	°C	140.5	55.0
Outlet pressure	bar	5.1	31.1

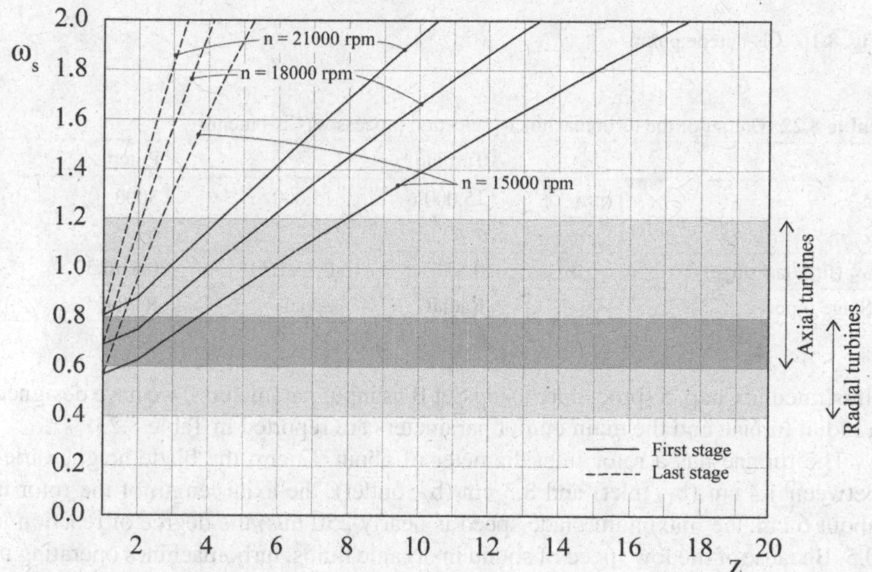

Fig. 8.15 Cis-butene turbine

highlighting those related to the first and last stage of the axial turbine; as previously mentioned, both the multistage axial and the single-stage radial turbines will be analyzed below.

After choosing the turbomachine type, the rotational speed and the number of stages, the turbomachinery preliminary design may start. By applying the procedure

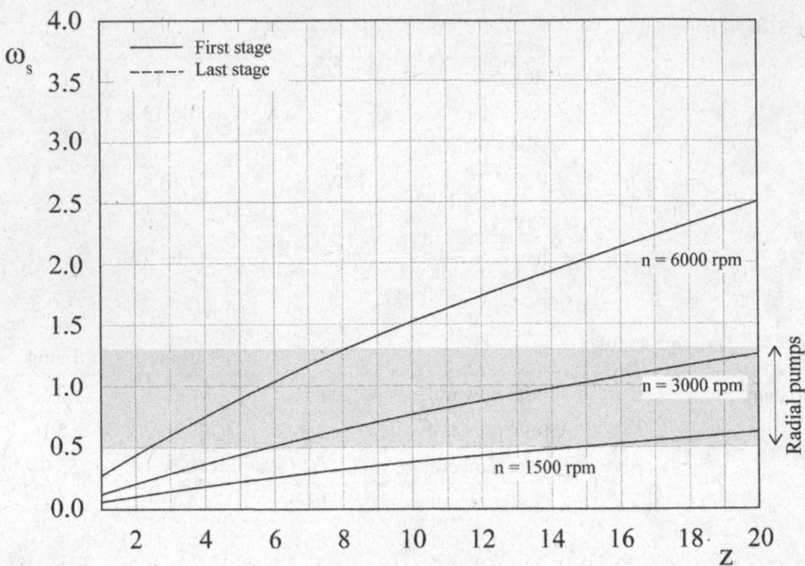

Fig. 8.16 Cis-butene pump

Table 8.22 Output of the turbomachinery selection processes (Cis-butene)

		Turbine		Pump
n	rpm	15,000	15,000	3000
Z	–	1	3	6
ω_S (first/last stage)		0.57	0.73/1.31	0.51/0.50
Stage type	–	Radial	Axial	Radial

illustrated in Chap. 5 (procedure using Set B as input parameters), we have designed a radial turbine and the main output parameters are reported in Table 8.23.

The turbine has a rotor inlet diameter of about 32 cm; the blade height varies between 1.4 cm (b_1, inlet) and 8.3 cm (b_2, outlet); the axial length of the rotor is about 6 cm; the maximum blade speed is nearly 250 m/s; the degree of reaction is 0.5. Because of the low speed of sound in organic fluids, turbomachines operating in ORC are characterized by high Mach numbers. In this case, we have a Mach number higher than one at the stator outlet; this value can however be considered acceptable. In terms of overall performance, the fluid flow transfers a power of 1.9 MW to the blades with a polytropic efficiency of about 83%. Such efficiency is decidedly lower (about 5 percentage points) than the assumption made in Sect. 8.1.4.

In order to increase this efficiency and reduce the Mach number, a multistage axial turbine solution has been considered, as previously illustrated. We have designed a 3-stage axial turbine by applying the procedure shown in Chap. 3. Table 8.24 reports the main results concerning the first and last stage. The input parameters and the

Table 8.23 Radial turbine design: main output (Cis-butene)

		Turbine
Input [f(ω_s)]		
ψ_{is}	–	1.152
α_1	°	74.57
ξ	–	0.962
Assumptions		
α_{2M}		0.00
δ_h		0.185
Output parameters		
Dimensionless coefficients		
ψ		0.990
φ		0.284
Kinematics		
δ_t	–	0.706
R	–	0.508
β_1	°	−1.99
β_{2M}	°	57.48
u_1	m/s	251.3
Thermodynamics		
p_1	bar	13.21
T_1	°C	97.11
p_2	bar	5.10
T_2	°C	63.09
Geometry		
D_0	cm	41.9
D_1	cm	32.0
D_{2t}	cm	22.6
b_1	cm	1.42
b_2	cm	8.33
$D_{hyd,R}$	cm	2.71
$L_{hyd,R}$	cm	12.92
$D_{hyd,S}$	cm	0.99
$L_{hyd,S}$	cm	5.26
$N_{B,R}$	–	13
$N_{B,S}$	–	25
Dimensionless numbers		
$Ma_{1,S}$	–	1.32
Ma_2	–	0.90
$Re_{1,S}$		8.08×10^6
Re_2		4.81×10^6
Performances		
P_W	MW	1.91
η_{TT}	%	85.98
η_p	%	83.17

Table 8.24 Axial turbine design: main output (Cis-butene)

		Turbine	
		First stage	Last stage
Input parameters			
ψ	–	0.80	0.890
φ	–	0.40	0.40
R	–	0.60	0.555
Output parameters			
Kinematics			
α_1	°	63.43	65.80
α_2	°	0.00	0.00
β_1	°	−26.57	−15.36
β_2	°	68.20	68.20
u	m/s	160.3	160.3
Thermodynamics			
p_1	bar	22.53	6.944
T_1	°C	125.0	72.72
p_2	bar	15.82	5.10
T_2	°C	106.9	61.43
Geometry			
D_M	cm	20.41	20.41
$h_{B,S}$	cm	1.09	4.19
$h_{B,R}$	cm	1.62	5.58
C_R	cm	1.65	1.92
C_S	cm	1.56	1.80
$C_{a,R}$	cm	0.82	1.02
$C_{a,S}$	cm	1.02	1.10
$N_{B,R}$	–	49	43
$N_{B,S}$	–	51	46
Dimensionless numbers			
$Ma_{1,S}$		0.812	0.745
Ma_2		0.900	0.813
$Re_{1,S}$		1.13×10^7	4.94×10^6
Re_2		1.04×10^7	4.49×10^6
Performances			
P_W	MW	2.00	
η_{TT}	%	89.99	
η_p	%	87.89	

kinematic parameters are referred to the mean diameter. The work coefficient, the flow coefficient and the degree of reaction have been chosen ensuring a reasonable compromise between efficiency and turbine geometry.

As can be seen, in the axial turbine the Mach number is lower than one and the efficiency is significantly higher than that of the radial configuration, even if the first-stage geometry shows a critical aspect due to small blade heights which could require a design revision.

The turbine has a mean diameter of about 20 cm; the blade height varies between 1.1 cm (first stage) and 5.6 cm (last stage); the axial length is about 10 cm, considering an axial clearance equal to about 30% of the axial chord; the maximum blade tip speed reaches 190 m/s; the degree of reaction varies from 0.34 (hub) to 0.68 (tip). In terms of the overall performance, the fluid flow transfers a power of 2 MW to the blades with a polytropic efficiency of almost 88%. Such efficiency is, therefore, very similar to the assumption made in Sect. 8.1.4.

Finally, by applying the procedure illustrated in Chap. 7 (procedure using Set B as input parameters), we have designed a centrifugal multi-stage pump and the main output parameters are reported in Table 8.25.

The pump has an impeller outlet diameter of about 25 cm; the blade height varies between about 3 cm (b_1 first stage) and 2 cm (b_2 last stage); the maximum blade speed does not exceed 40 m/s; the degree of reaction is 0.73. In terms of the overall performance, the pump transfers a power of 154 kW from the blades to the fluid with an hydraulic efficiency of about 89%. Such efficiency is, therefore, higher than the assumption made in Sect. 8.1.4.

Table 8.25 Pump design: main output (Cis-butene)

		Pump	
		First stage	Last stage
Input parameters [f(ω_s)]			
ψ_{is}	–	0.492	0.493
δ_t	–	0.552	0.551
δ_b	–	0.081	0.081
Assumptions			
α_1		0.00	0.00
δ_h		0.30	0.30
Output parameters			
Dimensionless coefficients			
ψ		0.554	0.554
φ		0.131	0.131
Kinematics			
ξ	–	0.658	0.657
R	–	0.732	0.732
α_2		81.16	81.19
β_1	°	72.90	72.92
β_2	°	79.08	79.10
u_2	m/s	38.97	38.96
Operation			
p_2	bar	8.44	30.11
ΔH	m	76.19	76.22
$NPSH_R$	m	7.45	7.43
Geometry			
D_{1t}	cm	13.69	13.68
D_2	cm	24.82	24.82
D_3	cm	33.95	33.93
b_1	cm	3.12	3.12
b_2	cm	2.02	2.02
SF	–	0.842	0.842
$D_{hyd,R}$	cm	2.51	2.51
$L_{hyd,R}$	cm	26.74	26.79
$D_{hyd,vaned}$	cm	2.44	2.44
$L_{hyd,vaned}$	cm	13.97	13.96
$N_{B,R}$	–	5	5
$N_{B,S}$	–	8	8
Dimensionless numbers			
Re_1		1.93×10^6	1.90×10^6
$Re_{2,S}$		2.30×10^6	2.27×10^6
Performances			
P_W	MW	0.154	
η_{hyd}	%	88.89	

References

Belmonte AA, Sebastian A, Aguilar J, Romaero M (2016) Performance comparison of different thermodynamic cycles for an innovative central receiver solar power plant. SolarPACES

Bennet J, Wilkes J, Allison T, Pelton R, Wygant K (2017) Cycle modeling and optimization of an integrally geared sCO_2 compander. In: Proceedings of ASME turbo expo (2017)

Beserati SM, Goswami DY (2014) Analysis of advanced supercritical carbon dioxide power cycles with a bottoming cycle for concentrating solar power applications. ASME J Solar Eng

Beserati SM, Goswami DY (2017) Supercritical CO_2 and other advanced power cycles for concentrating solar thermal (CST) systems. In: Advances in concentrating solar thermal research and technology. Elsevier, Amsterdam

Dostal V, Hejzlar P, Driscoll MJ (2006a) High-performance supercritical carbon dioxide cycle for next-generation nuclear reactors. Nucl Tecnhol

Dostal V, Hejzlar P, Driscoll MJ (2006b) The supercritical carbon dioxide power cycle: comparison to other advanced power cycles. Nucl Technol

Fleming D, Holschuh T, Conboy T, Pasch J, Wright S, Rochau G (2012) Scaling considerations for a multi-megawatt class supercritical CO_2 brayton cycle and path forward for commercialization. In: Proceedings of the ASME turbo expo (2012)

Kalra C, Sevincer E, Brun K, Hofer D, Moore J (2014) Development of high efficiency hot gas turbo-expander for optimized CSP supercritical CO_2 power block operation. In: The 4th international symposium—supercritical CO_2 power cycles

Kusterer K, Braun R, Moritz N, Lin G, Bohn D (2012) Helium Brayton cycles with solar central receivers: thermodynamic and design considerations. In: Proceedings of ASME turbo expo (2012)

Kusterer K, Braun R, Moritz N, Sugimoto T, Tanimura K (2013) Comparative study of solar thermal Brayton cycles operated with helium or argon. In: Proceedings of ASME turbo expo

Lemmon EW, Bell IH, Huber ML, McLinden MO. REFPROP—reference fluid thermodynamic and transport properties. NIST Standard Reference Database 23, version 10.0

Macadam S, Follet W, Kutin M, Subbaraman G (2019) Supercritical CO_2 power cycle projects at GTI. In: 3rd European supercritical CO_2 conference

Moore JJ, Towler M, Mortzheim J, Cich S, Hofer D (2018) Testing of a 10 MWe supercritical CO_2 turbine. In: 47th turbomachinery & 34th pump symposia

Pelton R, Jung S, Allison T, Smith N (2018) Design of a wide-range centrifugal compressor stage for supercritical CO_2 power cycles. ASME J Eng Gas Turbines Power

Stein WH, Buck R (2017) Advanced power cycles for concentrated solar power. Solar Energy

Turchi CS, Ma Z, Dyreby J (2012) Supercritical carbon dioxide power cycle configurations for use in concentrating solar power systems. In: Proceedings of ASME turbo expo

Turchi CS, Ma Z, Neises TW, Wagner MJ (2013) Thermodynamic study of advanced supercritical carbon dioxide power cycles for concentrating solar power systems. J Solar Energy Eng

Wilkes J, Allison T, Schmitt J, Bennett J, Wygant K, Pelton R, Bosen W (2016) Application of an integrally geared compander to an sCO2 recompression brayton cycle. In: The 5th international symposium—supercritical CO_2 power cycles

Printed in the United States
by Baker & Taylor Publisher Services

Printed in the United States
by Baker & Taylor Publisher Services